CISM COURSES AND LECTURES

The series presents lecture notes, monographs, edited works and proceedings in the field of Mechanics, Engineering, Computer Science and Applied Mathematics.
Purpose of the series is to make known in the international scientific and technical community results obtained in some of the activities organized by CISM, the International Centre for Mechanical Sciences.

INTERNATIONAL CENTRE FOR MECHANICAL SCIENCES

COURSES AND LECTURES - No. 426

ADVANCED NUMERICAL APPLICATIONS AND PLASTICITY IN GEOMECHANICS

EDITED BY

VAUGHAN D. GRIFFITHS
COLORADO SCHOOL OF MINES

GIANCARLO GIODA
POLITECNICO DI MILANO

Springer-Verlag Wien GmbH

This volume contains 162 illustrations

Originally published by Springer-Verlag Wien New York in 2001
SPIN 10837221

In order to make this volume available as economically and as rapidly as possible the authors' typescripts have been reproduced in their original forms. This method unfortunately has its typographical limitations but it is hoped that they in no way distract the reader.

ISBN 978-3-211-83337-7 ISBN 978-3-7091-2578-6 (eBook)
DOI 10.1007/978-3-7091-2578-6

PREFACE

Numerical application of plasticity to geomechanics is an area of research that has grown rapidly since its origins in the late 1960s. While early work was mainly confined to academic research, it is gratifying to see that the new growth area for some of these methodologies lies in geotechnical practice.

There are at least two well-established conference circuits, one international association and various journals that emphasize numerical and analytical approaches to solving geomechanics problems. It is no coincidence though, that the evolution of numerical application of plasticity to geomechanics, has run in parallel with the development of the finite element method and other numerical methods, and the extraordinary increase in power and speed of digital computers.

This text contains papers that were presented as part of a course of the same name offered at the Centre International des Sciences et Mécaniques (CISM) in Udine, Italy in late June 2000. The material spans a remarkable range of topics, from theoretical developments involving novel algorithms and constitutive models to practical applications involving prediction of stresses and deformations in tunnel linings.

The contributions are arranged in alphabetic order according to the first author, however if the reader wishes to move in an approximate route from theory to practice, he or she might start with the paper by Woodward in which a review is presented on different numerical models for geomaterials and their numerical implementation. The paper by Benvenuti and Loret continues the constitutive theme with a focus on damaged solids. The growing interest in coupled problems of varying complexity is reflected in the papers by Loret, Gajo and Hueckel and Loret and Khalili, with potential applications to swelling soils and environmental geotechnics. Modeling consolidation behavior of nonlinear geomaterials in theory and practice is discussed in the paper by Dluzewski, and the impact of statistically defined soil properties in classical problems of slope stability and bearing capacity in the paper by Griffiths. The paper by Cividini covers applications of numerical modeling and plasticity to several ground improvement techniques including freezing, and the paper by Cividini, Gioda and Sterpi applies elasto-plastic models to practical tunneling problems.

It is hoped that the reader will find this text a valuable reference to current trends in theory and practice relating to numerical applications and plasticity in geomechanics.

Vaughan D. Griffiths
Giancarlo Gioda

CONTENTS

Page

Internal Length-Scales in Damaged Solids: a Lagrange Multiplier Approach

Elena Benvenuti[1] and Benjamin Loret[2]

[1] Dipartimento di Ingegneria. Università degli Studi di Ferrara, Via G. Saragat 1, 44100 Ferrara, Italia,
[2] Laboratoire Sols, Solides, Structures, B.P. 53X, 38041 Grenoble Cedex, France

Abstract. Like softening in elastic-plastic solids and fluid-saturated porous media, damage in brittle solids is likely to induce instabilities. The corresponding equations of motion loose their hyperbolicity and the computations depend pathologically on the finite element mesh. Non-local operators are introduced to bring an internal length into the governing equations. For elastic-damaging solids, the associated computational aspects are addressed in a unified framework, which views the non-local operators as constraints and introduces them into the energy through a Lagrange multiplier: the resulting matrix systems, although larger, are symmetric.

1 Introduction

Quasi-brittle materials generally exhibit a load-carrying capacity which decreases for increasing values of the strain or they may show snap-back. This behaviour is accompanied by a change in the nature of the equations of motion which loose their hyperbolicity. Simultaneously, the strain field tends to localize into a volume as small as possible: limitations in real materials are imposed by the microstructure. In finite element simulations, quasi-static problems involving no internal lengths will display localized zones spuriously bounded by the element size. This pathological dependency with respect to the mesh can be alleviated if an internal length, a macroscopic representative of the microstructure, is introduced in the constitutive equations. Then energy can dissipate in a finite volume. In a dynamic context, alternatives are available, as a Duvaut-Lions viscoplasticity produces in the field equations a length as product of an elastic wave-speed times the relaxation time, Loret and Prévost [1990].

This qualitative description applies to both damaging and plastic materials. On the practical side, many problems remain open. For example, the choice of the variable(s) which should contain/introduce the length-scale is by no means obvious. Also, there exist several ways of involving the length-scale, e.g. second order gradient approach, integral approach ...

In Loret, Simões and Martins [1999],[2000], the presence of a (plastic) internal length is shown to allow the mechanical information to travel some distance, precluding various instabilities. The analysis is performed for fluid-saturated elastic-plastic materials and delineates the influence of the three lengths present in the equations: one is attached to the motion, the second to the diffusion process and the third to the plastic behaviour. Clearly, a length-scale emanating from a diffusive property is unlikely to be able to prevent plastic instabilities. That is, a regularizing operator should be applied to the proper variable to be effective in restoring hyperbolicity: since for solids and fluid-saturated materials the source of the instability resides in the mechanical constitutive equations, the regularization should address these equations.

Here, we address computational aspects of elastic-damaging solids. Different formulations (gradient and integral models) are embodied in a unified framework. The averaging, or non-local operators, which introduce the length-scale into the constitutive equations are viewed as constraints and brought into the energy through a Lagrange multiplier. The resulting equations to be solved follow from the stationarity of the augmented energy. Finite element simulations of academic structures are presented to highlight the pros and cons of the usual methods (e.g. implicit gradient) versus their lagrangian approach. The latter provides matrix systems which, although larger, are symmetric.

Notation: Compact or index tensorial notation will be used throughout. Tensor quantities are identified by boldface letters. **I** is the second order identity tensor (Kronecker delta). Symbols '$\cdot$' and ':' between tensors of various orders denote their inner product with single and double contraction respectively. Time derivative of any time-dependent quantity a is denoted either by a superimposed dot, namely $\dot{a}$.

2 Local Elastic Damage

2.1 Experimental Observations on Rocks

Damage is defined as the progressive degradation of elastic properties as displayed by Fig. 1. The experiment has been performed on a cylindrical sample a 38mm in diameter and with an aspect ratio of 2. The material is initially anisotropic due to its mode of deposition. The behaviour is not completely reversible in the axial strain, but there is a strong ratchetting effect in the lateral strain. In fact, the lateral moduli are seen clearly to decrease during cycles, as consequence of the formation of macroscopic cracks essentially aligned with the vertical (loading) direction. The decrease would be still more important if the loading were aligned with the bedding plane: this situation would ease the opening of initial imperfections between deposition layers.

2.2 Basic Entities for Local Isotropic Damage

The constitutive equations for isotropic elastic damage involves four basic ideas.

(L1): First, a scalar measure of the strain undergone by the material called *equivalent strain* is defined. Among the most common propositions is the one due to Mazars and Pijaudier-Cabot [1989],

$$\epsilon_{\mathrm{eq}}(\boldsymbol{\epsilon}) = [\, \sum_{i\in[1,\mathrm{n}]} \langle \epsilon_i \rangle^2 \,]^{\frac{1}{2}} \,, \tag{2.1}$$

where the ϵ_i's are the eigenstrains, n the dimension of the problem, and the operator $\langle \cdot \rangle$ denotes the positive part of its argument. Another proposal by Peerlings and

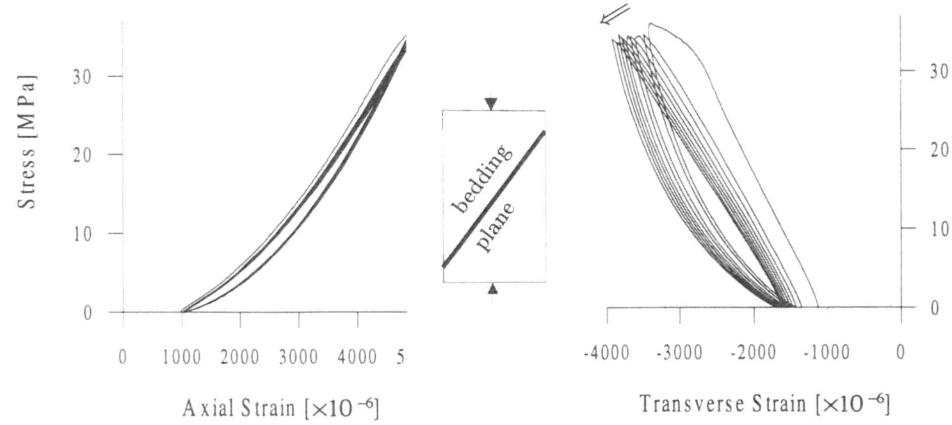

(a) The axial strain-axial stress shows a very small hysteresis loop.

(b) The lateral strain and consequently the lateral elastic modulus decrease during the cycles.

Figure 1 Experimental stress-strain data on sandstone by N. Gatelier [2000]. The behaviour is transversely isotropic due to both the deposition mode of the material and to the loading: cracks form more easily parallel to the bedding planes. The compression direction is vertical while the bedding direction makes an angle equal to 30° with the vertical. The decrease of lateral elastic moduli is stronger than that of the axial moduli.

de Borst [1998] is aimed at introducing the effect of confinement on the behaviour of polymers: it involves two material parameters, namely Poisson's ratio ν and $r = \sigma_{fc}/\sigma_{ft}$ the ratio between the compressive and tensile strengths (if $r \to \infty$, no failure occurs in compression),

$$\epsilon_{eq}(\epsilon) = \frac{r-1}{2r(1-2\nu)} I_1 + \Big[\Big[\frac{r-1}{2r(1-2\nu)} I_1\Big]^2 + \frac{1}{2r} \frac{1}{(1+\nu)^2} J_2 \Big]^{\frac{1}{2}}, \tag{2.2}$$

where the isotropic invariants of the strain I_1 and J_2 are defined as

$$I_1 = \operatorname{tr} \epsilon = \sum_{i \in [1,n]} \epsilon_i\,, \quad J_2 = 3 \operatorname{tr} \epsilon^2 - \operatorname{tr}^2 \epsilon = \sum_{i,j \subset [1,n]} (\epsilon_i - \epsilon_j)^2\,. \tag{2.3}$$

(L2): The second ingredient consists in using the maximum value up to the present time of the equivalent strain, say κ, to define a damage function $F(\epsilon_{eq}, \kappa)$,

$$F(\epsilon_{eq}, \kappa) = \epsilon_{eq} - \kappa \begin{cases} \leq 0 & : \text{no damage evolution,} \\ = 0 & : \text{damage evolves.} \end{cases} \tag{2.4}$$

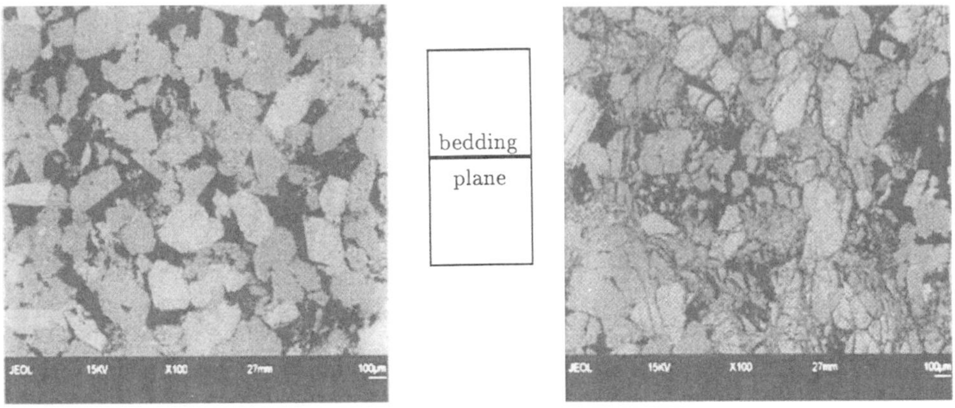

(a) Microstructure before mechanical loading.

(b) Microstructure after mechanical loading.

Figure 2 Scanning Electron Microscope view of the microstructure of Vosges sandstone, (a) before, (b) after mechanical loading (courtesy N. Gatelier). The compression direction is vertical and the bedding plane horizontal.

(L3): Still, the entity *damage* has to be defined: the idea is to view damage as a scalar D that varies from 0 for a sound material to 1 for a completely damaged material. The yield κ serves as an internal measure that controls D, for example Pijaudier-Cabot and Bažant [1987],

$$D(\kappa) = \begin{cases} 0 & \text{if } \kappa \leq \kappa_0, \\ 1 - \dfrac{\kappa_0}{\kappa}[1 - \alpha + \alpha \exp(-\beta(\kappa - \kappa_0))] & \text{otherwise,} \end{cases} \tag{2.5}$$

which involves three nondimensional material constants κ_0, α and β. Another proposal by Geers [1997] defines total damage for a finite value of maximum equivalent strain and involves two parameters κ_0 and κ_c,

$$D(\kappa) = \begin{cases} 0 & \text{if } \kappa \leq \kappa_0, \\ 1 - \left(\dfrac{\kappa_0}{\kappa}\right)^{\alpha} \left(\dfrac{\kappa_c - \kappa}{\kappa_c - \kappa_0}\right)^{\beta} & \text{if } \kappa_0 \leq \kappa \leq \kappa_c, \\ 1 & \text{otherwise.} \end{cases} \tag{2.6}$$

(L4): The fourth ingredient consists in interpreting damage as the entity that decreases the sound elastic stiffness $\mathbb{E}$, defined by Young's modulus E and Poisson's ratio ν (no direct connection exists at this point with possible changes of the microstructure). In the single scalar damage formulation, only Young's modulus is affected, a not really physical assumption, and, for the so-called equivalent strain model, damage affects the sound elastic stiffness linearly,

$$\boldsymbol{\sigma} = (1 - D)\,\mathbb{E} : \boldsymbol{\epsilon}\,. \tag{2.7}$$

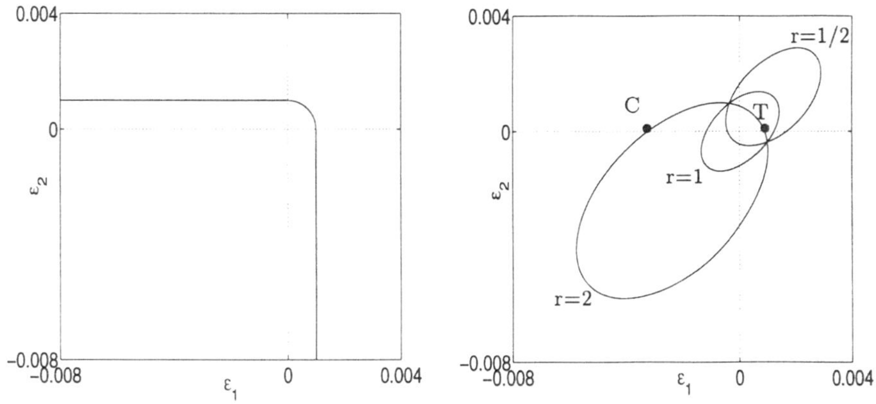

(a) model (2.1) (b) model (2.2) for $\nu = 0.2$

Figure 3 Contours of equivalent strain $\epsilon_{eq} = 0.1\%$ in plane strain; (a) model (2.1) for a material that does not undergo damage in compression; (b) model (2.2) for $\nu = 0.2$ and different r for a material more (resp. less) resistant in compression than in traction for $r > 1$ (resp. $r < 1$), but damageable for both loadings. The ratio between the uniaxial damage strains in traction and compression for model (2.2), points T and C shown here for $r = 1$, is R_t/R_c, with R_t and R_c given in the legend of next figure.

For the local damage models, a boundary value problem can now be formulated. Let us consider a body Ω, subject to body forces $\mathbf{b}$ in Ω, to prescribed displacements $\mathbf{u}^d$ on a non vanishing part Γ_u of the boundary $\partial\Omega$ and to tractions $\mathbf{p}^d$ on the complementary part Γ_p of the boundary The boundary value problem can be stated as follows:

Proposition 2.2.1 *Local BVP: Strong form.*
Solve for the displacement field $\mathbf{u}(\mathbf{x})$ *that satisfies the field equations,*

$$\begin{aligned} \operatorname{div} \boldsymbol{\sigma} + \mathbf{b} = \mathbf{0} &\quad \text{in } \Omega\,, \\ \boldsymbol{\sigma} \cdot \mathbf{n} = \mathbf{p}^d &\quad \text{on } \Gamma_p\,, \\ \mathbf{u} = \mathbf{u}^d &\quad \text{on } \Gamma_u\,, \end{aligned} \tag{2.8}$$

where the stress tensor $\boldsymbol{\sigma}$ *depends on the displacement through the local constitutive behaviour (2.7) where* $D = D(\kappa(\epsilon_{eq}))$.

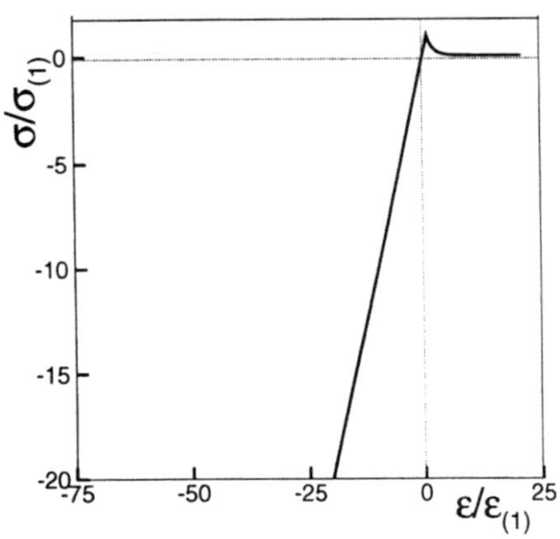

(a) Uniaxial response of model (2.1)

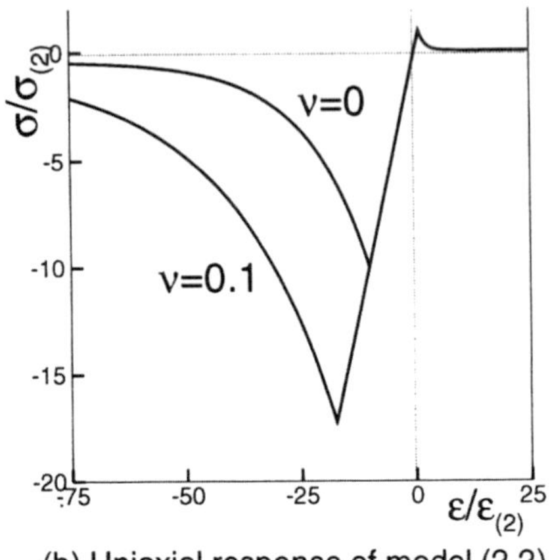

(b) Uniaxial response of model (2.2)

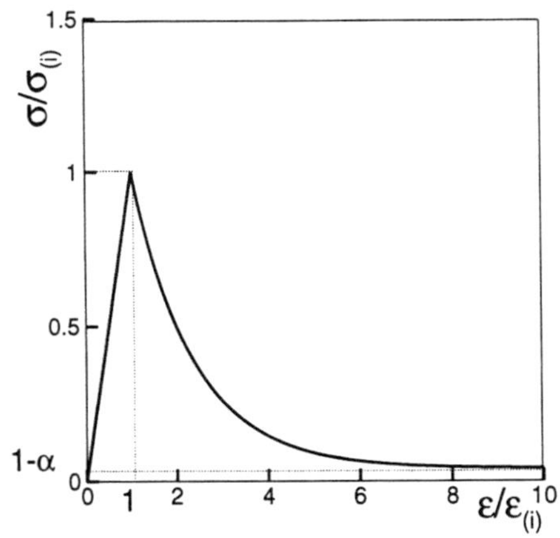

(c) Tensile uniaxial response of both models

Figure 4 Normalized uniaxial stress-strain relation using (2.2),(2.5),(2.7), with $r = 10$, $\alpha = 0.96, \beta = 350$. For both models, damage initiates at peak stress, that is there is no smooth transition between a positive and a negative slope $d\sigma/d\epsilon$. Let $\sigma_0 \equiv E\,\kappa_0$ be a normalizing stress. For model (2.1), tensile damage initiates at $\epsilon_{(1)} = \kappa_0$ and $\sigma_{(1)} = E\,\kappa_0$. For model (2.2), both these values are divided by $R_t = r_1 + (r_1^2 + r_2)^{1/2}$ with $r_1 = (r-1)/(2\,r\,(1-2\,\nu))$ and $r_2 = 1/(r\,(1+\nu)^2)$, that is $\epsilon_{(2)} = \kappa_0/R_t$ and $\sigma_{(2)} = E\,\kappa_0/R_t$. The asymptotic value of the tensile stress at large strains is, in both cases $i = 1, 2$, $\sigma/\sigma_{(i)} = 1-\alpha$. For model (2.1), there is no compressive damage while, for model (2.2), the compressive damage strain is $\epsilon_{(2)} = \kappa_0/R_c$ and the compressive damage stress $\sigma_{(2)} = E\,\kappa_0/R_c$, with $R_c = r_1 - (r_1^2 + r_2)^{1/2} < 0$. The asymptotic value of the compressive stress at large strains is again $\sigma/\sigma_{(2)} = 1 - \alpha$.

3 Non-Local Elastic Damage

3.1 Introduction of a Length-Scale

When the field equations $(2.8)_1$ loose hyperbolicity, which is quite likely when the stress-carrying capacity decreases, introduction of one or several internal length(s) may restore well-posedness. The usual approach delineates three steps consisting in

(NL1): choosing a *local* variable, of scalar or tensorial nature, say $\xi = \xi(\mathbf{x})$. For isotropic damage, the field of choice is naturally the equivalent strain, but in other situations, like anisotropic plastic damage, the possibilities are many;

(NL2): defining its *averaged* or *non-local* approximation $\bar{\xi} = \bar{\xi}(\mathbf{x})$, through the smoothing operator $\mathcal{N}$ with support or *control volume* $V(\mathbf{x})$ around $\mathbf{x}$ has a size that provides a length-scale L,

$$\bar{\xi}(\mathbf{x}) = \mathcal{N}(\xi)(\mathbf{x}); \tag{3.1}$$

(NL3): replacing the local variable $\xi(\mathbf{x})$ by its non-local approximation $\bar{\xi}(\mathbf{x})$ in the field and constitutive equations.

Practically, two types of non-local operators are used. They are either of the *integral type*,

$$\underline{\text{integral}} \text{ operator } \mathcal{N}_{\text{int}} \;:\; \xi(\mathbf{x}) \longmapsto \bar{\xi}(\mathbf{x}) = \begin{cases} \dfrac{1}{V_\psi} \displaystyle\int_{V(\mathbf{x})} \psi(\mathbf{x}-\mathbf{s})\,\xi(\mathbf{s})\,dV(\mathbf{s}) = \\ \dfrac{1}{V_\psi} \displaystyle\int_{V(\mathbf{0})} \psi(-\mathbf{s})\,\xi(\mathbf{x}+\mathbf{s})\,dV(\mathbf{s}), \end{cases} \tag{3.2}$$

with $V_\psi = \int_{V(\mathbf{x})} \psi(\mathbf{x}-\mathbf{s})\,dV(\mathbf{s}) = \int_{V(\mathbf{0})} \psi(-\mathbf{s})\,dV(\mathbf{s})$ in order to leave invariant constant fields. The function averaging or weighting ψ is aimed at smoothing out the irregularities of the local field. It is usually isotropic in space, and so $\psi(-\mathbf{s}) = \psi(\mathbf{s})$, and bell-shaped as it has to put more weight on the points close to the control point $\mathbf{x}$. A typical candidate for the weight function is the normalized Gauss function

$$\psi(\mathbf{y}) = \frac{1}{(2\,\pi)^{\text{n}/2}} \exp\Big(-\frac{\|\,\mathbf{y}\,\|^2}{2\,L_{\text{int}}^2}\Big), \tag{3.3}$$

where $L = L_{\text{int}}$ is a length of interaction between material points. Assuming for simplicity $V(\mathbf{x}) = \mathbb{R}^{\text{n}}$, we have $V_\psi = (\sqrt{2}\,L_{\text{int}})^{\text{n}}$. In actual computations, the control volume $V(\mathbf{x})$ around $\mathbf{x}$ may be finite to save on computation time.

The second type of non-local operator consists in approximating the field $\xi(\mathbf{x})$ through an even order Taylor expansion: if one views this approach as an expansion of the integral approach where the weight ψ is isotropic, then odd order derivatives should vanish. In fact, the first formulations of the *explicit type*, had the drawback of differentiating the local field $\xi(\mathbf{x})$ which may be irregular. The simplest second order gradient operator, with $\Delta = \sum_{i\in[1,\text{n}]} \partial^2/\partial x_i^2$ the laplacian, involves a parameter b with the dimension length^2, namely e.g. Mühlhaus and Aifantis [1991],

$$\underline{\text{explicit}}\ \underline{\text{gradient}}\ \text{operator}\ \mathcal{N}_{\text{ge}}\ :\ \xi(\mathbf{x}) \longmapsto \begin{cases} \bar{\xi}(\mathbf{x}) = \xi(\mathbf{x}) + b\,\Delta\xi(\mathbf{x}) & \text{in } \Omega\,, \\ \nabla\xi\cdot\mathbf{n} = 0 & \text{on } \partial\Omega\,. \end{cases} \tag{3.4}$$

An expansion of the Gaussian above yields $b = L_{\text{int}}^2/4$ for one-dimensional problems $\mathrm{n} = 1$. To avoid differentiating the local field, a modification of the scheme referred to as *implicit second gradient* approximation has been proposed by Peerlings et al. [1996],

$$\underline{\text{implicit}}\ \underline{\text{gradient}}\ \text{operator}\ \mathcal{N}_{\text{gi}}\ :\ \xi(\mathbf{x}) \longmapsto \begin{cases} \bar{\xi}(\mathbf{x}) = \xi(\mathbf{x}) + b\,\Delta\bar{\xi}(\mathbf{x}) & \text{in } \Omega\,, \\ \nabla\bar{\xi}\cdot\mathbf{n} = 0 & \text{on } \partial\Omega\,. \end{cases} \tag{3.5}$$

To account for the higher order derivative introduced in $(3.4)_1$, $(3.5)_1$ an additional boundary condition has to be specified, usually as in $(3.4)_2$, $(3.5)_2$ with $\mathbf{n}$ unit normal to $\partial\Omega$ at $\mathbf{x}$. This Neumann condition is reminiscent of the condition used for elastic-plastic non-local materials by Mühlhaus and Aifantis [1991] for the plastic strain, namely $\nabla\epsilon^{\mathrm{pl}}(\mathbf{x})\cdot\mathbf{n} = 0$ at the boundary of the plastic zone. There is however no compelling physical justification, except of course its simplicity, see Jirásek [1999].

A comparison between explicit and implicit gradient formulations (for plasticity) by Pamin et al. [1999] shows that the amount of spurious stress oscillations is reduced by the implicit approach.

The boundary value problem for non-local elastic damaging materials becomes:

Proposition 3.1.1 *Non Local BVP: Strong form.*

Solve for the displacement field $\mathbf{u}(\mathbf{x})$ *<u>and</u> the non-local field* $\bar{\xi}(\mathbf{x})$ *that satisfy the field equations* [1],

$$\begin{aligned} \operatorname{div}\boldsymbol{\sigma} + \mathbf{b} = \mathbf{0} \quad & \text{in } \Omega\,, \\ \boldsymbol{\sigma}\cdot\mathbf{n} = \mathbf{p}^d \quad & \text{on } \Gamma_{\mathrm{p}}\,, \\ \mathbf{u} = \mathbf{u}^d \quad & \text{on } \Gamma_{\mathrm{u}}\,, \\ \bar{\xi}(\mathbf{x}) = \mathcal{N}(\xi(\mathbf{x})) \quad & \text{in } \Omega\,, \\ [\nabla\bar{\xi}\cdot\mathbf{n} = 0 \quad & \text{on } \partial\Omega]\,, \end{aligned} \tag{3.6}$$

where the stress tensor $\boldsymbol{\sigma}$ *depends, inter alia, on the displacement through the non-local constitutive behaviour (2.7) where* $D = D(\kappa(\bar{\epsilon}_{\text{eq}}))$ *if, as it is typical, the field which is made non-local is the equivalent strain,*

$$\xi(\mathbf{x}) = \epsilon_{\text{eq}}(\mathbf{x}), \quad \forall\mathbf{x}\in\Omega\,, \tag{3.7}$$

and $\epsilon_{\text{eq}} = \epsilon_{\text{eq}}(\mathbf{u})$ *through a relation of the form (2.1) or (2.2).*

---

[1] The boundary condition $(3.6)_5$ applies for the implicit gradient operator only. For explicit gradient operator, change $(3.6)_5$ to $\nabla\xi\cdot\mathbf{n} = 0$.

4 A Lagrange Approach to Non-Local Elastic Damage

4.1 A Lagrange Formulation for the Implicit Gradient Model

In a Lagrange approach, the non-local operator $\mathcal{N}$ is viewed as a constraint, which is enforced in a weak form. In fact, since the elastic tensor $\mathbb{E}$ displays the major symmetry, the weak form of (3.6) is equivalent to find, for kinematically admissible displacement, the minimum of the functional

$$\mathcal{F}(\mathbf{u},\bar{\epsilon}_{\mathrm{eq}}) = \int_\Omega \frac{1}{2}(1-D(\bar{\epsilon}_{\mathrm{eq}}))\,\boldsymbol{\epsilon} : \mathbb{E} : \boldsymbol{\epsilon}\, d\Omega - \int_\Omega \mathbf{u}\cdot\mathbf{b}\, d\Omega - \int_{\Gamma_p} \mathbf{u}\cdot\mathbf{p}^d\, d\Gamma\,, \tag{4.1}$$

subject to the pointwise constraint,

$$\mathcal{G}_{\mathrm{gi}}(\mathbf{u},\bar{\epsilon}_{\mathrm{eq}})(\mathbf{x}) = \bar{\epsilon}_{\mathrm{eq}}(\mathbf{x}) - \epsilon_{\mathrm{eq}}(\mathbf{x}) - b\,\Delta\bar{\epsilon}_{\mathrm{eq}}(\mathbf{x}) = 0\,,\ \forall \mathbf{x}\in\Omega, \quad \nabla\bar{\epsilon}_{\mathrm{eq}}\cdot\mathbf{n} = 0,\ \forall\mathbf{x}\in\partial\Omega\,. \tag{4.2}$$

This problem is replaced by the search of the stationary points of the Lagrangian

$$\mathcal{L}_{\mathrm{gi}}(\mathbf{u},\bar{\epsilon}_{\mathrm{eq}},\lambda) = \mathcal{F}(\mathbf{u},\bar{\epsilon}_{\mathrm{eq}}) + \int_\Omega \lambda\, \mathcal{G}_{\mathrm{gi}}(\mathbf{u},\bar{\epsilon}_{\mathrm{eq}})\, d\Omega\,, \tag{4.3}$$

where λ denotes the Lagrange multiplier. Integrating $\mathcal{G}(\mathbf{u},\bar{\epsilon}_{\mathrm{eq}})(\mathbf{x})$ by part using the Green-Gauss formula and the boundary condition $(4.2)_2$, the Lagrangian becomes:

Proposition 4.1.1 *BVP: Lagrange multiplier approach for the implicit gradient model.*
Find the displacement field $\mathbf{u}(\mathbf{x})$ *satisfying* $(3.6)_3$*, the non-local field* $\bar{\epsilon}_{\mathrm{eq}}(\mathbf{x})$ *and the lagrangian field* $\lambda(\mathbf{x})$ *that realize the stationary points of the functional*

$$\begin{aligned}\mathcal{L}_{\mathrm{gi}}(\mathbf{u},\bar{\epsilon}_{\mathrm{eq}},\lambda) = &\int_\Omega \frac{1}{2}(1-D(\bar{\epsilon}_{\mathrm{eq}}))\,\boldsymbol{\epsilon} : \mathbb{E} : \boldsymbol{\epsilon}\, d\Omega - \int_\Omega \mathbf{u}\cdot\mathbf{b}\, d\Omega - \int_{\Gamma_p} \mathbf{u}\cdot\mathbf{p}^d\, d\Gamma \\ &+ \int_\Omega \lambda\,(\bar{\epsilon}_{\mathrm{eq}} - \epsilon_{\mathrm{eq}}) + b\,\nabla\lambda\cdot\nabla\bar{\epsilon}_{\mathrm{eq}}\, d\Omega\,.\end{aligned} \tag{4.4}$$

The finite element discretization of the three unknown fields $(\mathbf{u},\bar{\epsilon}_{\mathrm{eq}},\lambda)$ may use different interpolation functions for the displacement, equivalent strain and lagrangian multiplier, formally $(\mathbf{U},\overline{\mathbf{E}}_\epsilon,\boldsymbol{\Lambda})$ denoting the respective nodal vectors:

$$\mathbf{u}(\mathbf{x}) = \mathbf{N}_{\mathrm{u}}(\mathbf{x})\,\mathbf{U}\,, \qquad \bar{\epsilon}_{\mathrm{eq}}(\mathbf{x}) = \mathbf{N}_\epsilon(\mathbf{x})\,\overline{\mathbf{E}}_\epsilon\,, \qquad \lambda(\mathbf{x}) = \mathbf{N}_\lambda(\mathbf{x})\,\boldsymbol{\Lambda}\,, \tag{4.5}$$

with associated (symmetrized) gradients,

$$\boldsymbol{\epsilon}(\mathbf{x}) = \mathbf{B}_{\mathrm{u}}(\mathbf{x})\,\mathbf{U}\,, \quad \nabla\bar{\epsilon}_{\mathrm{eq}}(\mathbf{x}) = \mathbf{B}_\epsilon(\mathbf{x})\,\overline{\mathbf{E}}_\epsilon\,, \quad \nabla\lambda(\mathbf{x}) = \mathbf{B}_\lambda(\mathbf{x})\,\boldsymbol{\Lambda}\,. \tag{4.6}$$

The discretized functional becomes (the element dependance is not displayed)

$$\begin{aligned}\mathbf{L}_{\mathrm{gi}}(\mathbf{U},\overline{\mathbf{E}}_\epsilon,\boldsymbol{\Lambda}) = &\int_\Omega \frac{1}{2}(1-D)\,\mathbf{U}^T\,\mathbf{B}_{\mathrm{u}}^T\,\mathbb{E}\,\mathbf{B}_{\mathrm{u}}\,\mathbf{U} - \mathbf{U}^T\mathbf{N}_{\mathrm{u}}^T\,\mathbf{b}\, d\Omega - \int_{\Gamma_p} \mathbf{U}^T\mathbf{N}_{\mathrm{u}}^T\,\mathbf{p}^d\, d\Gamma \\ &+ \int_\Omega \overline{\mathbf{E}}_\epsilon^T\,(\mathbf{N}_\epsilon^T\,\mathbf{N}_\lambda + \mathbf{B}_\epsilon^T\,b\,\mathbf{B}_\lambda)\,\boldsymbol{\Lambda} - (\mathbf{N}_\lambda\,\boldsymbol{\Lambda})\,\epsilon_{\mathrm{eq}}\, d\Omega\,,\end{aligned} \tag{4.7}$$

where $D{=}D(\bar\epsilon_{\mathrm{eq}})$, $\bar\epsilon_{\mathrm{eq}} = \mathbf{N}_\epsilon(\mathbf{x})\,\overline{\mathbf{E}}_\epsilon$ and $\epsilon_{\mathrm{eq}}{=}\epsilon_{\mathrm{eq}}(\mathbf{B}_{\mathrm{u}}(\mathbf{x})\,\mathbf{U})$. Requiring its stationarity provides three sets of non-linear equations (the same notations are used for the element unknowns and the global unknowns):

$$\begin{aligned}
\mathbf{R}_u &= \frac{\partial \mathbf{L}_{\mathrm{gi}}}{\partial \mathbf{U}} = \int_\Omega (1-D)\,\mathbf{B}_{\mathrm{u}}^T \mathbb{E}\,\mathbf{B}_{\mathrm{u}}\,\mathbf{U} - (\mathbf{N}_\lambda\,\boldsymbol{\Lambda})\,\frac{\partial \epsilon_{\mathrm{eq}}}{\partial \mathbf{U}} - \mathbf{N}_{\mathrm{u}}^T\,\mathbf{b}\,d\Omega - \int_{\Gamma_p} \mathbf{N}_{\mathrm{u}}^T\,\mathbf{p}^d\,d\Gamma = \mathbf{0}\,,\\
\mathbf{R}_\epsilon &= \frac{\partial \mathbf{L}_{\mathrm{gi}}}{\partial \overline{\mathbf{E}}_\epsilon} = \int_\Omega -\frac{1}{2}\frac{\partial D}{\partial \overline{\mathbf{E}}_\epsilon}\,\mathbf{U}^T\mathbf{B}_{\mathrm{u}}^T\mathbb{E}\,\mathbf{B}_{\mathrm{u}}\,\mathbf{U} + (\mathbf{N}_\epsilon^T\,\mathbf{N}_\lambda + \mathbf{B}_\epsilon^T\,b\,\mathbf{B}_\lambda)\,\boldsymbol{\Lambda}\,d\Omega = \mathbf{0}\,,\\
\mathbf{R}_\lambda &= \frac{\partial \mathbf{L}_{\mathrm{gi}}}{\partial \boldsymbol{\Lambda}} = \int_\Omega \overline{\mathbf{E}}_\epsilon^T\,(\mathbf{N}_\epsilon^T\,\mathbf{N}_\lambda + \mathbf{B}_\epsilon^T\,b\,\mathbf{B}_\lambda) - \epsilon_{\mathrm{eq}}\,\mathbf{N}_\lambda\,d\Omega = \mathbf{0}\,.
\end{aligned} \tag{4.8}$$

This non-linear system is solved iteratively through Newton iterations (or more refined techniques), namely by expanding at iteration $i+1$ the residuals

$$\mathbf{R}_a^{i+1} \simeq \mathbf{R}_a^i + \left(\frac{\partial \mathbf{R}_a}{\partial \mathbf{U}}\right)^i \delta\mathbf{U}^{i+1} + \left(\frac{\partial \mathbf{R}_a}{\partial \overline{\mathbf{E}}_\epsilon}\right)^i \delta\overline{\mathbf{E}}_\epsilon^{i+1} + \left(\frac{\partial \mathbf{R}_a}{\partial \boldsymbol{\Lambda}}\right)^i \delta\boldsymbol{\Lambda}^{i+1} = \mathbf{0}\,, \quad a = u, \epsilon, \lambda\,, \tag{4.9}$$

and solving for the variation $\delta\mathbf{X}^{i+1}$ of the global unknown vector $\mathbf{X} = (\mathbf{U}, \overline{\mathbf{E}}_\epsilon, \boldsymbol{\Lambda})$. For the full Newton method without line search, $\mathbf{X}^{i+1}{=}\mathbf{X}^i{+}\delta\,\mathbf{X}^{i+1}$. In matrix form, the augmented Lagrange system

$$\mathbf{K}^i\,\delta\,\mathbf{X}^{i+1} = -\mathbf{R}^i \iff \begin{bmatrix} \mathbf{K}_{uu} & \mathbf{K}_{u\epsilon} & \mathbf{K}_{u\lambda} \\ \mathbf{K}_{\epsilon u} & \mathbf{K}_{\epsilon\epsilon} & \mathbf{K}_{\epsilon\lambda} \\ \mathbf{K}_{\lambda u} & \mathbf{K}_{\lambda\epsilon} & \mathbf{0} \end{bmatrix}^i \delta \begin{bmatrix} \mathbf{U} \\ \overline{\mathbf{E}}_\epsilon \\ \boldsymbol{\Lambda} \end{bmatrix}^{i+1} = - \begin{bmatrix} \mathbf{R}_u \\ \mathbf{R}_\epsilon \\ \mathbf{R}_\lambda \end{bmatrix}^i \tag{4.10}$$

has the following block components,

$$\begin{aligned}
\mathbf{K}_{uu} &= \int_\Omega (1-D)\,\mathbf{B}_{\mathrm{u}}^T\mathbb{E}\,\mathbf{B}_{\mathrm{u}} - \lambda\,\frac{\partial^2 \epsilon_{\mathrm{eq}}}{\partial \mathbf{U}^2}\,d\Omega,\\
\mathbf{K}_{u\epsilon} &= \mathbf{K}_{\epsilon u}^T = \int_\Omega -(1-D)^{-1}\,\frac{\partial D}{\partial \bar\epsilon_{\mathrm{eq}}}(\mathbf{B}_{\mathrm{u}}^T\boldsymbol{\sigma})\,\mathbf{N}_\epsilon\,d\Omega,\\
\mathbf{K}_{u\lambda} &= \mathbf{K}_{\lambda u}^T = \int_\Omega -\frac{\partial \epsilon_{\mathrm{eq}}}{\partial \mathbf{U}}\,\mathbf{N}_\lambda\,d\Omega,\\
\mathbf{K}_{\epsilon\epsilon} &= \int_\Omega -\frac{1}{2}(1-D)^{-1}\,\frac{\partial^2 D}{\partial \bar\epsilon_{\mathrm{eq}}^2}(\mathbf{U}^T\mathbf{B}_{\mathrm{u}}^T\,\boldsymbol{\sigma})\,\mathbf{N}_\epsilon^T\,\mathbf{N}_\epsilon\,d\Omega,\\
\mathbf{K}_{\epsilon\lambda} &= \mathbf{K}_{\lambda\epsilon}^T = \int_\Omega \mathbf{N}_\epsilon^T\,\mathbf{N}_\lambda + \mathbf{B}_\epsilon^T\,b\,\mathbf{B}_\lambda\,d\Omega,
\end{aligned} \tag{4.11}$$

and residuals,

$$\begin{aligned}
\mathbf{R}_u &= \int_\Omega \mathbf{B}_\mathrm{u}^T\,\boldsymbol{\sigma} - \lambda\,\frac{\partial \epsilon_\mathrm{eq}}{\partial \mathbf{U}} - \mathbf{N}_\mathrm{u}^T\,\mathbf{b}\;d\Omega - \int_{\Gamma_p} \mathbf{N}_\mathrm{u}^T\,\mathbf{p}^d\;d\Gamma,\\
\mathbf{R}_\epsilon &= \int_\Omega -\frac{1}{2}(1-D)^{-1}\,\frac{\partial D}{\partial \overline{\epsilon}_\mathrm{eq}}(\mathbf{U}^T\mathbf{B}_\mathrm{u}^T\,\boldsymbol{\sigma})\,\mathbf{N}_\epsilon\;d\Omega + \mathbf{K}_{\epsilon\lambda}\,\boldsymbol{\Lambda},\\
\mathbf{R}_\lambda &= \mathbf{K}_{\lambda\epsilon}\,\overline{\mathbf{E}}_\epsilon - \int_\Omega \epsilon_\mathrm{eq}\,\mathbf{N}_\lambda\;d\Omega.
\end{aligned} \tag{4.12}$$

The same set of equations can be obtained from the virtual work principle and assuming identical interpolation functions for the variations and unknowns.

While three unknown fields have to be computed here in contrast to the two-field method, see Section 5.1 below, the **matrix system** (4.10) is **symmetric**.

4.2 A Lagrange Formulation for the Integral Model

The Lagrange approach applies as well for the integral model. The modification with respect to the preceding scheme lies in the constraint associated to the non-local operator which is now,

$$\mathcal{G}_\mathrm{int}(\mathbf{u},\overline{\epsilon}_\mathrm{eq})(\mathbf{x}) = \overline{\epsilon}_\mathrm{eq}(\mathbf{x}) - \frac{1}{V_\psi}\int_{V(\mathbf{x})} \psi(\mathbf{x}-\mathbf{s})\,\epsilon_\mathrm{eq}(\mathbf{s})\,dV(\mathbf{s}), \quad \forall \mathbf{x}\in\Omega\,. \tag{4.13}$$

The Lagrangian problem

$$\mathcal{L}_\mathrm{int}(\mathbf{u},\overline{\epsilon}_\mathrm{eq},\lambda) = \mathcal{F}(\mathbf{u},\overline{\epsilon}_\mathrm{eq}) + \int_\Omega \lambda\;\mathcal{G}_\mathrm{int}(\mathbf{u},\overline{\epsilon}_\mathrm{eq})\,d\Omega\,, \tag{4.14}$$

can be stated in the following way:

Proposition 4.2.1 *Non Local BVP: Lagrange multiplier approach for the integral model.*

Find the displacement field $\mathbf{u}(\mathbf{x})$ *satisfying* $(3.6)_3$, *the non-local field* $\overline{\epsilon}_\mathrm{eq}(\mathbf{x})$ *and the lagrangian field* $\lambda(\mathbf{x})$ *that realize the stationary points of the functional*

$$\begin{aligned}
\mathcal{L}_\mathrm{int}(\mathbf{u},\overline{\epsilon}_\mathrm{eq},\lambda) = &\int_\Omega \frac{1}{2}(1-D(\overline{\epsilon}_\mathrm{eq}))\,\boldsymbol{\epsilon}:\mathbb{E}:\boldsymbol{\epsilon}\,d\Omega - \int_\Omega \mathbf{u}\cdot\mathbf{b}\,d\Omega - \int_{\Gamma_p}\mathbf{u}\cdot\mathbf{p}^d\,d\Gamma\\
&+\int_\Omega \Big(\overline{\epsilon}_\mathrm{eq}(\mathbf{x}) - \mathcal{N}_\mathrm{int}(\epsilon_\mathrm{eq})(\mathbf{x})\Big)\,\lambda\;d\Omega(\mathbf{x})\,.
\end{aligned} \tag{4.15}$$

Note that no boundary condition is required, unlike for the gradient models. The discretized functional becomes (the element dependance is not displayed),

$$\begin{aligned}
\mathbf{L}_\mathrm{int}(\mathbf{U},\overline{\mathbf{E}}_\epsilon,\boldsymbol{\Lambda}) = &\int_\Omega \frac{1}{2}(1-D)\,\mathbf{U}^T\,\mathbf{B}_\mathrm{u}^T\,\mathbb{E}\,\mathbf{B}_\mathrm{u}\,\mathbf{U} - \mathbf{U}^T\mathbf{N}_\mathrm{u}^T\,\mathbf{b}\;d\Omega - \int_{\Gamma_p}\mathbf{U}^T\mathbf{N}_\mathrm{u}^T\,\mathbf{p}^d\;d\Gamma\\
&+\int_\Omega \Big(\overline{\mathbf{E}}_\epsilon^T\mathbf{N}_\epsilon^T - \mathcal{N}_\mathrm{int}(\epsilon_\mathrm{eq})\Big)(\mathbf{N}_\lambda\,\boldsymbol{\Lambda})\;d\Omega\,,
\end{aligned} \tag{4.16}$$

where $D{=}D(\bar{\epsilon}_{eq})$ and $\epsilon_{eq}{=}\epsilon_{eq}(\mathbf{B}_u(\mathbf{s})\,\mathbf{U})$. Its stationarity provides three sets of non-linear equations (the same notations are used for the element unknowns and the global unknowns):

$$\mathbf{R}_u = \frac{\partial \mathbf{L}_{\rm int}}{\partial \mathbf{U}} = \int_\Omega (1-D)\,\mathbf{B}_{\rm u}^T \mathbb{E}\,\mathbf{B}_{\rm u}\,\mathbf{U} - (\mathbf{N}_\lambda\,\boldsymbol{\Lambda})\,\mathcal{N}_{\rm int}\Big(\frac{\partial \epsilon_{\rm eq}}{\partial \mathbf{U}}\Big) - \mathbf{N}_{\rm u}^T\,\mathbf{b}\,d\Omega - \int_{\Gamma_p} \mathbf{N}_{\rm u}^T\,\mathbf{p}^d\,d\Gamma = \mathbf{0},$$
$$\mathbf{R}_\epsilon = \frac{\partial \mathbf{L}_{\rm int}}{\partial \overline{\mathbf{E}}_\epsilon} = \int_\Omega -\frac{1}{2}\frac{\partial D}{\partial \overline{\mathbf{E}}_\epsilon}\,\mathbf{U}^T \mathbf{B}_{\rm u}^T \mathbb{E}\,\mathbf{B}_{\rm u}\,\mathbf{U} + \mathbf{N}_\epsilon^T\,(\mathbf{N}_\lambda\,\boldsymbol{\Lambda})\,d\Omega = \mathbf{0}\,,$$
$$\mathbf{R}_\lambda = \frac{\partial \mathbf{L}_{\rm int}}{\partial \boldsymbol{\Lambda}} = \int_\Omega \Big(\overline{\mathbf{E}}_\epsilon^T \mathbf{N}_\epsilon^T - \mathcal{N}_{\rm int}(\epsilon_{\rm eq})\Big)\,\mathbf{N}_\lambda\,d\Omega\,. \tag{4.17}$$

The Newton iterations to solve this non-linear problem involve the following block components,

$$\begin{aligned}
\mathbf{K}_{uu} &= \int_\Omega (1-D)\,\mathbf{B}_{\rm u}^T \mathbb{E}\,\mathbf{B}_{\rm u} - \lambda\,\mathcal{N}_{\rm int}\Big(\frac{\partial^2 \epsilon_{\rm eq}}{\partial \mathbf{U}^2}\Big)\,d\Omega,\\
\mathbf{K}_{u\epsilon} = \mathbf{K}_{\epsilon u}^T &= \int_\Omega -(1-D)^{-1}\,\frac{\partial D}{\partial \bar{\epsilon}_{\rm eq}}\,(\mathbf{B}_{\rm u}^T \boldsymbol{\sigma})\,\mathbf{N}_\epsilon d\Omega,\\
\mathbf{K}_{u\lambda} = \mathbf{K}_{\lambda u}^T &= \int_\Omega -\mathcal{N}_{\rm int}\Big(\frac{\partial \epsilon_{\rm eq}}{\partial \mathbf{U}}\Big)\,\mathbf{N}_\lambda\,d\Omega,\\
\mathbf{K}_{\epsilon\epsilon} &= \int_\Omega -\frac{1}{2}(1-D)^{-1}\,\frac{\partial^2 D}{\partial \bar{\epsilon}_{\rm eq}^2}\,(\mathbf{U}^T \mathbf{B}_{\rm u}^T\,\boldsymbol{\sigma})\,\mathbf{N}_\epsilon^T\,\mathbf{N}_\epsilon\,d\Omega,\\
\mathbf{K}_{\epsilon\lambda} = \mathbf{K}_{\lambda\epsilon}^T &= \int_\Omega \mathbf{N}_\epsilon^T\,\mathbf{N}_\lambda\,d\Omega,
\end{aligned} \tag{4.18}$$

and residuals,

$$\begin{aligned}
\mathbf{R}_u &= \int_\Omega \mathbf{B}_{\rm u}^T\,\boldsymbol{\sigma} - \lambda\,\mathcal{N}_{\rm int}\Big(\frac{\partial \epsilon_{\rm eq}}{\partial \mathbf{U}}\Big) - \mathbf{N}_{\rm u}^T\,\mathbf{b}\,d\Omega - \int_{\Gamma_p} \mathbf{N}_{\rm u}^T\,\mathbf{p}^d\,d\Gamma,\\
\mathbf{R}_\epsilon &= \int_\Omega -\frac{1}{2}(1-D)^{-1}\,\frac{\partial D}{\partial \bar{\epsilon}_{\rm eq}}\,(\mathbf{U}^T \mathbf{B}_{\rm u}^T\,\boldsymbol{\sigma})\,\mathbf{N}_\epsilon\,d\Omega + \mathbf{K}_{\epsilon\lambda}\,\boldsymbol{\Lambda},\\
\mathbf{R}_\lambda &= \mathbf{K}_{\lambda\epsilon}\,\overline{\mathbf{E}}_\epsilon - \int_\Omega \mathcal{N}_{\rm int}(\epsilon_{\rm eq})\,\mathbf{N}_\lambda\,d\Omega.
\end{aligned} \tag{4.19}$$

By construction, the Newton matrix is **symmetric**.

5 Computational Aspects of Non-Local Damage

In order to highlight the computational aspects of the Lagrange approach, we first return to the basic two-field approach.

5.1 The Two-Field Approach

The most straightforward idea to solve the non-local boundary value problem stated in Proposition 3.1.1 is to consider a two-field approach, namely displacements and equivalent strains, Peerlings et al. [1996] for the implicit gradient method. The weak form of the non-local boundary value problem results from the principle of virtual work. The first equation of the weak form can be obtained by multiplying $(3.6)_1$ by variations of the displacement that vanish on Γ_u and integration by part over Ω using the natural boundary condition $(3.6)_2$. A second equation is obtained by multiplying $(3.6)_4$ by variations of the equivalent strain, integration by part and, for the implicit gradient operator, use of the boundary condition $(3.6)_5$. Discretization of the unknowns and variations by a Galerkin method yields the following sets of non-linear equations,

$$\begin{aligned} \mathbf{R}_u &= \int_\Omega (1-D)\,\mathbf{B}_\mathrm{u}^T \mathbb{E}\,\mathbf{B}_\mathrm{u}\,\mathbf{U} - \mathbf{N}_\mathrm{u}^T\,\mathbf{b}\,d\Omega - \int_{\Gamma_p} \mathbf{N}_\mathrm{u}^T\,\mathbf{p}^d\,d\Gamma = \mathbf{0}\,, \\ \mathbf{R}_\epsilon &= \int_\Omega (\mathbf{N}_\epsilon^T\,\mathbf{N}_\epsilon + \mathbf{B}_\epsilon^T\,b\,\mathbf{B}_\epsilon)\,\overline{\mathbf{E}}_\epsilon - \epsilon_\mathrm{eq}\,\mathbf{N}_\epsilon^T\,d\Omega = \mathbf{0}\,. \end{aligned} \tag{5.1}$$

The full Newton method,

$$\mathbf{K}^i\,\delta\,\mathbf{X}^{i+1} = -\mathbf{R}^i \iff \begin{bmatrix} \mathbf{K}_{uu}\ \mathbf{K}_{u\epsilon} \\ \mathbf{K}_{\epsilon u}\ \mathbf{K}_{\epsilon\epsilon} \end{bmatrix}^i \delta \begin{bmatrix} \mathbf{U} \\ \overline{\mathbf{E}}_\epsilon \end{bmatrix}^{i+1} = - \begin{bmatrix} \mathbf{R}_u \\ \mathbf{R}_\epsilon \end{bmatrix}^i \tag{5.2}$$

has the following block components,

$$\begin{aligned} \mathbf{K}_{uu} &= \int_\Omega (1-D)\,\mathbf{B}_\mathrm{u}^T \mathbb{E}\,\mathbf{B}_\mathrm{u}\,d\Omega, \\ \mathbf{K}_{u\epsilon} &= \int_\Omega -(1-D)^{-1}\,\frac{\partial D}{\partial \overline{\epsilon}_\mathrm{eq}}\,(\mathbf{B}_\mathrm{u}^T\boldsymbol{\sigma})\,\mathbf{N}_\epsilon\,d\Omega, \\ \mathbf{K}_{\epsilon u} &= \int_\Omega -\mathbf{N}_\epsilon^T\,\frac{\partial \epsilon_\mathrm{eq}}{\partial \mathbf{U}}\,d\Omega, \\ \mathbf{K}_{\epsilon\epsilon} &= \int_\Omega \mathbf{N}_\epsilon^T\,\mathbf{N}_\epsilon + \mathbf{B}_\epsilon^T\,b\,\mathbf{B}_\epsilon\,d\Omega, \end{aligned} \tag{5.3}$$

and residuals,

$$\begin{aligned} \mathbf{R}_u &= \int_\Omega \mathbf{B}_\mathrm{u}^T\,\boldsymbol{\sigma} - \mathbf{N}_\mathrm{u}^T\,\mathbf{b}\,d\Omega - \int_{\Gamma_p} \mathbf{N}_\mathrm{u}^T\,\mathbf{p}^d\,d\Gamma, \\ \mathbf{R}_\epsilon &= \mathbf{K}_{\epsilon\epsilon}\,\overline{\mathbf{E}}_\epsilon - \int_\Omega \epsilon_\mathrm{eq}\,\mathbf{N}_\epsilon^T\,d\Omega. \end{aligned} \tag{5.4}$$

5.2 The Lagrange Multiplier Approach with a Penalty Method

Let us now return to the Lagrange approach. In order to preserve the symmetry of the non-linear system to be iteratively solved, one may substitute to (or use in conjunction with) the Lagrange approach a penalty approach,

$$\mathcal{L}_\mathrm{int}(\mathbf{u},\overline{\epsilon}_\mathrm{eq},\lambda) = \mathcal{F}(\mathbf{u},\overline{\epsilon}_\mathrm{eq}) + \int_\Omega \lambda\,\mathcal{G}_\mathrm{int}(\mathbf{u},\overline{\epsilon}_\mathrm{eq})\,d\Omega + \frac{1}{2}\gamma\Big(\int_\Omega \mathcal{G}_\mathrm{int}(\mathbf{u},\overline{\epsilon}_\mathrm{eq})\,d\Omega\Big)^2. \tag{5.5}$$

The penalty parameter $\gamma > 0$ should be sufficiently large, but it is bounded above by a stability condition The penalty method alone does not involve additional unknowns.

5.3 Solution of the Lagrange Equations by Uzawa's Algorithm

Instead of solving the system (4.10) at once for the three fields, one may use Uzawa's algorithm. The idea is, at each iteration, to solve first the system obtained by decoupling the full three-field system (in fact coupling comes through the residual): freezing the Lagrange multiplier to $\boldsymbol{\Lambda}^i$, one solves for $\mathbf{W} \equiv (\mathbf{U}, \overline{\mathbf{E}}_\epsilon)^T$ by requiring stationarity of $\partial \mathbf{L}_{\mathrm{gi}}/\partial \mathbf{W} \equiv (\mathbf{R}_u, \mathbf{R}_\epsilon)^T$ with respect to $\mathbf{W}$ itself,

$$\begin{bmatrix} \mathbf{R}_u \\ \mathbf{R}_\epsilon \end{bmatrix}^{i+1} = \frac{\partial}{\partial \mathbf{W}} \begin{bmatrix} \mathbf{R}_u \\ \mathbf{R}_\epsilon \end{bmatrix}^{i} \delta \mathbf{W}^{i+1} + \begin{bmatrix} \mathbf{R}_u \\ \mathbf{R}_\epsilon \end{bmatrix}^{i} \Leftrightarrow \begin{bmatrix} \mathbf{K}_{uu} & \mathbf{K}_{u\epsilon} \\ \mathbf{K}_{\epsilon u} & \mathbf{K}_{\epsilon\epsilon} \end{bmatrix}^{i} \delta \begin{bmatrix} \mathbf{U} \\ \overline{\mathbf{E}}_\epsilon \end{bmatrix}^{i+1} = - \begin{bmatrix} \mathbf{R}_u \\ \mathbf{R}_\epsilon \end{bmatrix}^{i} . \tag{5.6}$$

In these expressions, the field of Lagrange multipliers is replaced by its most recent values; an initial value must be given at the first iteration $i = 1$. The updated values of the basic fields $\mathbf{W}$,

$$\mathbf{W}^{i+1} = \mathbf{W}^i + \delta \mathbf{W}^{i+1} , \tag{5.7}$$

are used in the subsequent computations. The second system to solve is obtained by linearizing $\mathbf{R}_\lambda = \mathbf{0}$, eqns (4.8) or (4.17), accounting for the implicit dependance of $\mathbf{W}$ on the Lagrange field, namely,

$$\mathbf{R}_\lambda^{i+1} = \mathbf{R}_\lambda^i + \left(\frac{\partial \mathbf{R}_\lambda}{\partial \boldsymbol{\Lambda}}\right)_{|\boldsymbol{\Lambda}^i} \delta \boldsymbol{\Lambda} = \mathbf{0} \Longleftrightarrow \left(\frac{\partial \mathbf{R}_\lambda}{\partial \mathbf{W}} \frac{\partial \mathbf{W}}{\partial \boldsymbol{\Lambda}}\right)_{|\boldsymbol{\Lambda}^i} \delta \boldsymbol{\Lambda} = -\mathbf{R}_\lambda^i . \tag{5.8}$$

Using $\mathbf{R}_\lambda = \partial \mathbf{L}_{\mathrm{gi}}/\partial \boldsymbol{\Lambda}$, eqn $(4.8)_3$, the stationarity of $\partial \mathbf{L}_{\mathrm{gi}}/\partial \mathbf{W}$ for both $\mathbf{W}$ and $\boldsymbol{\Lambda}$ varying, i.e. $d(\partial \mathbf{L}_{\mathrm{gi}}/\partial \mathbf{W}) = \mathbf{0}$, provides,

$$\frac{\partial \mathbf{W}}{\partial \boldsymbol{\Lambda}} = - \left[\frac{\partial^2 \mathbf{L}_{\mathrm{gi}}}{\partial \mathbf{W} \partial \mathbf{W}}\right]^{-1} \frac{\partial \mathbf{R}_\lambda}{\partial \mathbf{W}} . \tag{5.9}$$

Therefore, omitting the iteration index, with $\mathbf{Z}$ the square matrix of size (nb. of Lagrange multipliers)^2

$$\mathbf{Z} = [\mathbf{K}_{\lambda u} \ \mathbf{K}_{\lambda\epsilon}] \begin{bmatrix} \mathbf{K}_{uu} & \mathbf{K}_{u\epsilon} \\ \mathbf{K}_{\epsilon u} & \mathbf{K}_{\epsilon\epsilon} \end{bmatrix}^{-1} \begin{bmatrix} \mathbf{K}_{u\lambda} \\ \mathbf{K}_{\epsilon\lambda} \end{bmatrix} , \tag{5.10}$$

the increment of Lagrange multiplier is given as

$$\delta \boldsymbol{\Lambda} = \mathbf{Z}^{-1} \ \mathbf{R}_\lambda . \tag{5.11}$$

The updated field $\boldsymbol{\Lambda}$ is used in subsequent computations, that is for the next iteration.

Although the above computations entail inversions of systems of sizes associated either to the basic unknowns $\mathbf{W}$ or to the Lagrange multipliers $\boldsymbol{\Lambda}$ only, they are heavy. Instead, one might couple the Lagrange method with the penalty method, as mentioned in the previous section. Then, for large values of the penalty γ in eqn (5.5), namely

$$\mathcal{L}_{\mathrm{int}}(\mathbf{W}, \boldsymbol{\Lambda}) = \mathcal{F}(\mathbf{W}) + \boldsymbol{\Lambda} \, \mathbf{R}_\lambda + \frac{1}{2} \gamma \, (\mathbf{R}_\lambda)^T \mathbf{R}_\lambda , \tag{5.12}$$

one may approximate the inverse of matrix $\mathbf{Z}$ by the dominating penalty term, yielding in place of (5.11),

$$\delta \boldsymbol{\Lambda} \simeq \gamma \mathbf{R}_\lambda \,. \tag{5.13}$$

Then, with respect to the two-field approach, the Lagrange formulation not only does not entail substantially more computations than the two-field method, it does better since it saves half of the storage of the Newton matrix which is now symmetric!

6 Finite Element Simulations

Numerical tests are described now for both the two-field implicit gradient and its three-field lagrangian version.

For the implicit gradient, mesh-independence of the load-displacement curve and field variables is tested on a notched concrete slab loaded in tension under plane stress condition, Sect. 6.1. Next, the lagrangian approach is used to solve the standard inhomogeneous bar subjected to tension at its free end, Sect. 6.2. The equivalence of the two- and three-field approaches is checked and discussed on this example.

Difficulties of convergence are traced to ill-conditioning and bad scaling of the system of non-linear equations. In order to follow the decreasing load-carrying capacity of the structures, arc-length techniques are used: for simplicity, the only variables entering these schemes may be the displacements, the influence of non-local equivalent strains and lagrangians being not considered. However, incorporation of equivalent strains in the arc-length methods was found either to improve the convergence, Sect. 6.1, or to be necessary to obtain convergence, Sect. 6.2. The arc-length schemes used are the *orthogonal residual method*, Hededal and Krenk [1995], and an ad-hoc variant, the *indirect displacement control*, de Borst [1996]: in the latter, which is efficient but problem dependent (that is dependent on the geometry, loading and boundary conditions), only the displacement variables that can be speculated to increase monotonously enter the arc-length control.

6.1 A Notched Concrete Slab with the Implicit Gradient

As a plane-stress example, the concrete slab experimentally studied by Hassanzadeh [1991] has been analyzed through the two-field gradient model, where the gradient parameter b has been set equal to 1 mm^2, Fig. 5. The equivalent strain is defined by eqn (2.2) and the damage D by eqn (2.5). Whereas in the experiment of Hassanzadeh, the specimen has been first loaded under deformation control in pure tension up to the peak load, and then subjected to a constant ratio between normal and tangential displacement, in this numerical example, a uniform distribution of traction load has been considered applied at the top of the specimen, whereas the nodes at the bottom are fixed. Moreover, in the computations, the thickness has been set to 1 mm.

Because of the symmetry of the specimen, only one half of it has been studied. Three meshes, denoted in the following by (a), (b) and (c), have been adopted, all defined by eight-noded serendipity elements for the displacement field and four-noded bilinear elements for the averaged equivalent strain field. For simplicity, both fields have been integrated with a four-point integration scheme.

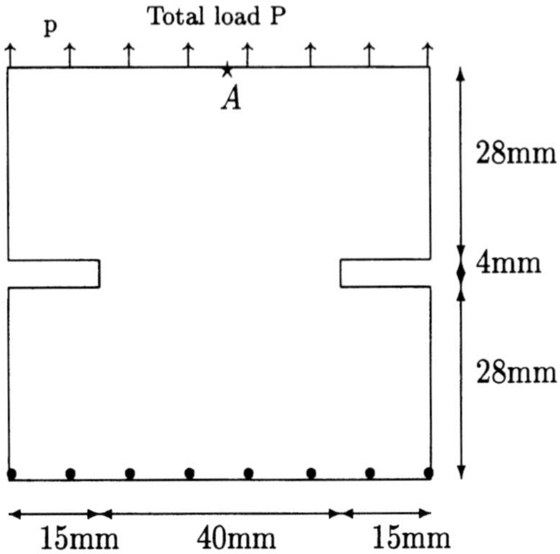

(a) Geometry of the notched concrete slab (fictitious thickness: 1mm). Linear load density p(N/mm).

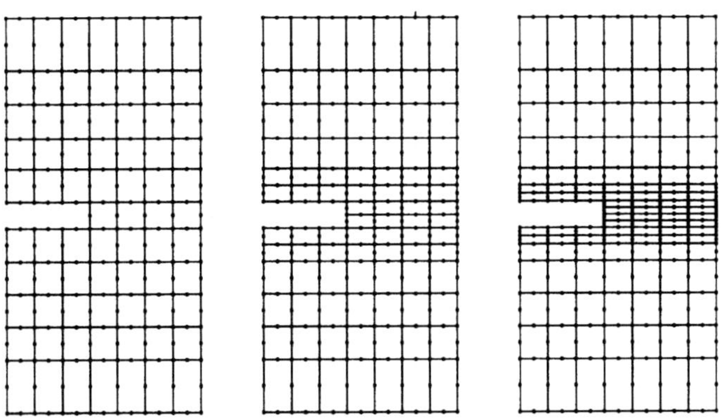

(b) Finite element meshes exploiting the symmetry of the problem.

Figure 5 Notched concrete slab subjected to uniform distribution of tensile load on the upper platens and with fixed lower platens. Geometry and meshes used to solved the problem. The non-local material behaviour is defined by the parameters given in the text. The meshes use 8-node elements for displacement and 4-node elements for equivalent strain.
The material parameters are: $r = 10$, $\alpha = 0.96$, $\beta = 350$, $\kappa_0 = 10^{-4}$, Young's modulus E= 32900 N/mm^2, Poisson's ratio $\nu = 0.2$.

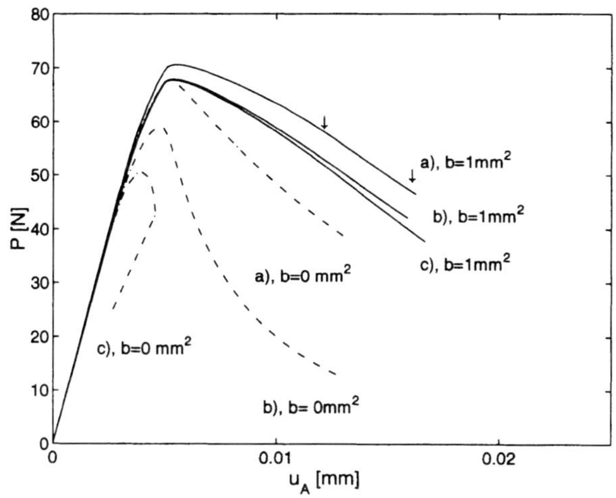

Figure 6 Load versus displacement of point A for the meshes of the concrete slab described by Fig. 5: a) coarse mesh, b) medium mesh and c) fines mesh. The arrows show the displacements at which contours are drawn in the figures below. The local damage analysis $b = 0$ clearly introduces a pathological mesh-dependence. For $b = 1$ mm^2, convergence of the results occurs and shows that a satisfactory mesh size should be smaller than 1 mm, see Fig. 5, that is smaller than the internal length $L = \sqrt{b}$.

The damage zone, strictly related to the value of the averaged (non-local) equivalent strain, tends to localize at the notch, as it is well visible in the plots of the contours of the averaged equivalent strain, taken at two displacements along the descending branch of the load-displacement curve, Fig. 7. These contours also show convergent values as the mesh is refined.

This trend is confirmed from the plots showing the resultant total load versus the vertical displacement of the point A placed on the vertical symmetry axis, Fig. 6. In addition, as an illustration of the effectiveness of the gradient regularization, results corresponding to $b = 0$, that is to a local material, are shown for comparison. Note that, for a given load, the displacement decreases as mesh is refined: a qualitative explanation of this phenomenon will be provided in next section in relation with the 1D-bar.

Covergence of the iterative procedure depends on both the search direction and the arc-length procedure. For simplicity, initially only the displacements were included in the arc-length procedure. Even at the incipience of damage, in the raising range of the load, where a priori the arc-length method is not required, the full Newton-Raphson method did not converge, even if the tangent stiffness was checked to be invertible. The reason for this lack of convergence is traced to the non monotonous behaviour, precisely at the incipience of damage, of the derivative of the damage function $D = D(\kappa)$ which is involved in the tangent matrix in $\mathbf{K}_{u\epsilon}$, eqn (5.3). One possibility to recover convergence is to perform line-searches. Since these are costly, we have preferred to abandon the quadratic convergence rate of the Newton-Raphson method by deleting the out-of-diagonal term $\mathbf{K}_{u\epsilon}$, which incidentally vanishes for an elastic behaviour. So the iterations (5.2) are

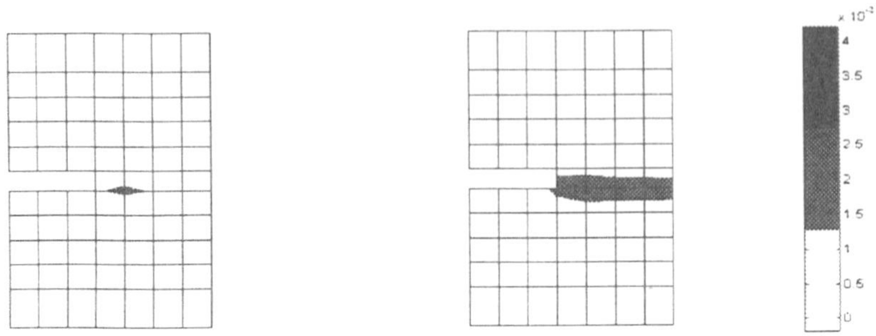

(a) Coarse mesh: displacement u_A=0.012 mm and 0.016 mm

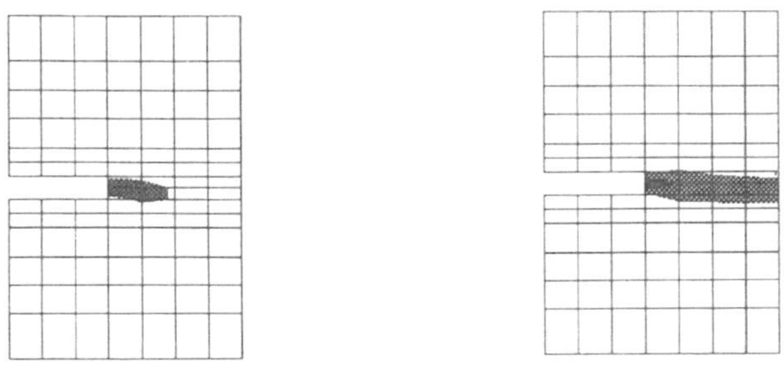

(b) Medium mesh: idem

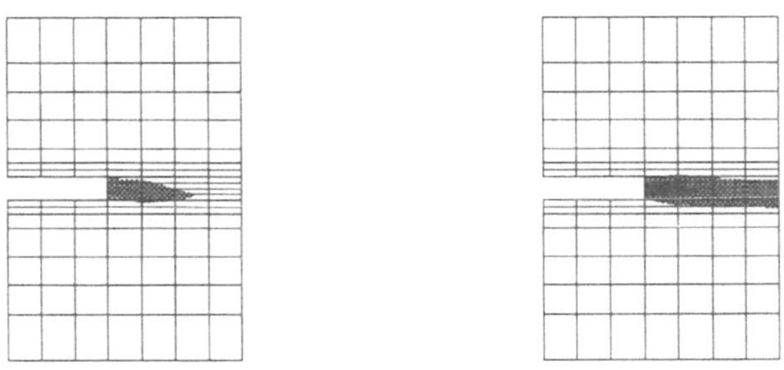

(c) Fine mesh: idem

Figure 7 Notched concrete slab subjected to a tensile load along the upper side and with fixed lower side, see Fig. 5. Coarse, medium and fine meshes. Contours of the non-local equivalent strain $\bar{\epsilon}_{eq}$ for two values of the displacement at point A.

replaced by

$$\begin{bmatrix} \mathbf{K}_{uu} & \mathbf{0}_{u\epsilon} \\ \mathbf{K}_{\epsilon u} & \mathbf{K}_{\epsilon\epsilon} \end{bmatrix}^i \delta \begin{bmatrix} \mathbf{U} \\ \overline{\mathbf{E}}_\epsilon \end{bmatrix}^{i+1} = - \begin{bmatrix} \mathbf{R}_u \\ \mathbf{R}_\epsilon \end{bmatrix}^i . \tag{6.1}$$

The above remarks pertain to the case where only the displacements are included in the arc-length procedure. If the equivalent strains are included as well, the Newton-Raphson matrix **K** was found to lead to convergence.

6.2 The 1D-Bar with the Lagrange Approach

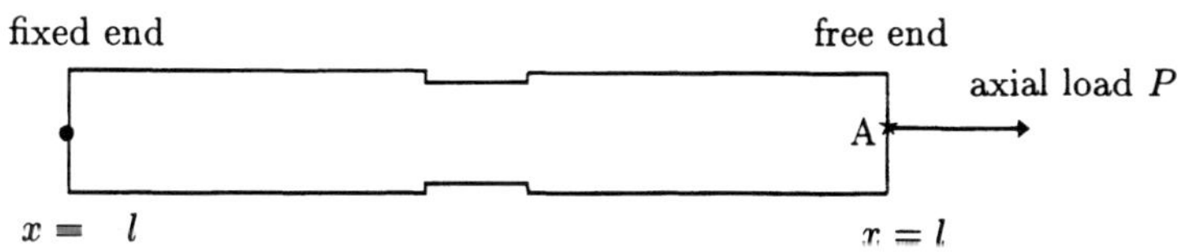

Figure 8 1D-bar fixed at its left end $x = -l$ and subjected to an axial load P at the right end A at $x = l = 50\,\text{mm}$. Young's modulus is equal to 20 000 N/mm^2. An inhomogeneity is introduced through a thiner section of length 10 mm around its center: there the section is equal to 9 mm^2 instead of 10 mm^2 elsewhere. The equivalent strain is defined by eqn (2.1) and the damage law $D = D(\kappa)$ by eqn (2.6) with $\alpha = 10$, $\beta = 0.8$, $\kappa_0 = 2\,10^{-4}$ and $\kappa_c = 0.0125$. The non-local damage model is defined by the implicit gradient model with $b = 1\,\text{mm}^2$ and analyzed through the lagrangian approach.

The Lagrange approach is now used to solve a 1-dimensional bar containing an inhomogeneity in the form of a reduced section where damage localizes, Fig. 8. The equivalent strain is defined by eqn (2.1) and the damage law $D = D(\bar{\kappa})$ by eqn (2.6) with $\kappa_0 = 2\,10^{-4}$ and $\kappa_c = 0.0125$ as in Geers [1997] in order to compare the two-field approach (used by Geers) with the three-field approach used here. In a 2D context, the partial derivative of ϵ_{eq} with respect to $\boldsymbol{\epsilon}$ is required. The eigenvalues are $\epsilon_i = (\epsilon_{11} + \epsilon_{22} + s_i\sqrt{\Delta})/2$ with $s_i = (-1)^{i+1}$, $i = 1, 2$, and $\Delta = (\epsilon_{11} - \epsilon_{22})^2 + 4\,\epsilon_{12}^2$. So

$$\frac{\partial \epsilon_i}{\partial \boldsymbol{\epsilon}} = \frac{s_i}{2\sqrt{\Delta}} \begin{bmatrix} \epsilon_{11} - \epsilon_{22} + s_i\sqrt{\Delta} & 2\,\epsilon_{12} \\ 2\,\epsilon_{12} & -\epsilon_{11} + \epsilon_{22} + s_i\sqrt{\Delta} \end{bmatrix} . \tag{6.2}$$

so that

$$\frac{\partial \epsilon_{\text{eq}}}{\partial \boldsymbol{\epsilon}} = \frac{1}{\epsilon_{\text{eq}}} \left(< \epsilon_1 > \frac{\partial \epsilon_1}{\partial \boldsymbol{\epsilon}} + < \epsilon_2 > \frac{\partial \epsilon_2}{\partial \boldsymbol{\epsilon}} \right). \tag{6.3}$$

Notice that the derivatives $\partial\epsilon_i/\partial\boldsymbol{\epsilon}$ are not continuous for $<\epsilon_i>$ around 0, but $\partial\epsilon_{\text{eq}}/\partial\boldsymbol{\epsilon}$ is continuous. In a 1D context, some simplifications arise as $\epsilon_{\text{eq}} =< \epsilon >$, and so, in tension, $\partial\epsilon_{\text{eq}}/\partial\epsilon = 1$, and $\partial^2\epsilon_{\text{eq}}/\partial\epsilon^2 = 0$.

The load-displacement curves converge as the mesh is refined, Fig. 9. Even the coarse mesh, for which the element size 2.5mm is larger than the internal length $L = \sqrt{b}$=1mm, provides results quite close from the converged ones. The results are quite close as well

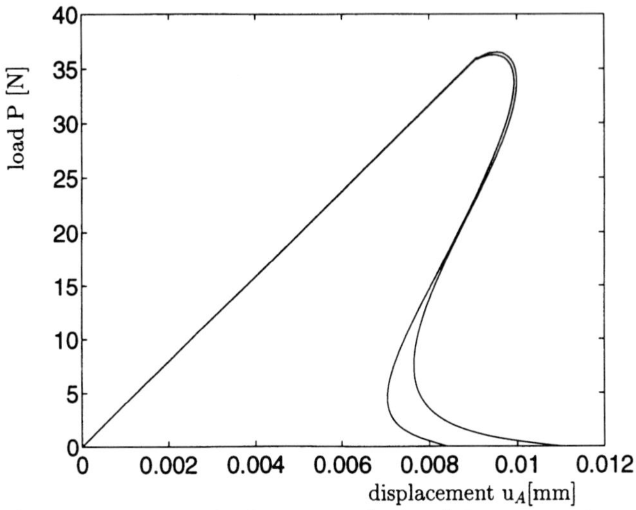

Figure 9 Idem fig. 8. Load versus displacement of point A for two meshes containing 40 and 80 elements. Even the coarse mesh for which the element size 2.5mm is larger than the internal length $L = \sqrt{b}$=1mm provides results quite close from the converged ones.

to those obtained by the two-field approach, Fig. 10, although some differences appear in the residuals $\mathbf{R}_u$ due to the additional term in $(4.12)_1$ with respect to $(5.4)_1$: this term is proportional to $D'(\kappa) = dD/d\kappa$ and it is important at the incipience of damage, as can be seen on Fig. 10. Note that, on the other hand, the residuals $\mathbf{R}_\epsilon$, eqns $(5.4)_2$ and $(4.12)_3$, are identical.

The derivative $D'(\kappa)$ in the residuals has been estimated numerically as $(\,D(\kappa+t) - D(\kappa-t)\,)/2t$, with $t = 10^{-4}$. In the arc-length procedure, it has been found necessary to incorporate the equivalent strains to obtain convergence. If only displacements are included in the arc-length procedure, even a line search is not able to ensure convergence.

The localization of the damage in the central region with smaller section is apparent from Fig. 11. The Lagrange multiplier, which appears as a generalized stress (dimension N/mm^2) is non zero only where damage occurs. This can be checked from eqn $(4.12)_2$. Indeed, first, $\partial D/\partial\kappa$ vanishes in the elastic domain where $\kappa \leq \kappa_0$, eqn (2.6); second, $\mathbf{K}_{\epsilon\lambda}$, eqn $(4.11)_5$, is invertible; the result then follows from the vanishing of the residual $\mathbf{R}_\epsilon$, eqn $(4.12)_2$. At the inception of damage, $\partial D/\partial\kappa = \alpha/\kappa_0 + \beta/(\kappa_c - \kappa_0)$, the strain ϵ is equal to κ_0, and so, from eqn $(4.12)_2$, the lagrangians in the damage zone are of the order of the stress, that is $1/2\,E\,\kappa_0^2\,\partial D/\partial\kappa$. More details can be found in Benvenuti [2000].

As noted earlier in relation with the concrete slab in tension, refining the mesh yields a decreasing displacement at the extremity of the structure where the load is applied. Indeed, let us idealize the slab and inhomogeneous bar as purely symmetric 1-dimensional samples and let us consider their upper part, Fig. 12. A zone in the center, of length l_d, undergoes damage and has a strain ϵ_d, while the remaining part of the sample, of length $l_e = l - l_d$, undergoes elastic unloading (in the post-peak domain) and has strain ϵ_e. The

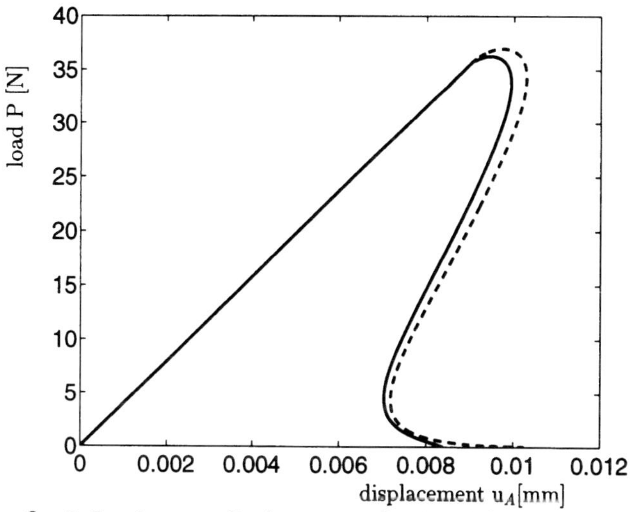

Figure 10 Idem fig. 8. Load versus displacement of point A for the lagrangian approach, continuous curve, and the two-field approach, dash curve, both for the mesh with 80 elements.

displacement of the extremity of the sample with respect to its center is $u_A = \epsilon_d\, l_d + \epsilon_e\, l_e$. In the post-peak domain, $\Delta\epsilon_e = \Delta P/ES$, S section of the sample, is negative.

When the mesh is refined, the damage length l_d decreases since the spreading over the element at the damage-elastic transition can shrink. Therefore l_e increases, and consequently u_A decreases. In fact, the decrease can be so large as to trigger snap-back (which, for a given inhomogeneity around the center, is more likely to occur for a long sample than for a short one), since l_e becomes larger.

References

Bažant, Z.P. and Lin, F.B. (1988). Non-local yield limit degradation. *Int. J. Num. Meth. Engineering*, **26**, 1805-1823.

Benvenuti, E., Loret B. and A. Tralli (2000). A lagrangian approach for non-local damaging materials. SIMAI 2000, V Congresso Nazionale della Societá Italiana di Matematica Applicata e Industriale, Ischia Porto, 5-9 June 2000, to appear.

Benvenuti, E., Loret B. and A. Tralli (2000). On implicit gradient models for quasi-brittle materials. IGF 2000, XV Congresso Nazionale del Gruppo Italiano Frattura, Bari, 3-5 Maggio 2000, 89-96

Benvenuti, E., Borino G. and A. Tralli (2000). A thermodynamically consistent non local formulation for elasto-damaging materials: theory and computations. ECCOMAS 2000, European Congress on Computational Methods in Applied Sciences and Engineering, Barcelona 11-14 September 2000, on CD-Rom.

Benvenuti, E. (2000). Enhanced non-local numerical models for damaging materials. Ph'd thesis, Università degli Studi di Ferrara, Italia.

Bowen, R.M (1976). Theory of mixtures. *Continuum Physics*, vol. 3, 1-127, A.C. Eringen ed., Academic Press, New York.

de Borst, R. (1986). Non-linear Analysis of Frictional Materials. Ph'd thesis, Delft University of Technology.

Gatelier, N. (2000). Etude expérimentale et modélisation du comportement fragile d'un grès des Vosges. Thèse de Doctorat, Université Joseph Fourier, Grenoble, en préparation.

Geers, M.G.D. (1997). Experimental analysis and computational modeling of damage and fracture. Ph'd thesis, Eindhoven University of Technology.

Hassanzadeh, M. (1991). Behaviour of fracture process zone in concrete influenced by simultaneously applied normal and shear displacements. Ph'd thesis, Lund University of Technology.

Hededal,O. and S. Krenk (1995). *Femlab-Matlab Toolbox for the Finite Element Method*, version 1.0, Aalborg University.

Jirásek, M. (1999). Computational aspects of non-local models. ECCM'99, München, Aug. 31-Sept. 3, Conference Proceedings on CD-Rom.

Loret, B. and J.-H. Prévost (1990). Dynamic strain localization in elasto-visco-plastic solids. *Computer Methods in Applied Mechanics and Engineering*, **83**, 247-273.

Loret, B., Simões, F.M.F. and Martins, J.A.C. (1999). Flutter instability and ill-posedness in solids and fluid-saturated porous media. CISM course on *Material Instabilities in Elastic and Plastic Solids*, Udine, edited by H. Petryk, Springer Verlag, Wien, 99 pages.

Loret, B., Simões, F.M.F. and Martins, J.A.C. (2000). Regularization of flutter ill-posedness in fluid-saturated porous media, *Continuum thermomechanics: the art and science of modeling matter's behaviour. Paul Germain's anniversary volume*, R. Drouot, G. Maugin and F. Sidoroff eds., Kluwer Academic Publishers, Dordrecht, The Netherlands.

Mazars, J. and G. Pijaudier-Cabot (1989). Continuum damage theory. Application to concrete. *J. of Engng. Mechanics Div.*, Transactions of the ASCE, **11**, 345-363.

Mühlhaus, H.B. and E. Aifantis (1991). A variational principle for gradient plasticity. *Int. J. Solids Structures*, **28**, 845-857.

Pamin, J. Askes, H. and R. de Borst (1999). An element-free Galerkin method for plasticity. ECCM'99, München, Aug. 31-Sept. 3, Conference Proceedings on CD-Rom.

Peerlings, R.H.J, de Borst, R., Brekelmans, W.A.M. and J.H.P. de Vree (1996). Gradient-enhanced damage for quasi-brittle materials. *Int. J. Num. Meth. Engineering*, **39**, 3391-3403.

Peerlings, R.H.J. and R. de Borst (1998). Gradient-enhanced damage modelling of concrete fracture. *Mech. Cohesive Frictional Materials*, **3**, 323-342.

Pijaudier-Cabot, G. and Z. Bažant (1987). Non-local damage theory. *J. of Engng. Mechanics Div.*, Transactions of the ASCE, **10**, 1512-1533.

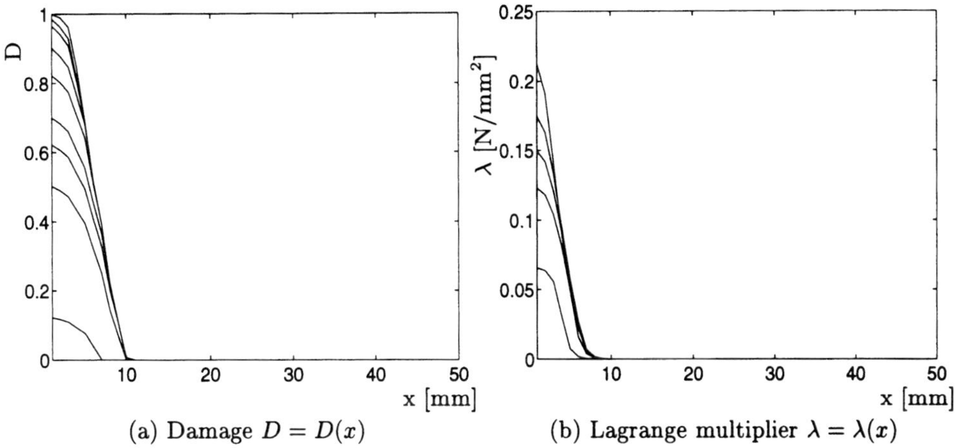

Figure 11 Idem fig. 8. Evolution along the right half $x \in [0, l]$ of the damage parameter D and of the lagrangian multiplier λ for the finest mesh analysis. The lagrangian multiplier is non zero only where damage occurs.

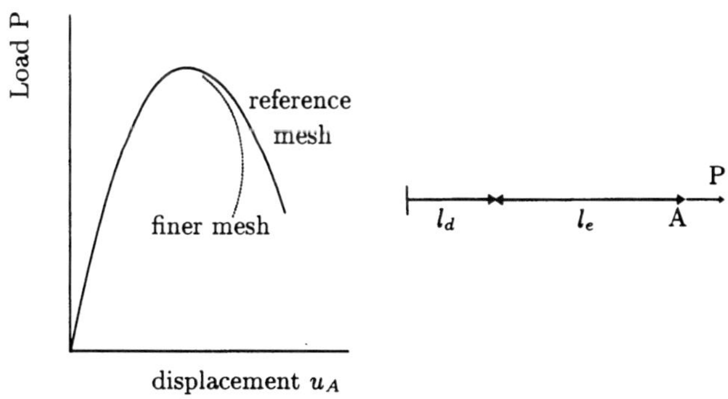

Figure 12 In the post-peak domain, mesh refinement on essentially 1-dimensional sample decreases the displacement at the end where load is applied and may trigger snap-back, as it implies effects similar to an increase of the length of the sample.

Laboratory Investigation and Numerical Modelling of Soil Improvement Techniques

Annamaria Cividini

Department of Structural Engineering, Politecnico di Milano, Milan, Italy

Abstract: A discussion is presented of the use of the laboratory investigation and of the finite element modelling for the analysis of geotechnical problems involving techniques of soil improvements. Among the variety of possible applications, those related to the low pressure grouting of granular soils, to the artificial freezing of grounds and to reinforced earth structures are discussed in some details. The analyses of cases related to actual design problems are also presented and commented upon.

1 Introduction

The use of particular techniques aimed at improving the mechanical, or hydraulic, characteristics of soil or rock masses is nowadays a common practice in geotechnical engineering.

Treatments of this kind have been known, and applied, since ancient times. For instance, layers of cohesive soil reinforced with mats of vegetable fibres were used in Mesopotamia for the construction of temples, or Ziqqurat, since the XXII century BC. The tower of Babel, described in the Bible, is believed to be an example of them.

These techniques, whose range of applications dramatically increased with the developments of modern technology, can be subdivided into two main categories depending on their temporary or permanent effects. Among the temporary ones the electro-osmosis, the artificial ground freezing, the temporary lowering of the water table by pumping wells, etc. can be mentioned. Among those that lead to permanent improvements of the characteristics of the geotechnical medium, the most popular ones are the deep and shallow compaction, the permanent drainage systems, the low and high pressure grouting, the soil pre-loading, the reinforced earth, the thermal treatments, etc.

The design of relevant geotechnical works involving these techniques requires a proper modelling, either physical or numerical, of the improvement process and/or of the subsequent behaviour of the treated soil. In fact, because of their intrinsic limits, closed form analytical solutions can hardly provide an adequate insight into these complex problems.

Here the discussion is focused on the numerical modelling, through the finite element method, of the following well-established procedures for soil improvement: the low pressure grouting of granular soils, the artificial freezing of saturated soils and the so-called earth reinforced structures.

From the computational mechanics standpoint, the analysis of the mentioned improvement techniques involves various relevant aspects. They are related, in particular, to the solution of elasto-plastic (strain softening) problems, to the analysis of seepage and consolidation, to the non-linear transfer of heat and to the use of homogenisation procedures for the analysis of in-homogeneous, non-isotropic materials.

In the following, the main characteristics of the finite element analysis of the mentioned problems are presented, as well as some applications to practical cases.

In general terms, it should also be observed that any numerical analysis of complex geo-technical problems requires, from the one hand, the choice of the relevant material parameters and, from the other hand, a proper validation of the numerical results. To this purpose, the results of some laboratory tests, and their interpretation through the described numerical models, are also discussed and commented upon. This permits a deeper understanding of the advantages and shortcoming of the described solution procedures and some insight into their potential for application to actual design problems.

2 Low Pressure Grouting

Since about 1975 a series of industrial and agricultural changes led to an appreciable increase of the water table level in the city of Milan (northern Italy). The phreatic surface gradually reached an elevation higher than that of various underground facilities, which have been completed between 1960 and 1980, and that were not properly waterproofed since at that time the water table had a much lower level. As a consequence, some extensive leakage of water took place at various locations within these structures and, in particular, within subway tunnels.

One of the geotechnical provisions envisaged to limit these negative effects is based on the low pressure grouting of the granular soil surrounding them to decrease its permeability and to reduce the seepage flow toward the structures.

An experimental and numerical study concerning the mentioned provision is summarised in the following, which was carried out in the framework of a research project supported by the Milan Metro (Metropolitana Milanese). First, on the basis of a series of laboratory tests, an empirical law is derived that governs the variation with time of the permeability of a granular soil subjected to the seepage flow of silicate solutions (Cividini, 2000). This law is subsequently adopted in some numerical examples to get an insight into the effectiveness of the treatment and of the time necessary to complete it.

2.1 Laboratory Investigation

The laboratory investigation was carried out on samples of granular soil that were compacted in layers of small thickness within a "permeameter". This consists of a Plexiglas pipe (internal diameter of 10cm, height of the samples between 42 and 45 cm), within which the fluid flow takes place in the vertical direction. The hydraulic head difference was generated by means of a container filled with water, or with the silicate solution, placed at a proper elevation.

The grouting stage of the tests was performed by connecting the base of the sample to the tank containing the silicate solution and recording during time the volume of fluid leaving the sample and the pore pressure at different elevations within it. Depending on the adopted grouting solution, the total duration of each test ranged between 20 and 30 days.

Since the chemical properties play a major role in the effectiveness of the grouting treatment (e.g. Nonveiller, 1989), three different solutions of sodium silicate have been used during the laboratory tests. They consist of solution of SiO_2 and of Na_2O with different ratios between the two components. Phosphoric acid was added to them, since the time required for the formation of the silica gel is influenced by the solution pH.

Some typical experimental results, i.e. the increase during time of the total volume of fluid leaving the sample and the corresponding variation of the average coefficient of permeability for the entire sample, are reported in Figure 1. These results show that, for the tested granular soil, the grouting through diluted solutions of sodium silicate may lead to a reduction of the coefficient of permeability of two to three orders of magnitude.

A main scope of this experimental investigation was to derive an empirical law expressing the variation of the coefficient of permeability with time, to be adopted in the numerical analyses. To this purpose, the linear laws depicted in Figure 2 were adopted as an engineering approximation of the test results. Here k_i and k_g denote, respectively, the initial (natural) coefficient of permeability and that of the grouted soil, t_g represents the time necessary for the formation of the silica gel and $k_{min}(t)$ is the coefficient of permeability corresponding to the maximum concentration of the solution.

On these bases, the time dependent value of k_{min} can be expressed as follows

$$k_{\min}(t) = k_i \cdot \left(k_g / k_i\right)^{t/t_g} \quad , \quad 0 \le t \le t_g \quad . \tag{1}$$

The assessment of validity of this empirical law was based on the numerical simulation of the experimental results in Figure 2, as discussed in Section 2.3.

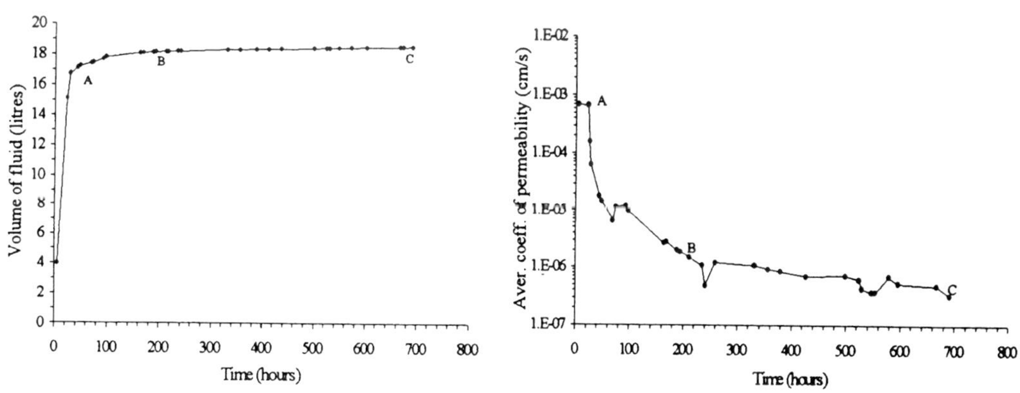

Figure 1. Volume of fluid leaving the sample during time and corresponding variation of the average coefficient of permeability of the sample during time. Points A, B and C denote the beginning and the end of grouting and the end of the test. (Seepage tests performed with ratio of 1:3 between silicate and sodium solutions, with 0.3 moles per litre of solution and 0.03 moles per litre of phosphoric acid, pH=11.09).

2.2 Finite Element Formulation

The grouting process can be modelled as the seepage flow of an incompressible liquid, containing a given substance as a solute, through a saturated porous medium. The mass of substance per unit volume of the liquid is expressed by its concentration c.

The macroscopic flux of the substance is customarily subdivided into three parts, referred to as advective, dispersive and diffusive fluxes (see e.g. Bear and Verruijt, 1987), that depend on different physical causes.

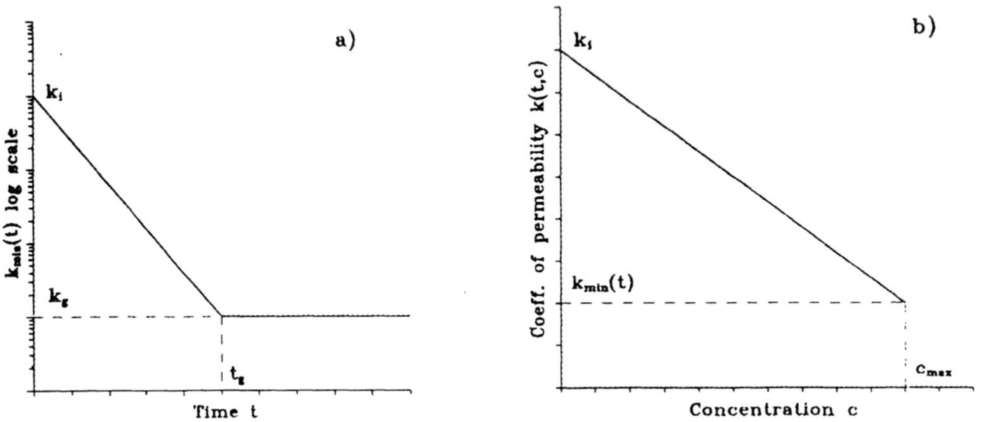

Figure 2. Adopted variation of the coefficient of permeability of the soil with time a) and concentration b).

The advective flux is related to the macroscopic velocity (in Darcy's sense) of the liquid flow. The dispersive flux depends on local fluctuations of the velocity field, and can be related to the gradient of the concentration. The diffusive flux is due to the molecular diffusion of the substance within the liquid caused by the random movements of the fluid molecules. This last part of the flux could be incorporated into that related to the mechanical dispersion (see e.g. Rowe and Booker, 1985; 1997). Considering that the diffusive flux depends on a coefficient of molecular diffusion, which has a very small order of magnitude for most cases of interest in geomechanics, its contribution is often negligible with respect to those of the first two components and it was not considered in the analyses.

The continuity equation for the advective-diffusive flux is obtained by imposing that the accumulation rate of the solute produced by seepage and by dispersive flux compensates the variation of the mass of substance contained within the porous element. This, in turn, is due to the variation of the concentration and/or porosity, n, during the time increment, and to the contribution of possible internal sources having an imposed flux q_c.

The finite element formulation of the problem governed by the above mentioned continuity equation can be based on a procedure quite similar to that adopted for seepage or diffusion analyses. The procedure, whose details have been presented by Cividini (2000) and are omitted here for briefness, leads to the following matrix form of the solving equations,

$$\underline{\underline{A}}_1 \cdot \underline{c} + \underline{\underline{A}}_2 \cdot \underline{c} + \underline{\underline{A}}_3 \cdot \dot{\underline{c}} = \underline{a}_1 \quad , \tag{2}$$

where

$$\underline{\underline{A}}_1 = \int_V \underline{\underline{B}}^T \underline{\underline{K}} \cdot \underline{\underline{B}} \cdot \underline{h} \cdot \underline{f}^T dV \quad , \tag{3a}$$

$$\underline{A}_2 = n \cdot \int_V \underline{B}^T \underline{D} \cdot \underline{B} dV \quad , \tag{3b}$$

$$\underline{A}_3 = n \cdot \int_V \underline{f} \cdot \underline{f}^T dV \quad \text{and} \quad \underline{a}_1 = -n \cdot \int_V \underline{f} \cdot q_c dV \quad . \tag{3c,d}$$

In eqs.(2) and (3), $\underline{c}$ and $\underline{h}$ are the nodal concentration and hydraulic head vectors, $\underline{K}$ and $\underline{D}$ are the permeability and diffusivity matrices, $\underline{B}$ is the matrix of the derivatives of the shape functions $\underline{f}$.

It is worthwhile observing that the finite element evaluation of the nodal hydraulic heads is governed by the following equation,

$$\underline{A}_4 \cdot \underline{h} = \underline{a}_2 \quad , \tag{4}$$

where

$$\underline{A}_4 = \int_V \underline{B}^T \underline{K} \cdot \underline{B} dV \quad \text{and} \quad \underline{a}_2 = \int_V \underline{f} \cdot q_w dV \quad . \tag{5a,b}$$

The compressibility of the soil skeleton is neglected in the above equations, i.e. the porosity n does not vary with time; q_w represents the imposed flow due to possible internal sources.

Among the various numerical schemes, proposed in the literature for the time integration of advective-dispersive flow problems (see e.g. Wendland and Schmid, 2000), here a two-step time integration is adopted. The first step consists in determining the hydraulic head vector $\underline{h}$ by means of eq.(4). If the coefficients of permeability are constant, and if the boundary conditions do not vary with time, the assumptions of incompressibility of the soil skeleton and of the liquid phase lead to a time independent distribution of the hydraulic head. Hence, under the above assumptions, only the time dependent change of the concentration vector $\underline{c}$ has to be evaluated during the second step of the analysis through an iterative procedure.

This step is subdivided into a series of time increments Δt during which a linear variation of vector $\underline{c}$ is assumed. Since $\underline{A}_1$ is in general non-symmetric, its contribution can be moved to the right hand side of eq.(2) to preserve the symmetric nature of the system matrix. This leads to the following recursive solution equation for $\underline{c}_{t+\Delta t}$,

$$[\vartheta \Delta t \cdot \underline{A}_2 + \underline{A}_3] \cdot (\underline{c}_{t+\Delta t})_i =$$

$$= \Delta t \cdot \underline{a}_1 - [(1-\vartheta)\Delta t \cdot \underline{A}_2 - \underline{A}_3] \cdot \underline{c}_t \; - \Delta t \cdot \underline{A}_1 \cdot [\vartheta (\underline{c}_{t+\Delta t})_{i-1} + (1-\vartheta)\underline{c}_t] \quad , \quad 0 \le \vartheta \le 1 \quad , \tag{6}$$

where i denotes the iteration number.

Note that the evaluation of the nodal pore pressures and concentrations are "uncoupled", with a consequent saving in size of the relevant arrays of the finite element code.

The extension of this integration technique to the case of non-constant permeability (due e.g. to the time dependent variation of the viscosity of the liquid-solute mixture) is rather straightforward. Since the variation with time of the coefficient of permeability leads to a time dependent distribution of the hydraulic head, an iterative procedure should be employed to evaluate both vectors $\underline{h}$ and $\underline{c}$ at the end of each time increment Δt.

For the first iteration the nodal concentrations at time $t+\Delta t$ are set equal to those at time t and the permeability of the soil elements reached by the grouting mixture is determined through the adopted empirical law. Then, a first approximation of the nodal hydraulic heads and concentrations at time $t+\Delta t$ is determined following the described two-step scheme. The time increment is repeated, updating the values of the coefficient of permeability on the basis of the computed values of the concentration at time $t+\Delta t$. The iterations are repeated until further changes of the final nodal variables become negligible.

When convergence has been reached, the iterative procedure is applied to the solution for the subsequent time increment. As to the time required for calculation, a rapid convergence was observed for all examined cases, leading to a computational effort comparable to that necessary to solve non-linear or unconfined seepage problems.

2.3 Application

First, the numerical simulation of the laboratory tests was carried out for validating the empirical law expressed by eq.(1). Figure 3 reports the pore pressure distributions recorded along the sample height at the beginning of the test and during the grouting process, before the formation of the silica gel has been completed.

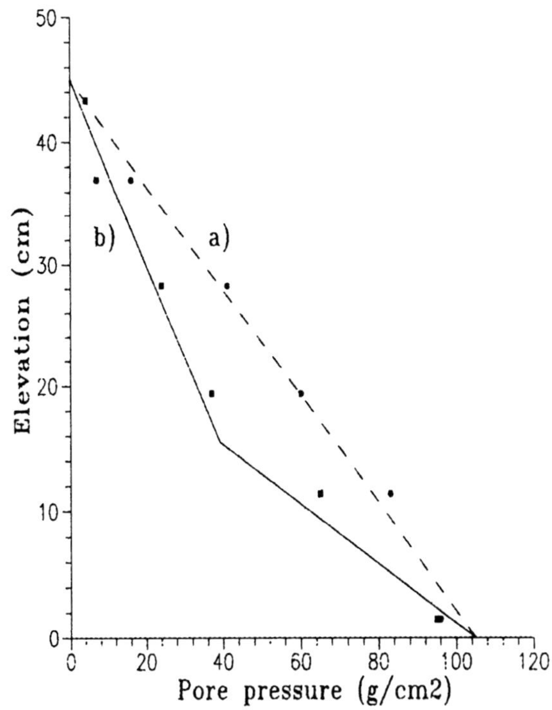

Figure 3. Pore pressure values measured along the sample height (dots) and numerical interpretation of the experimental results (dashed and solid lines): a) before grouting, b) during the test, before the complete formation of the silica gel.

The comparison between the experimental data and the numerical results shows that the adopted relationship between permeability, time and concentration leads to an acceptable approximation of the experimental data and, hence, that it can be adopted for the numerical analysis of in situ treatments.

The described finite element approach was subsequently adopted in the numerical simulation of a grouting treatment in which the solution is injected from the tip of a single pipe. The analyses were carried out in axisymmetric regime, adopting the mentioned empirical law for expressing the time dependent variation of the permeability of the grouted soil. A soil portion of 10m x10m was discretised into 4 node, isoparametric finite elements. At the beginning of calculation this zone is subjected to a constant hydraulic head distribution of 10m. The hydraulic head at the pipe tip is kept equal to 15m. The left vertical side of the mesh (where the pipe is located and that represents a line of symmetry of the problem) and its bottom side are impervious, while the hydraulic head is imposed on the remaining sides. The problem was solved by considering a reduction of two orders of magnitude of the coefficient of permeability at the end of grouting, with respect to the initial permeability of the natural soil.

The results of calculations are summarised in Figure 4 that shows the contour lines of the hydraulic head and of the concentration for different time values. They indicate that the formation of the silica gel, and the parallel reduction of the coefficient of permeability, tends to limit the increase during time of the size of the treated zone.

On this basis it could be concluded that a further delay between the beginning of grouting and the initiation of the silica gel formation seems necessary to increase the spreading of the solution and, hence, to improve the effectiveness of the treatment. This information could be useful in the choice of the chemical characteristics of the grouting solution and in the evaluation of the overall cost of the treatment.

3 Artificial Freezing of Saturated Sand

The artificial freezing of ground is based on the circulation of brine or liquid nitrogen into pipes installed within the soil mass. This techniques is adopted for improving the shear resistance of coarse and medium fine soils during the excavation of tunnels, or shafts, located below the water table (see e.g. Andersland et al., 1989; Frivik, 1980; Andersland and Ladanyi, 1994).

The relatively high cost of this treatment could be limited by "optimising" its design with respect to some relevant parameters, such as the distance of the pipes, their diameter, the temperature of the cooling fluid, etc. Considering the complex, non-linear nature of the heat transfer and freezing processes, an experimental investigation was carried to provide the necessary bases for the development of a finite element program for freezing analysis.

In the following, first the results of some laboratory freezing tests are summarised (Gioda et al., 1994). Then the main characteristics of a finite element approach are illustrated for heat transfer analyses with phase change, customarily referred to as Stefan problem. Some comments are also presented on the technique that was adopted to account for the latent heat effects during the water/ice phase transition.

Finally, the finite element procedure is adopted in the interpretation of the laboratory tests and in a parametric study, concerning the effects of the diameter of the pipes and of the coolant temperature on the spreading with time of the frozen zone.

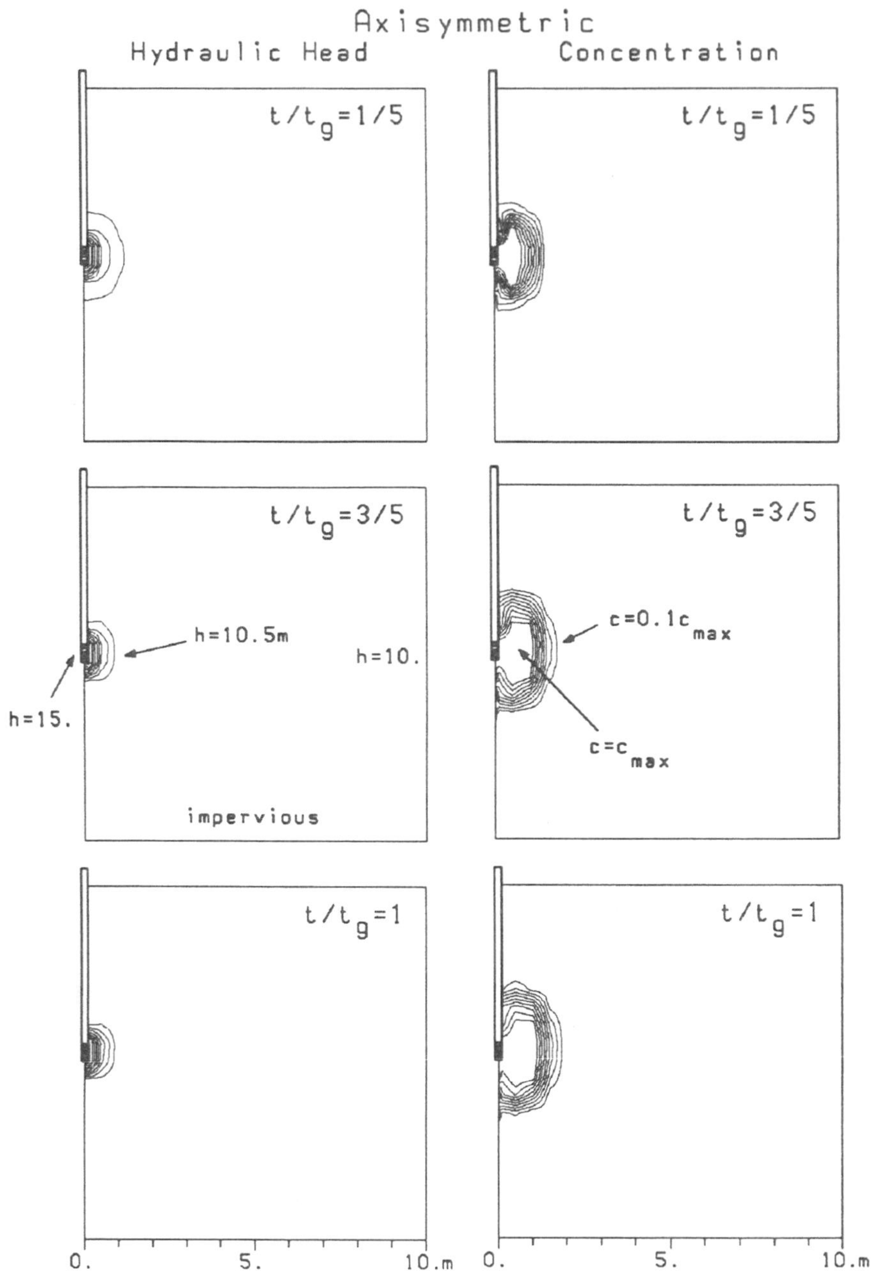

Figure 4. Contour lines of the hydraulic head and of the concentration for different time values. Contour line intervals: hydraulic head 0.5m, concentration 0.1.

3.1 Freezing Tests

The experimental set up shown in Figure 5 consists of a thermally insulated cylinder, having internal diameter and height of 1 m, filled with saturated sand. A steel pipe, installed at the centre of the container, is used for the circulation of liquid nitrogen during the test. The cylindrical geometry of the container and its insulation (consisting of polyurethane foam) lead to axisymmetric and planar heat flow conditions during the tests. This permits an easier interpretation of the experimental results by means of a one dimensional finite element model.

Nine thermal transducers were placed within the sand during its deposition. They were situated at the level of the mid-plane of the container in order to reduce the influence of its upper and lower bases. The transducers were located along three radii, at an angle of 120° from each other, and at a distance of 150 mm, 270 mm and 390 mm from the cylinder axis. Three additional transducers were used to measure the temperatures of the outer part of the insulation, of the nitrogen at its outlet from the steel pipe and the room temperature. The transducers were connected to a personal computer for the real time acquisition of the temperatures, and for their graphical representation.

Various tests have been performed using freezing pipes with different external diameters. In order to facilitate the subsequent interpretation of the experimental data, the same temperature (-183°C) was imposed at the outlet of the nitrogen for all tests, by a suitable regulation of its flux.

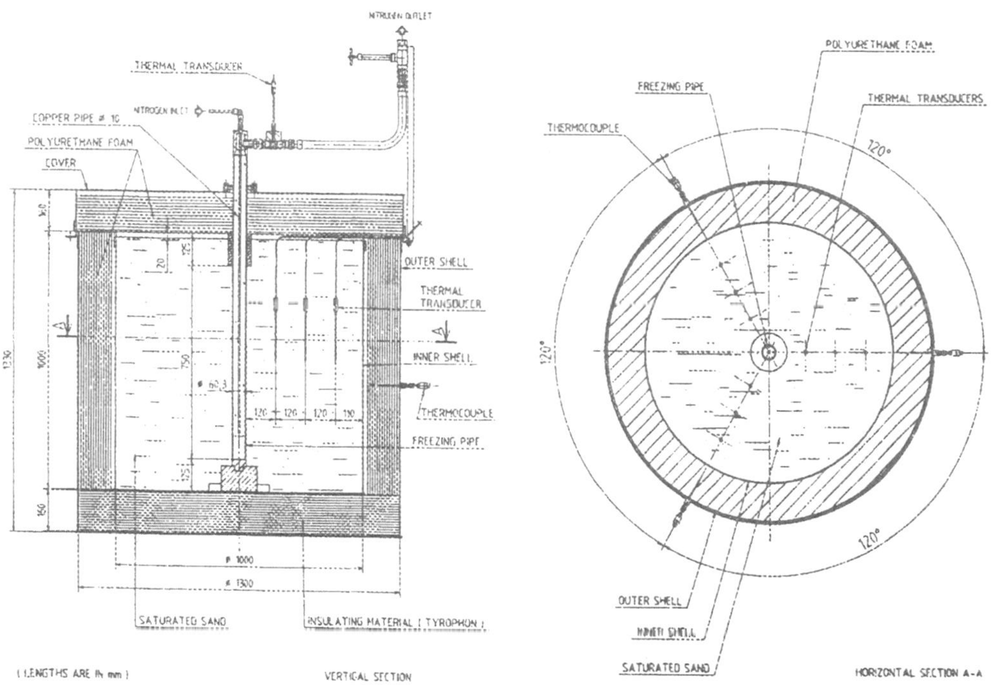

a) b)

Figure 5. a) Vertical section and b) horizontal section through A-A in (a) of the equipment used for laboratory freezing tests (All dimensions are in mm).

The measured changes of temperature with time are shown, for one of these tests, by the dashed lines in Figure 6. The diagrams a, b and c, corresponding to the temperatures within the sand, were obtained by averaging the readings of the transducers located at the same distance from the freezing pipe. The maximum difference between the temperatures measured at the same distance from the pipe, on different radii, was less than 1.5°C. This indicates that the heat flow within the container was reasonably close to the assumed axisymmetric conditions.

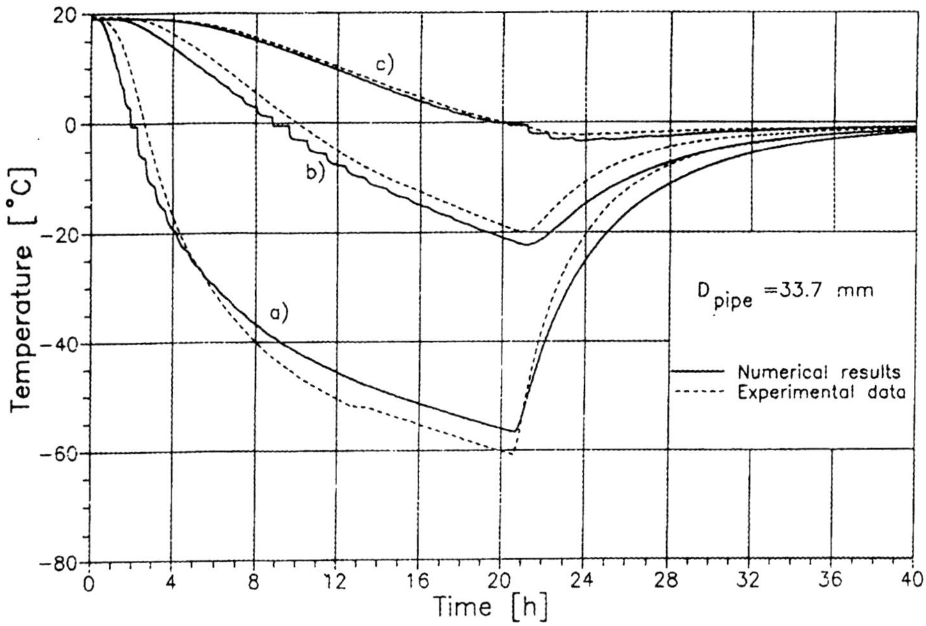

Figure 6. Comparison between experimental (dashed lines) and numerical (solid lines) temperature vs. time data. Temperatures of the transducers located at (a) 150.mm, (b) 270.mm and (c) 390.mm from the cylinder axis.

The flux of liquid nitrogen was stopped when the temperature of the farthest transducers from the axis reached 0°C. The recorded data show that, after some hours, the temperature of all transducers reached the same value equal to about -1°C. This remained practically constant for several days, until the equipment was dismounted. This indicates that the polyurethane provided an adequate thermal insulation of the container.

3.2 Finite Element Formulation and Boundary Conditions

The following main assumptions are adopted in the numerical analysis (Gioda, 1993):

- The soil is perfectly saturated.
- The thermal properties of soil are not influenced by the stress and strain variation caused by freezing.

- The effects of the possible water seepage flow and of the consequent heat convection can be ignored.
- The migration of water towards the phase change zone is neglected, which is a reasonable assumption for coarse soils, like sand and gravel.

The third assumption rules out the movement of fluid from the problem at hand. This seems reasonable; in fact a significant seepage flow should be avoided in order to permit an effective freezing treatment. In addition, this assumption allows using the same governing equation for heat transfer both above and below the freezing temperature, which permits a simpler formulation of the solution procedure.

The effects of the water/ice phase transition are introduced in the calculations through a technique that views the latent heat as a nodal "load" which is activated when the melting/freezing temperature is reached

The case without phase change will be first considered. The finite element formulation for time dependent, heat transfer problems can be derived by discretising the governing equations, and the relevant boundary conditions, rewritten in a suitable integral form. The step-by-step time integration is based here on the assumption that the time dependent variables have a constant rate during a time increment $\Delta t>0$. Noting that the material parameters and the coefficients characterising the boundary conditions are, in general, functions of temperature and, hence, of time, the following iterative equation is reached that leads to the nodal temperatures at time $t+\Delta t$,

$$\left[\frac{1}{2}\Delta t(\underline{M}^{i-1}_{t+\Delta t}+\underline{M}_t)+(\underline{N}^{i-1}_{t+\Delta t}+\underline{N}_t)\right]\underline{T}^i_{t+\Delta t}=$$

$$\Delta t(\underline{f}^{i-1}_{t+\Delta t}+\underline{f}_t)+\left[-\frac{1}{2}\Delta t(\underline{M}^{i-1}_{t+\Delta t}+\underline{M}_t)+(\underline{N}^{i-1}_{t+\Delta t}+\underline{N}_t)\right]\underline{T}^i_t \quad . \tag{7}$$

Here t denotes the time, $\underline{T}$ is the nodal temperature vector, $\underline{M}$ and $\underline{N}$ are the thermal conductivity and the thermal capacity matrices, $\underline{f}$ is the vector of the nodal heat flow due to the internal heat generation per unit volume and to the imposed rate of heat flow on part of the contour. The suffix i in eq.(7) denotes the iteration number.

At the beginning of iterations for a given time increment the temperature and all relevant quantities at time t are known, and the temperature at time $t+\Delta t$ is extrapolated on the basis of its values computed at the end of the preceding time steps. The iterative process for the current time increment, based on eq.(7), ends when the difference between the temperature at time $t+\Delta t$ evaluated at two subsequient iterations decreases below a given lower limit.

As to the boundary conditions, only those concerning imposed values of temperature or heat flow, on chosen portions of the contour, have been considered in the calculations. In fact, convection and radiation boundary conditions are barely relevant for the artificial freezing of soil and were not been introduced into the calculations.

3.3 Phase Transition

In the case of freezing/melting processes, a moving surface $F(x,y,t)=0$ exists that has an a priori unknown shape and that separates liquid and solid phases. Similarly to other moving boundary problems, e.g. unconfined seepage (see Crank, 1984; Cividini and Gioda, 2000; etc), two different conditions, related respectively to the temperature and to its space derivatives, characterise this surface. The first one imposes that the temperatures of the frozen T_f and unfrozen T_u phases facing the moving surface are equal to the melting/freezing temperature T_{mf} of the material. The second one represents the relationship between the net rate of heat flow in the direction normal to the moving surface, the latent heat per unit volume and the velocity v_n of the surface itself.

Within the finite element method, two main approaches are known for the analysis of the Stefan problem. The first one, the so called "variable mesh" approach (see e.g. Zabras and Ruan, 1990), modifies the geometry of the grid during the iterative solution process until a part of its boundary approximates with the desired accuracy the shape of the moving surface at the chosen time. Approaches of this type become exceedingly complex if several moving surfaces exist that tend to join with each other when freezing propagates.

Since this situation is likely to occur when dealing with the artificial freezing of ground, an alternative class of approaches, the so called "fixed mesh" techniques (Voller et al., 1990), can be considered. They operate on grid of constant geometry, hence the nodal co-ordinates remain unchanged during the solution process, and allow the moving boundary to pass through the elements. Usually these procedures are less accurate than the preceding ones, but offer the advantage of handling several simultaneous moving surfaces without requiring particular provisions.

Due to the above considerations, a fixed mesh technique has been adopted for the purposes of this study.

Different procedures have been also proposed in the literature for introducing the latent heat effects. They require, for instance, a fictitious variation of the heat capacity or the use of the enthalpy as an additional variable. The procedure here adopted sees the latent heat L as a positive or negative source of heat, uniformly distributed on the volume V, which becomes active when the freezing/melting temperature is reached. The effect of this source is accounted for in the finite element analysis in integral terms, as customary in its context. At the beginning of calculations, the nodal vector $\underline{q}_L$ of total latent heat is evaluated as follows

$$\underline{q}_L = \sum_e \int_{V_e} \underline{b}_e L\rho \cdot dV \quad . \tag{8}$$

In eq.(8), the summation runs over the element of the mesh, the index e denotes the e-th element, V is the element volume, $\underline{b}$ is the vector of the interpolation functions for the temperature and ρ is the mass density.

When the temperature at a node reaches the phase transition value, the node is constrained so that no further variation of its temperature can occur. The corresponding increment of nodal heat $\Delta\underline{q}_i$ is evaluated for each time step, until the accumulated heat reaches the previously evaluated total value $\underline{q}_L$,

$$\sum_i \Delta\underline{q}_i = \underline{q}_L \quad . \tag{9}$$

When the condition expressed by eq.(9) is fulfilled, where the summation runs over the time increments, the node is released and its temperature is again allowed to vary with time. If during a time step the accumulated nodal heat overcomes the corresponding latent heat, the node is released and a heat increment is imposed to it, for that time step, equal to the difference between the previously accumulated heat and the total latent heat.

3.4 Numerical Analysis of Laboratory Tests

The details of the determination of the thermal properties of the soil introduced in the finite element calculations were discussed by Gioda et al. (1994). Here only a comparison between experimental and numerical results is represented in Figure 6. They show that the finite element analysis is able to provide an acceptable estimation of the heat flow within the sand and that the progress of the frozen boundary can be evaluated with reasonable accuracy.

Note that the stepwise pattern of the numerically evaluated diagrams, in the vicinity of the phase change temperature, depends on the previously described procedure for taking into account the latent heat effects.

3.5 Effect of Diameter and Distance of the Freezing Pipes

The diameter of the freezing pipes, and their relative distance, are among the geometrical parameters influencing the in situ freezing treatments. In order to investigate their effect, a parametric study was performed based on the described finite element program.

The finite element mesh adopted in the calculations represents a horizontal section of a portion of a grid of regularly spaced pipes. The temperature of the coolant is imposed at the external surface of the pipes, while the remaining boundaries of the mesh, account taken of the symmetry of the problem, are considered as impervious to heat. In the calculations the material parameters coincide with those used in the analysis of the laboratory test. The initial temperature of the soil is 20°C.

Some of the results of this parametric study are summarised in Figure 7 through the diagrams relating the progress of the frozen front, in terms of the non-dimensional quantity $x/(d/2)$, with time. Here d represents the distance between the axis of two adjacent pipes and x is the distance between the pipe axis and the corresponding 0°C isotherm. Consequently, the frozen fronts of two adjacent pipes join with each other when $x/(d/2)=1$.

Figure 7 shows two sets of curves corresponding, respectively, to -183°C and to -80°C. With reference to actual field conditions these temperatures refer to pipes close to, and far away from, the inlet of liquid nitrogen. The curves within each set correspond to pipes having external diameter of 88.9 mm, 60.3 mm e 33.7 mm. Similar diagrams were also obtained by changing the distance between two adjacent pipes (Gioda et al., 1994).

These results show that the numerical analysis is able to solve heat transfer problems related to ground freezing with an acceptable accuracy and that it could help the designer in choosing the characteristics of a freezing treatment and, perhaps, to "optimise" it.

4 Reinforced Earth Structures

Geosynthetic or metal grids embedded into layers of cohesionless soils form a composite material that is used in geotechnical practice for retaining structures, embankments, slope stabilisation, etc. A consequence of the growing use of this technology is the increasing interest for the

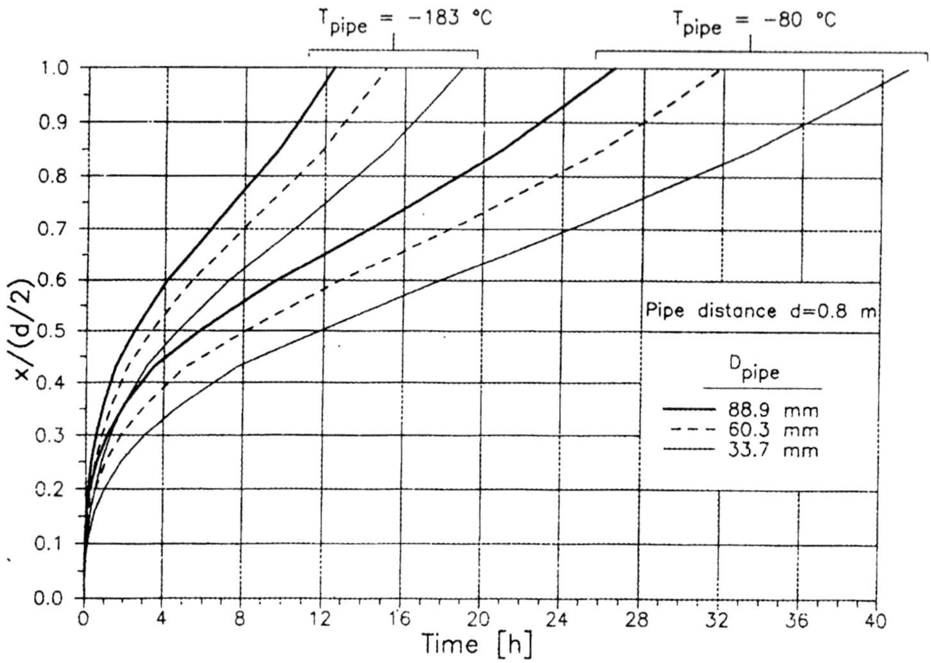

Figure 7. Variation with time of the distance x between the freezing pipe axis and the corresponding frozen front for various coolant temperatures T and pipe diameters D.

laboratory evaluation of the stress-strain response of reinforced sand samples, in view of the calibration of constitutive laws that can be adopted in the numerical analysis of such structures.

In the following some results are presented of an experimental study carried out on samples of sand reinforced with geotextiles and of their finite element interpretation. This study led to the proposal of an elasto-plastic constitutive law, characterised by a non-isotropic yield criterion, governing the behaviour of the composite material (Cividini et al., 1994).

An application to the finite element analysis of a vertical earth wall is subsequently presented. In particular, two alternative approaches are comparatively evaluated. They are referred to as "inhomogeneous" and "non-isotropic" schemes (Cividini et al., 1997). When the first scheme is adopted (e.g. Smith and Segrestin, 1992), the reinforced earth structure is discretised considering separately the reinforcements and the sand layers between them. On the contrary, in the "non-isotropic" case the reinforced medium is made equivalent to a continuous, non-isotropic, elasto-plastic composite material (e.g. De Buhan and Salencon, 1983, Cividini et al., 1994).

4.1 Experimental Investigation and Modelling of the Laboratory Tests

The ongoing experimental program includes axisymmetric triaxial and plane strain tests, aimed at evaluating the influence of the reinforcements on the overall load-displacement behaviour of the reinforced samples (Balduzzi and Milani, 1999; Lüthi, 2000). The results obtained so far allow identifying some basic features of the mentioned behaviour to be accounted for in the finite element analyses.

4.1.1 Reinforcement Characteristics

The laboratory tests were based on samples having a relatively small size. In particular, the cylindrical samples adopted for the standard triaxial tests have diameter of 7.cm and height of 14.cm. The prismatic samples for the plane-strain tests (Drescher et al., 1990; Cividini, 1997) have dimensions of 4.cm x 8.cm x 14.cm. This does not permit the use of the relatively thick geotextiles or geogrids actually used in field applications. Consequently, a thin polypropylene, spunlaid needled, nonwoven geotextile was chosen for the tests.

This choice leads to experimental results that cannot be directly used for the analysis of actual design problem. They permit, however, to define the main characteristics of a suitable numerical model of the composite material, which can be applied to the analysis of design problems, upon proper calibration of its parameters.

The modulus of elasticity and the tensile yield limit of the geotextile were evaluated on the basis of the force-elongation diagrams obtained from wide-width tensile tests. Note that these are conservative values, since the normal pressure exerted by the soil on the reinforcements is likely to increase the value of the mentioned mechanical parameters.

Some direct shear tests were also carried out to evaluate the shear resistance of the sand-reinforcement interface (Badiani and Zavanella, 1996). Their results indicate that, in the case of extensible reinforcements, the frictional resistance of the interface is close to that of the natural sand. This permits to avoid the use of particular joint elements in the finite element interpretation of the laboratory tests.

4.1.2 Axisymmetric Triaxial Tests on Natural and on Reinforced Sand

Some triaxial tests were performed on unreinforced sand samples adopting different values of the cell pressure. The samples were compacted, with a relative density of 70%, within a Plexiglas mould in subsequent layers of constant thickness. The "moist tamping" technique (Mulilis et al., 1975) was used to this purpose, which ensures a sufficiently uniform density distribution throughout the specimen. These tests led to the evaluation of the effective friction angle and of the average values of the elastic modulus of the sand, estimated from the initial part of the stress-strain curves.

The elastic and shear strength parameters of the reinforced sand were subsequently determined through consolidated, drained triaxial tests. It was observed that the elastic and frictional parameters of the reinforced sand are quite close to those of the "natural" material sand, however a non-negligible cohesion intercept is also present.

A relevant characteristic of the behaviour of reinforced samples under compression is that their shear resistance depends on the relative orientation of the reinforcements with respect to the principal stress directions. In fact, due to the negligible capacity of the reinforcements to carry compressive stresses, the shear resistance of the composite material practically coincides with that of the natural sand if the minor principal stress is parallel to the reinforcements. On the other hand, if the minor principal stress is normal to the reinforcements, the overall shear resistance is markedly increased. This observation indicates that a non-isotropic yield condition should be adopted for the homogeneous composite material "equivalent" to the reinforced sand.

4.1.3 Numerical Modelling of the Axisymmetric Triaxial Tests

Two numerical simulations of the above triaxial tests were carried out. The first one by discretising separately the sand layers and the reinforcements (inhomogeneous approach). The second one was based on the homogeneous approach and a non-isotropic elasto-plastic criterion (Cividini et al., 1994).

a) Inhomogeneous approach

An elastic perfectly plastic behaviour was adopted in the calculations, governed by Drucker-Prager yield condition, with a non associated flow rule, for the sand and by Huber-Hencky-von Mises criterion for the reinforcement.

In spite of this simple material model, in which the material parameters do not depend on the stress and strain history, an acceptable agreement between experimental and numerical results was obtained. This suggested the extension of the numerical analysis to the non-isotropic approach, which requires the choice of a material law describing the overall behaviour of the reinforced samples.

b) Non-isotropic Approach: a Yield Criterion for the Reinforced Sand.

The following modified form of Drucker-Prager failure condition was adopted for the non-isotropic composite material equivalent to the reinforced sand,

$$F(\underline{\sigma},\vartheta) = J_2 + \alpha(\vartheta)\cdot I_1 \cdot K(\vartheta) \qquad (10)$$

In eq.(10), F represents the yield condition, J_2 is the square root of the second invariant of the stress deviator and I_1 is the first stress invariant. The material parameters α, K (depending, respectively, on the frictional and cohesive characteristics), are function of the angle ϑ between the major principal stress and the reinforcements,

$$\alpha(\vartheta) = \left[(\alpha_M + \alpha_m) + (\alpha_M - \alpha_m)\cdot \cos(2\vartheta)\right] \ , \qquad (11a)$$

$$K(\vartheta) = \left[(K_M + K_m) + (K_M - K_m)\cdot \cos(2\vartheta)\right] \ . \qquad (11b)$$

The parameters α_M and K_M in eqs.(11) are determined from triaxial compression tests on sample with horizontal reinforcements while α_m and K_m, that characterise the samples with vertical reinforcements, are here assumed equal to those of the natural sand.

The above expression of the failure condition is particularly suitable for elasto-plastic analyses based on iterative solution procedures (for instance the initial stress method, Zienkiewicz et al., 1969). It is sufficient, in fact, to adopt at each iteration the values of the parameters α and K determined through eqs.(11) on the basis of the stress state calculated at the end of the previous iteration.

From the programming viewpoint, however, the failure condition expressed by eq.(10) involves some additional problems with respect to use of the standard Drucker-Prager criterion. In particular, the non-linear expression of eq.(10) in terms of the Cartesian stress components requires the numerical evaluation of the intersection between the yield surface and the incremental stress vector. A second difference concerns the gradient of the yield function that appears in the elasto-plastic constitutive matrix. In fact, since in eq.(10) the material parameters are functions of

ϑ, which in turn depends on the Cartesian stress components, the gradient of F contains a series of additional terms with respect to those of the standard yield criterion.

c) Numerical Validation

To verify the agreement between the results of the inhomogeneous and non-isotropic models, both of them were adopted in the numerical simulation of compression tests, in plane strain conditions, with various orientations of the reinforcements.

The results in Figure 8 show a reasonable agreement between the two approaches, even though some discrepancy can be observed. In fact, for the inhomogeneous analyses the stress difference at failure is practically constants when the angle ϑ exceeds about 60°. In this case, in fact, the reinforcements are subjected to compression and do not provide any appreciable contribution to the overall shear resistance. On the contrary, eq.(10) introduces a continuous variation with ϑ of the shear stress at failure, regardless the stresses carried by the reinforcements. This drawback, which could be eliminated by suitably modifying eq.(10), suggests extending the experimental investigation for verifying the effect of the reinforcement orientation.

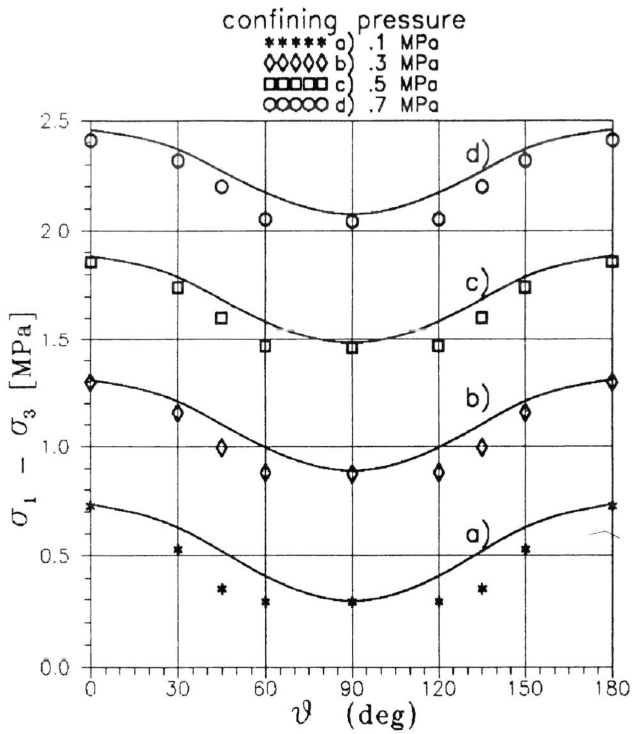

Figure 8. Variation of the compressive strength with the orientation of the reinforcements from inhomogeneous (dots) and non-isotropic (solid lines) analyses.

d) Non-isotropic Analysis of the Axisymmetric Triaxial Tests and Improvement of the Constitutive Model

The non-isotropic yield criterion was applied to the analysis of the triaxial tests carried out on reinforced sand samples, assuming an elastic perfectly plastic behaviour for the composite material. The comparison between experimental and numerical results shows that the assumption of perfectly plastic behaviour in the non-isotropic analyses does not provide an adequate approximation of the observed stress-strain curves. In particular, the axial strain for a given stress level is underestimated. This depends on the low axial stiffness of the reinforcements that was not accounted for in developing the mentioned homogeneous model.

To reduce this discrepancy, the non-isotropic calculation was modified introducing a simple strain hardening law. At the beginning of calculation the yield condition for the reinforced material coincides with that of the natural soil. When, in a finite element, the square root of the second invariant of the deviatoric plastic strains J_ε overcomes a chosen lower limit, the corresponding shear strength parameters (i.e. the cohesion in the present context) are linearly increased with J_ε up to those of the equivalent homogeneous material. This process ends when J_ε attains an upper limit.

The hardening analyses provided an approximation of the experimental results similar to that of the inhomogeneous calculations, upon a proper calibration of the upper and lower limits of J_2. This indicates that a hardening law has to be included into the non-isotropic model before attempting the application to actual design cases. Note that the hardening law depends on the stiffness of the reinforcements, as discussed in Section 4.2.

4.1.4 Plane Strain Tests.

As previously observed, the shear resistance of reinforced samples depends on the angle existing between the reinforcements and the principal stress directions. This effect was investigated by a series of plane strain compression tests (Cividini and Sterpi, 2000) with different values of the angle β between the horizontal direction and the reinforcements.

For briefness, only the results of six plane strain tests are described here. They were performed on samples reinforced with four layers of extensible geotextile, 3.cm apart. Three of them intend to investigate the effects of the lateral pressure on samples with horizontal reinforcements, while the remaining three concern samples with different orientations of the reinforcements (β=0°, 15°, 30°, 45°). The angle β was kept within the mentioned limits since the specimens with β=45° already show a shear resistance comparable to that of the natural (unreinforced) sand. To quantify the influence of the lateral stress, two values of the cell pressure (50.kPa and 200.kPa) were used for the samples with β=0° and β=45°.

Figure 9 reports the variation of the axial stress and of the volumetric strain versus the axial strain for the first three tests. The confining effect provided by the horizontal reinforcements leads to monotonous stress-strain behaviour. In addition, the specimens show a certain amount of dilatancy for low values of the confining pressure.

The influence of the reinforcement orientation on the axial stress-strain curves and on the volumetric changes is presented in Figure 10. A decrease of the overall stiffness and of the shear resistance is observed with increasing β, while only minor differences exist in the volumetric behaviour.

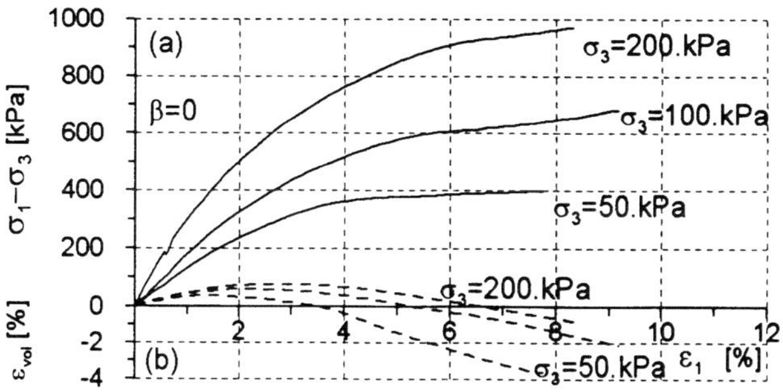

Figure 9. Plane strain compression tests on samples with horizontal reinforcements: (a) axial stress versus axial strain and (b) volume strain versus axial strain.

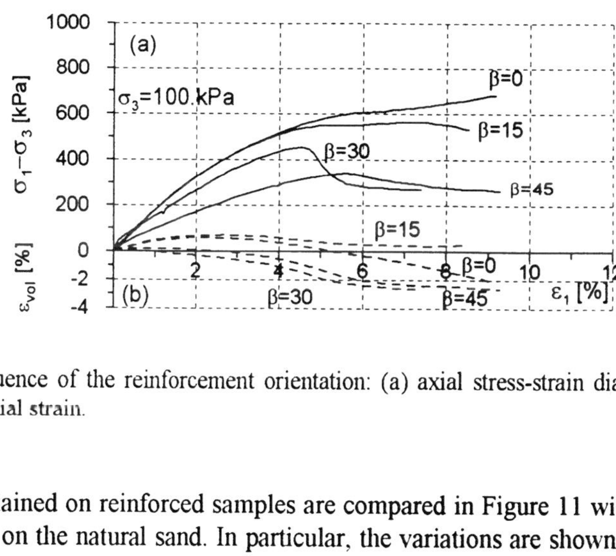

Figure 10. Influence of the reinforcement orientation: (a) axial stress-strain diagrams and (b) volume strains versus axial strain.

The results obtained on reinforced samples are compared in Figure 11 with those of the corresponding tests on the natural sand. In particular, the variations are shown of the first stress invariant, I_1, and of the square root J_2 of the second invariant of the deviatoric stress with the angle β. The data refer to both peak and end-of-test conditions. As previously observed, the shear resistance of the reinforced samples almost coincides with that of the unreinforced specimens when β approaches 45°. This implies that reinforced earth structures may give substantially different responses to external load increments, depending on the angle existing between the reinforcements and the compressive principal stress.

The experimental results indicate also that the overall behaviour of the composite material depends on a variety of mechanical and geometrical parameters, characterising both the basic materials and their assemblage. The numerical analyses discussed in the next Section intend to provide some insight into this behaviour and to single out those properties that have a major influence on overall response of the samples.

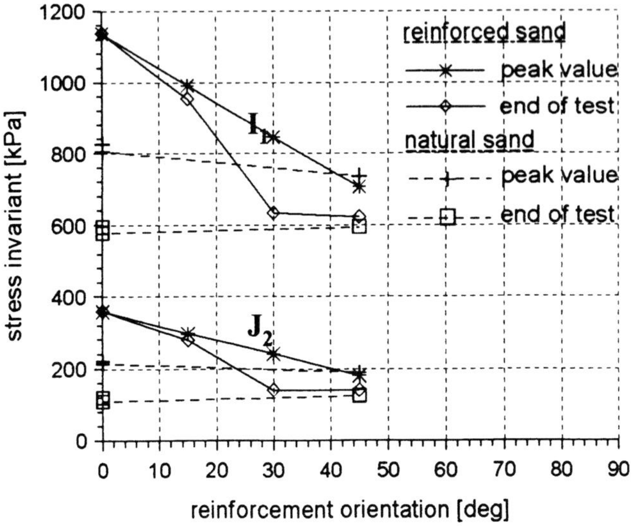

Figure 11. Values of the stress invariants in the reinforced specimens (solid lines) and in the natural sand (dashed lines), at the peak condition and at the end of the test.

4.1.5 Inhomogeneous Finite Element Analysis of Plane Strain Tests

The numerical simulation of the plane strain tests was carried out through an "inhomogeneous" finite element approach. Details of the performed analyses are reported in (Cividini and Sterpi, 2000), here only their main aspects are recalled.

An elastic-ideally plastic behaviour characterised by Huber-Hencky-von Mises criterion and a strain-softening model were adopted, respectively, for the reinforcements and for the sand. The strain-softening constitutive law (Cividini and Gioda, 1992) is based on the assumption that a peak yield condition exists (characterised by cohesion and friction angle) until the irreversible plastic strains attain a given limit. Then, with increasing strains, a decrease of the shear strength parameters occurs, until the ultimate condition is reached. In spite of the rather simple constitutive model adopted, the numerical results provided an acceptable approximation of the experimental data for the tests on natural sand, as shown in (Sterpi, 2000).

As to the reinforced samples, the numerical simulation was focused on two aspects (Dei Cas and Zanini, 2000). The first one concerns the modelling of the stress-strain behaviour for axial strain smaller than 4-5%. In fact, the behaviour under relative small strains is crucial for the evaluation of the stress-strain field of actual structures in working conditions, which is mainly governed by the elastic properties. The second aspect concerns the transition from ductile to strain softening behaviour, with increasing inclination of the reinforcements (cf. Figure 10).

Let discuss first the modelling under small strain conditions. A proper approximation of the experimental data (cf. Figures 12 and 13) was obtained considering the elastic behaviour of the geotextile as transversally isotropic. The values of the five elastic constants of the transversally isotropic material were calibrated through a back-analysis of the experimental data. It turned out that the influence of two of them is particularly relevant.

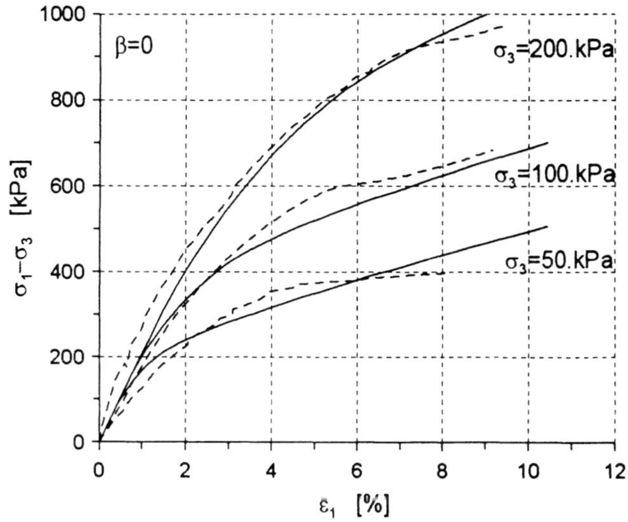

Figure 12. Comparison between the results of plane triaxial tests on horizontally reinforced samples (dashed lines) and of the elasto-plastic finite element analyses (solid lines).

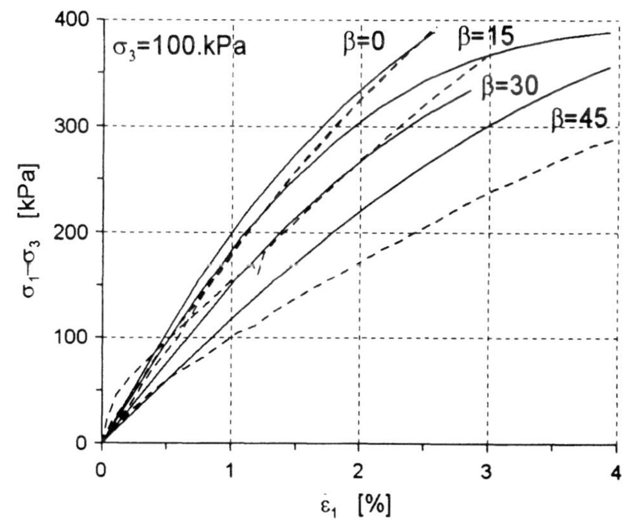

Figure 13. Specimens with inclined reinforcements: comparison between the results of plane strain tests (dashed lines) and of the finite element analyses (solid lines).

The first one is the elastic modulus E_v in the thickness direction. Its back-calculated value is consistent with that derived from the few data available from the laboratory determination of the nominal thickness of the geotextile.

The back-analysis of the second relevant parameter (the "out-of-plane" shear modulus G_{hv}) led to a relatively small value. Further investigation seems necessary to assess if such estimated low value is related solely to the geotextile properties or if, indirectly, it is influenced also by other physical or mechanical effects not considered in the numerical model. For instance, they can be due to the possible anisotropy of the sand layers or to the penetration of the sand particles within the geotextile, that in turn could modify its shear stiffness.

As to the second aspect of the numerical interpretation, it turned out that the strain softening isotropic model adopted for the sand layers is able to represent the marked influence of the reinforcement orientation on the stress-strain curves and on the failure mode (Figure 14). However, to reach an acceptable approximation of the experimental results on reinforced samples, the slope of the softening branch of the sand constitutive law should be lower than that used for interpreting the tests on the natural sand, and it should be related to the reinforcement inclination. Quite likely, this is due to a non-sufficient accuracy in modelling the mechanical behaviour of either sand or reinforcements.

The results presented in the preceding Sections suggest the following observations on the numerical modelling of tests on reinforced earth samples:

- When the non-isotropic approach is adopted, the elasto-plastic behaviour of the homogeneous, composite material ("equivalent" to the reinforced soil) requires the introduction of a non-isotropic yield condition and of a suitable strain hardening law.
- The above model provides an acceptable approximation of the results of axisymmetric triaxial tests with horizontal reinforcements. However, further improvements of this model seem necessary for an adequate simulation of the plane strain tests.
- The inhomogeneous approach provides a proper modelling of the plane strain tests if a strain softening law for the sand layers is adopted and the non-isotropic behaviour of the reinforcements is accounted for.

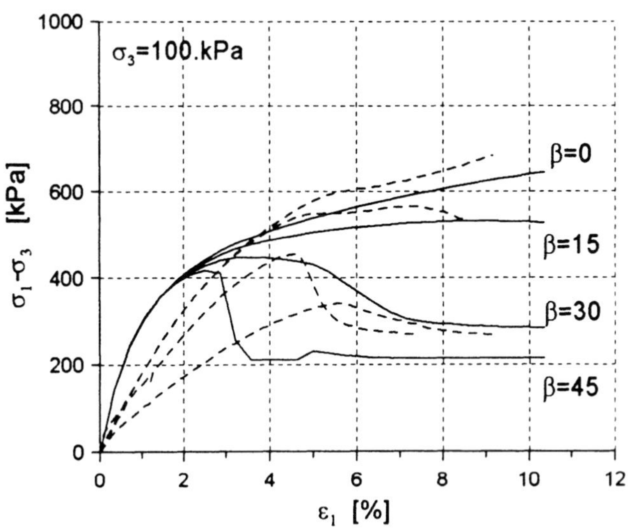

Figure 14. Specimens with inclined reinforcements: comparison between the results of plane strain tests (dashed lines) and of the strain-softening finite element analyses (solid lines).

4.2 Finite Element Analysis of a Reinforced Earth Structure

The design of reinforced earth structures is often based on the so-called limit equilibrium methods (see e.g. Ingold, 1982), which consider only the stress compatibility with respect to chosen yield criterion and the equilibrium conditions. These methods provide an estimation of the factor of safety of the structure, but cannot give any information on the strain and displacement distributions within it. This drawback can be overcome by adopting an elasto-plastic material model and a numerical solution technique. This permits, in fact, to evaluate the evolution of the strain, stress and displacement fields both during construction, through the simulation of the construction process, and in working conditions.

Here the results are summarised of a comparative evaluation of the mentioned inhomogeneous and non-isotropic finite element schemes, carried out for the case of a vertical earth wall (Smith and Segrestin, 1992; Cividini et al., 1997). In particular, the influence of the deformability of the reinforcements and of the three- or two-dimensional (plane strain) discretisation is investigated.

4.2.1 Finite Element Discretisation

The inhomogeneous and the non-isotropic approaches have been applied to the analysis of the construction process of a 7.5m high vertical wall considering two alternative kinds of reinforcements. The first one consists of the assemblage of oriented polyethylene grids (OPG). The second reinforcement consists of high adherence steel strips (HAS). Both OPG and HAS have vertical spacing of 0.75m, that corresponds to 10 reinforcement layers for the entire wall.

A first set of finite element analyses of this structure was based on elastic and elasto-perfectly plastic behaviour for the sand layers and the reinforcements, respectively. The analyses of both HAS and OPG cases were carried out using SoSIA (Soil Structure Interaction Analysis) finite element program (Cividini and Gioda, 1992), adopting both 2D, plane strain, and 3D models.

Figure 15 shows the 3D mesh used for the HAS case. This grid was used also for the OPG case, after modifying the length of the elements in the z direction to account for the different horizontal spacing of the reinforcements. Since the reinforcements are regularly spaced along the z direction, two vertical planes of symmetry were introduced that coincide with the centrelines of a "column" of reinforcements and of the adjacent "column" of sand. The grid for the 2D analyses is merely a vertical section (normal to the z-axis) of the 3D mesh.

The sand is discretised through 8 node isoparametric "brick" elements in the 3D mesh and through 4 node quadrilateral elements in the 2D calculations. The reinforcements are introduced by means of 4 node membrane elements and of 2 node truss elements in the 3D and 2D analyses, respectively. In the 2D case, the mesh used for the non-isotropic analyses does not contain the truss elements discretising the reinforcements.

Two details of the mesh, denoted as zones A and B in Figure 16, deserve some comments. The first one (zone A) concerns the thin layer of elastic elements, having the same characteristics of the sand, which was placed at the outer surface of the vertical wall. These elements are necessary in order to avoid the local failure of the sand between two adjacent reinforcements, caused by the lack of lateral confinement, which could induce a global failure of the wall. These thin elements should represent the actual surface "coating" installed on the vertical wall during construction.

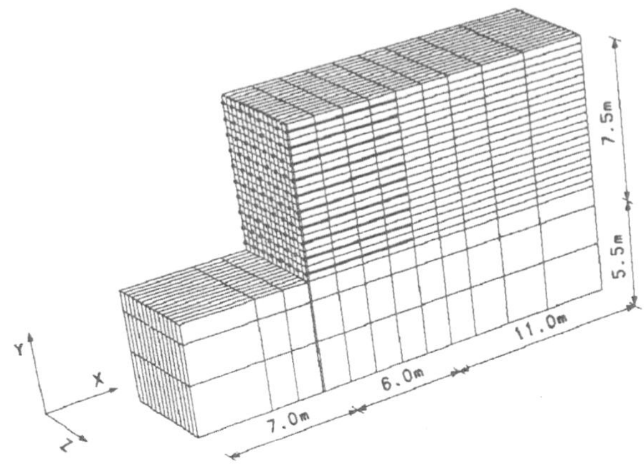

Figure 15. Finite element mesh used for the 3D analyses of the reinforced earth wall.

The second detail (zone B in Figure 16) concerns the use of "joint" elements placed at the end of the reinforcements. This was suggested in some studies to avoid that the reinforcements could "hang" on the sand beyond them. This provision does not seem necessary, and was not adopted here, since the elasto-plastic behaviour of the sand rules out any tensile or shear stress exceeding the limit imposed by its yield condition.

Both 2D and 3D analyses were initiated assuming an initial stress state in the soil mass representing the "foundation" of the reinforced structure. The wall construction was subdivided into 10 steps, each of which corresponds to the placement of one reinforcement and of one sand layer. Each step requires an elastic plastic analysis based on the new geometry of the mesh, considering the additional loads corresponding to the weight of the new sand layer.

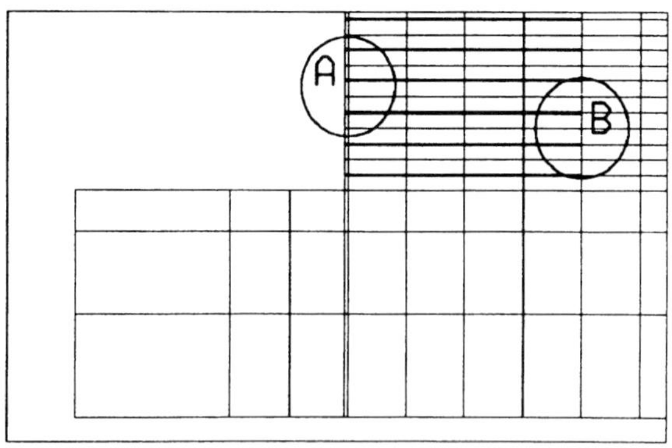

Figure 16. Details of the finite element mesh.

4.2.2 Determination of the Material Parameters for the Non-Isotropic Model

While the determination of the material constants to be used in the inhomogeneous analyses is straightforward, since they coincide with those of sand and reinforcements, a preliminary calculation is necessary to work out the shear strength parameters for the non-isotropic model. To this purpose, two plane strain analyses have been carried out, for the HAS and OPG cases, based on the inhomogeneous approach. They concern a rectangular portion of the reinforced earth structure, subjected by step to a vertical deformation up to failure. On the basis of the numerically evaluated ultimate load, the value of the K_M parameter in eqs.(11) is readily worked out in both HAS and OPG cases. The orthotropic elastic constants of the composite material were determined adopting the method proposed by Salamon (1968).

As observed in the interpretation of the triaxial tests (Sec. 4.1.3d), to obtain a fair agreement between inhomogeneous and non-isotropic results in the OPG case, it is necessary to account for the hardening behaviour of the non-isotropic composite material.

4.2.3 Comparison of the Finite Element Schemes in Modelling the Retaining Wall

Let consider now the results obtained in the analysis of the construction process of the reinforced structure, for both HAS and OPG cases, which are summarised in Figures 17 to 21. The analyses are based on a 3D inhomogeneous model and on 2D inhomogeneous and non-isotropic models. The hardening behaviour was accounted for in the 2D non-isotropic calculations.

a) Comparison of 2D and 3D modelling

The horizontal displacements at completion of the wall construction obtained by the 2D and 3D calculations are compared in Figure 17.

The difference between 2D and 3D results is negligible for the OPG case, while is about 10% in the HAS case. To explain this difference it should be considered that a relatively large distance exists between the high adherence steel strips (HAS) in the horizontal direction. As a consequence, the maximum displacements of the vertical sand "column" between two adjacent HAS "columns" obtained by the 3D calculation is slightly larger than the "average" horizontal displacement from the 2D calculations, in which the reinforcements are "smeared" in the direction normal to the mesh.

b) Comparison of the 2D schemes

The development of the horizontal displacements of the vertical wall during construction obtained by the 2D analyses is shown in Figures18a,b that refer, respectively, to the OPG and HAS cases. It can be observed that in both cases a non-linear relationship exists between the structure height and its lateral deformation. In fact, more than 50% of the total displacement develops during the last stage (about 25%) of construction.

For both OPG and HAS cases the non-isotropic, hardening 2D analyses lead to results quite similar to those of the corresponding 2D inhomogeneous analyses. This indicates that a proper non-isotropic model is able to provide adequate results, regardless the stiffness of the reinforcements.

Figure 19 shows the distribution along the wall of the "equivalent" pressure exerted by the reinforcements. This pressure is obtained by dividing the maximum axial force in the reinforcements at the end of construction by the portion of the total area of the wall corresponding to each of them. The data refer only to the inhomogeneous analyses, since the non-isotropic ones cannot provide a direct evaluation of the tensile force along the reinforcement.

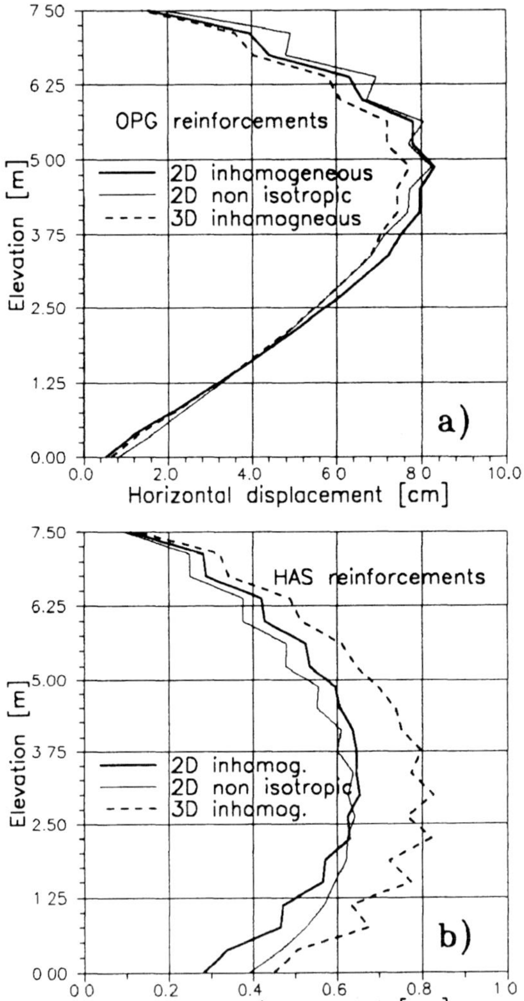

Figure 17. Horizontal displacements of the vertical wall at the end of construction from 2D and 3D analyses of OPG (a) and HAS (b) cases.

As already observed by Smith and Segrestin (1992), the equivalent pressure in the HAS case is close to the active earth pressure and is larger than that calculated in the OPG case. This is a consequence of the different behaviour within the reinforced zone in the HAS and OPG cases.

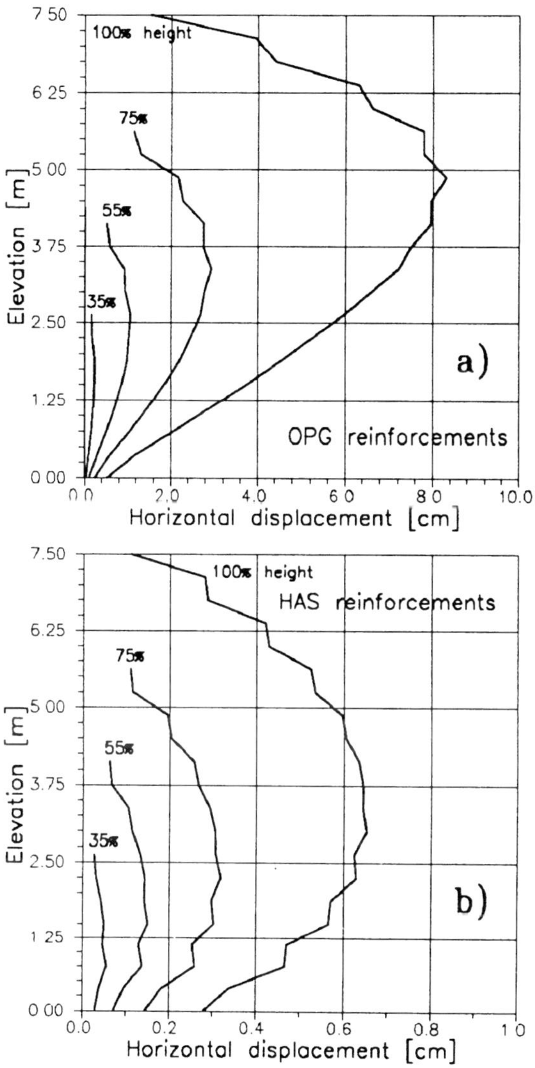

Figure 18. Horizontal displacements of the vertical wall at various stages of construction from 2D inhomogeneous analyses of OPG (a) and HAS (b) cases.

Finally, to get an insight into this effect, let consider the distribution of the irreversible deformation at the end of construction represented in Figures 20 and 21 by the square root of the second invariant of the deviatoric plastic strains.

The stiffness of the reinforcements has a marked influence on the overall behaviour of the reinforced structure. In particular, the use of deformable reinforcements lead to a global behaviour characterised by substantial hardening effects and by the development of relatively large plastic deformation within the structure, even under loads quite lower than the ultimate one. If stiff reinforcements are used, the behaviour of the structure is close to the elastic-perfectly plastic one, and appreciable plastic strains develop within it only approaching the limit load. This indicates that the behaviour of the structure with inextensible reinforcements under working loads is mainly governed by the elastic properties.

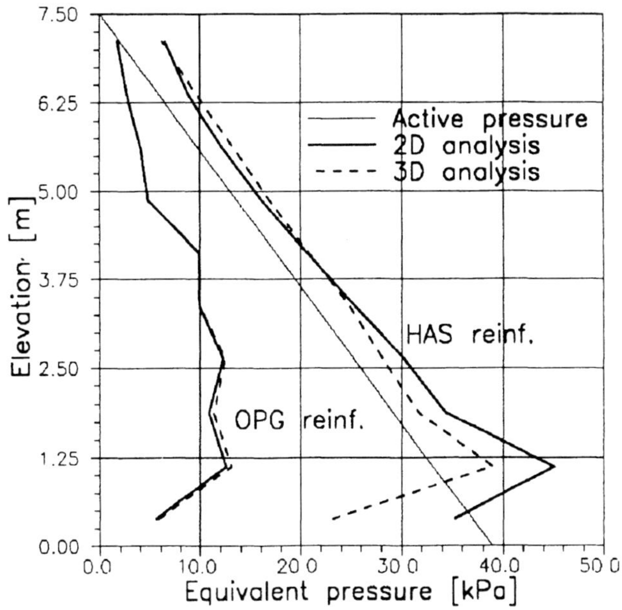

Figure 19. Distribution along the wall of the equivalent pressure of the reinforcements.

Concluding, both inhomogeneous and non-isotropic finite element approaches have advantages and shortcomings. For instance, those of the first group usually provide an accurate description of the stress and strain states in the reinforcements and sand layers, which is not possible with the non-isotropic scheme. On the other hand, they require relatively fine meshes, since the geometry of each layer must be considered and, hence, they lead to an increase of the computational effort.

The non-isotropic, homogeneous schemes do not require an accurate geometrical description of the layered system. However, the constitutive model for the composite material equivalent to the reinforced sand should have some complex characteristics, even under the assumption of elastic-perfectly plastic behaviour of sand and reinforcements, and neglecting the local problems due to the non linear interaction at their interface. In fact, it should be based on a non-isotropic yield condition and, in addition, a hardening behaviour should be introduced, strongly influenced by the characteristics (the stiffness, in particular) of the reinforcements. This implies a non-negligible complexity of the constitutive equations and a consequent increase of the programming effort with respect to the relatively simple inhomogeneous approach.

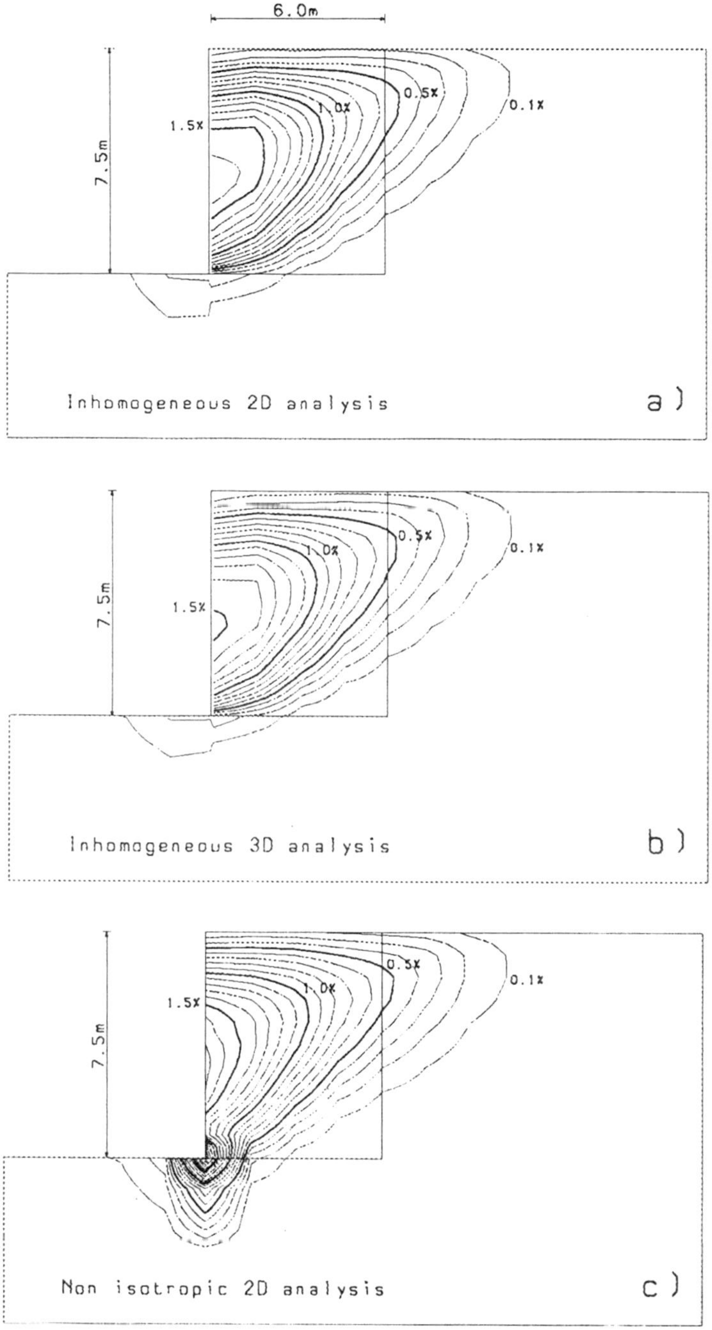

Figure 20. Vertical wall reinforced with oriented polyethylene grids (OPG). Contour lines of the square root of the second invariant of deviatoric plastic strains: a) inhomogeneous 2D analysis; b) inhomogeneous 3D analysis; c) non-isotropic 2D analysis.

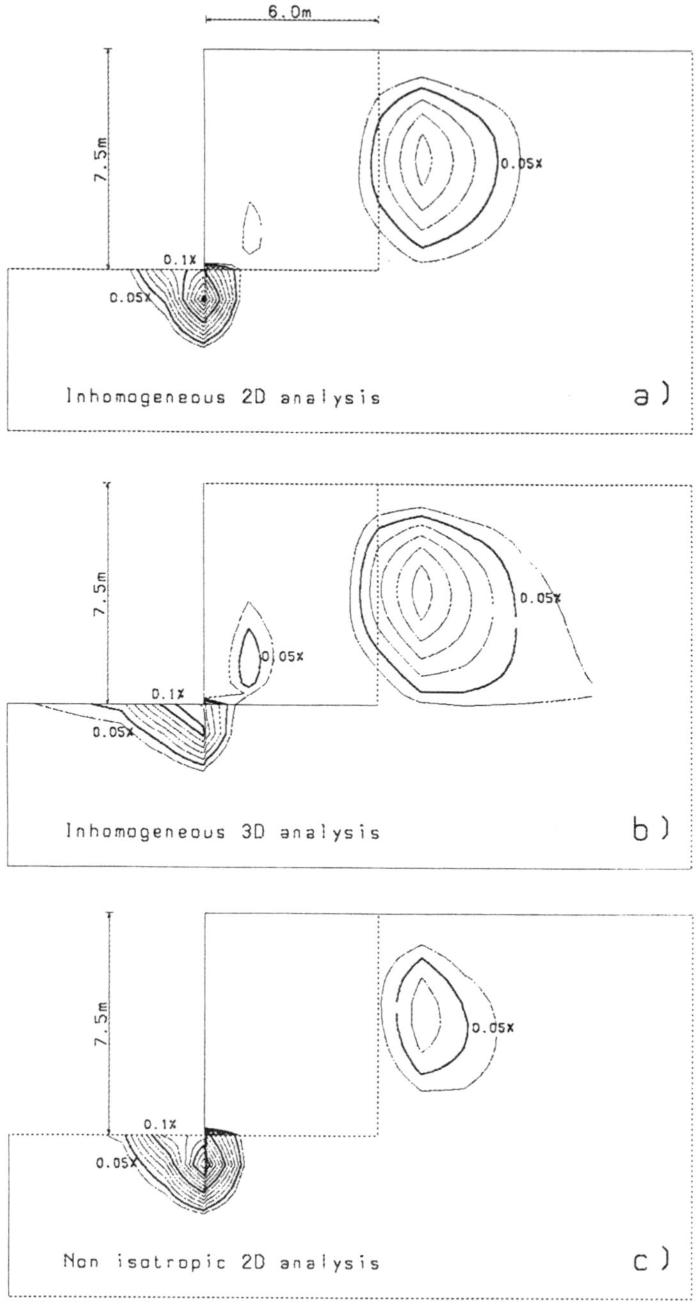

Figure 21. Wall reinforced with high adherence steel strips (HAS). Other characteristics as in Figure 20.

5 Concluding Remarks

A perhaps ambitious aim of this work was to show that the joined use of experimental and numerical techniques could contribute to the solution of complex geotechnical problems, and that they could represent a useful tool for the engineers during the design stage of the project.

To this purpose, some applications have been discussed of laboratory and numerical (finite element) procedures to the analysis of soil improvement techniques. In particular, three techniques of relatively common use in engineering practice have been considered, namely: the low pressure grouting of granular soils, the artificial freezing of grounds and the earth reinforcement through layers of metallic or synthetic materials.

Depending on the examined case, different conclusions have been reached on the advantages and shortcomings of the adopted experimental/numerical approach. In general terms, however, it was observed that the numerical analysis is able to provide a reasonable evaluation of the improvement process and/or of the subsequent behaviour of the treated soil. This requires the calibration of the relevant material parameters and the "validation" of the numerical procedure. These were based here on the interpretation and on the simulation of laboratory tests, properly chosen on the basis of the improvement technique under examination.

Acknowledgements

The financial support of the Ministry of University and Research of the Italian Government is gratefully acknowledged.

References

Andersland O.B., Gallavresi F., Goto S., Saareleinen S., Slunga E. (1989). General report of the specialty session on ground freezing. *Proc. XII ICSMFE, Rio de Janeiro.*

Andersland O.B., Ladanyi B. (1994). *An introduction to frozen ground engineering*. Chapman & Hall, NewYork.

Badiani B., Zavanella L. (1996). Studio sperimentale su un modello di opera in terra rinforzata, Laurea Thesis, Dept. of Civil Engineering, Universita' di Brescia.

Balduzzi G., Milani N. (1999). Indagine sperimentale del comportamento di una sabbia rinforzata, Laurea Thesis, Dept. of Structural Engineering, Politecnico di Milano.

Bear J., Verruijt A. (1987). *Modeling groundwater flow and pollution*. Dordrecht: D.Reidel.

Cividini A. (1997). Plane strain testing of strain softening soils. *Proc. 14th Int.Conf. on Soil Mechanics and Foundation Engineering*, Hamburg, Vol.1, 255-258.

Cividini A. (2000). A study of the low pressure grouting of granular soils. *Proc. of John Booker Memorial Symposium, Sydney*, Rotterdam: Balkema.

Cividini A., Donelli M., Sterpi D. (1994). On the mechanical behaviour of sand reinforced with geotextiles. *Proc.3rd Europ.Conf.Num.Meth.Geotech.Engin.*, Manchester, 299-304.

Cividini A., Gioda G. (1992). A finite element analysis of direct shear tests on stiff clays. *Int. Journal for Numerical and Analytical Methods in Geomechanics*, 16:869-886.

Cividini A., Gioda G. (2000). Finite element analysis of free surface seepage flows. *Chap. 20 in "Modeling in Geomechanics", (M.Zaman, G.Gioda, J.R.Booker Eds.)*, John Wiley & Sons Ltd, Chichester U.K., 505-524.

Cividini A., Gioda G., Sterpi D. (1997). An experimental and numerical study of the behaviour of reinforced sands, *(Invited paper) Proc. 9th Conf. of the Int. Association for Computer Methods and Advances in Geomechanics,* Wuhan, *People Republic of China*: 15-30.

Cividini A., Sterpi D. (2000). Plane strain tests on reinforced sand and their numerical modelling. *Proc. 2nd Europ. Conference on Geosynthetics and Exhibition*, Bologna, *Italy.*

Crank J. (1984). *Free and moving boundary problems.* Oxford University Press, Oxford, U.K.

De Buhan P., Salencon J. (1983). Determination of a macroscopic yield criterion for a multilayered material. *Proc. Int. Colloquium on Failure criteria for structured media,* Grenoble.

Dei Cas M., Zanini T. (2000). Analisi numerica di prove triassiali piane nelle deformazioni su sabbia rinforzata, Laurea Thesis, Dept. of Structural Engineering, Politecnico di Milano.

Drescher A., Vardoulakis I.G., Han C. (1990). A biaxial apparatus for testing soils, *Geotechnical Testing Journal,* 13, 226-234.

Frivik P.E. (1980). State-of-the-art report on ground freezing: thermal properties, modelling of processes and thermal design. *Proc. 2nd Int.Symp.on Ground Freezing, Trondheim.*

Gioda G. (1993). A finite element analysis of the artificial freezing of sand. *Seminar on the Impact* of *the Computational Mechanics to Engineering Problems*, Sydney.

Gioda G., Locatelli L., Gallavresi F. (1994). A numerical and experimental study of the artificial freezing of sand. *Canadian Geotechnical Journal*, 31(1).

Ingold T.S. (1982). An analytical study of geotextile-reinforced embankments. *Proc. 2nd Int.Conf. on Geotextiles*, Las Vegas.

Lüthi S. (2000). Comportamento di materiali geotecnici in prove triassiali piane ed assialsimmetriche, Laurea Thesis, Dept. of Structural Engineering, Politecnico di Milano.

Mulilis J.P., Chan C.K., Seed H.B. (1975). The effects of method of sample preparation on the cyclic stress-strain behaviour of sands. EERC 75/18, Univ. of California at Berkeley.

Nonveiller E. (1989). *Grouting theory and practice*. Elsevier.

Rowe R.K., Booker J.R. (1985). 1-D pollutant migration in soils of finite depth. *Journal of Geotechnical Engineering, ASCE* 111(4): 479-499.

Rowe R.K., Booker J.R. (1997). Recent advances in modeling of contaminant impact due to clogging. *Proc.9th Int.Conf. of International Association for Computer Methods and Advances in Geomechanics; Wuhan, People Republic of China*: 43-56.

Salamon M.D. (1968). Elastic moduli of a stratified rock mass. *Int.J. Rock Mech. Min.Sciences,* 5, 519-527.

Smith I.M., Segrestin P. (1992). Inextensible reinforcements versus extensible ties - FEM comparative analysis of reinforced or stabilized earth structures. In *"Earth Reinforcement Practice"* (H.Okhiai, S.Hayashi, J. Otani, Edts.), Balkema, pp.425-430.

Sterpi D. (2000). Influence of the kinematic testing conditions on the mechanical response of a sand. *Computers and Geotechnics*, 26, 23-41.

Voller V.R., Swaminathan C.R., Thomas B.G. (1990). Fixed grid techniques for phase change problems: a review. *Int.J.Numerical Methods in Engineering*, 30: 875-898.

Wendland E., Schmid G. (2000). A symmetrical streamline stabilization scheme for high advective transport. *International J.Numer.Anal. Methods in Geomechanics,* 24, 29-45.

Zabaras N., Ruan Y. (1990). Moving and deforming finite element simulation of Stefan problem, *Communications in Applied Numerical Methods*, 6: 495-506.

Zienkiewicz O.C., Valliappan S., King I.P. (1969). Elasto-plastic solutions of engineering problems. Initial-stress, finite element approach. *International Journal for Numerical Methods in Engineering*, 1, 75-100.

Application of Elasto-Plastic Analysis to Tunnelling

Annamaria Cividini, Giancarlo Gioda, and Donatella Sterpi

Department of Structural Engineering, Politecnico di Milano, Milano, Italy

Abstract: Some applications are presented of the elasto-plastic and visco-plastic stress analysis to tunnelling problems. First, the closed form solution is recalled of the so-called "characteristic line" of a circular opening, which is frequently adopted for a simplified evaluation of the load acting on the lining of deep tunnels. Subsequently, some examples are discussed, based on the finite element method, which illustrate a number of effects influencing the behaviour of underground openings. They concern, for instance, the mutual influence of parallel tunnels, the "squeezing" and the "strain softening" behaviour of the rock mass, the pore pressure changes induced by underground excavations in saturated media. The discussion is confined to static problems and to continuous rock/soil masses.

1. Introduction

The evaluation of the loads acting on the temporary and permanent supports of tunnels usually requires the solution of relatively complex stress-analysis problems. In fact, the pressure exerted by the rock/soil mass on the supporting structures depends on the interaction between them and the surrounding geotechnical medium, and can be evaluated by properly modelling the processes of excavation and construction of the tunnel.

This problem is here approached by recalling first the closed form solution for a circular tunnel, driven into an isotropic elasto-plastic medium, in plane-strain and axisymmetric conditions, which leads to the so called "characteristic curve" of the opening. This solution is frequently adopted in engineering practice for simplified analyses and permits to get some insight into the main characteristics of the mentioned interaction problem. It is recognised, however, that in the majority of cases the analysis is not amenable to such a simple solution scheme. This leads to a widespread use of numerical, computer oriented, methods.

Among the various numerical solution techniques, the finite element method is perhaps the most commonly adopted in tunnel engineering. In fact, it allows a relatively easy implementation of non-linear and time dependent constitutive laws for the rock/soil mass. In addition, it permits the simulation of the excavation and construction processes that lead to the completion of the opening.

The numerical aspects of the stress analysis of tunnels are here discussed with particular reference to the finite element method. The cases examined concern various situations of possible practical interest. In particular: the interaction between parallel tunnels; the effects of pore pressure, creep, and strain softening on the overall behaviour of the opening; the comparison between the results of three-dimensional and two-dimensional, plane strain calculations, etc.

All problems are treated in static conditions and the soil/rock mass is modelled as a continuous body, neglecting the local influence of possible discontinuities and joints.

2. Characteristic Curve of a Deep Circular Opening

Before discussing the derivation of the so-called "characteristic curve" of a deep tunnel, it is worthwhile to mention a solution reported by Poulos and Davis (1974) for a circular tunnel, embedded into a homogeneous and isotropic elastic medium.

The elastic medium is subjected to a uniform pressure distribution represented by a vertical stress p_0 and by a horizontal stress $K_0 \cdot p_0$, where K_0 is the coefficient of rock pressure at rest. Hence, the local effects of the rock own weight are neglected. The solution, developed in plane strain and axisymmetric conditions, provides the stress and strain distributions in the rock mass surrounding the tunnel, depending on its diameter, on the thickness of its lining, on K_0 and on the elastic parameters of both rock and lining.

The complete solution of this problem is not reported here, for sake of briefness. In fact, for the purpose of this discussion, it is sufficient to mention the expression of the normal pressure p_n exerted by the rock on the liner, when K_0 has a unit value and the lining is perfectly rigid,

$$p_n = 2p_0(1-\nu_r) \quad . \tag{1}$$

In the above equation, p_0 and ν_r represent, respectively, the in situ hydrostatic pressure and Poisson coefficient of the elastic rock mass.

Since Poisson ratio is smaller than 0.5, eq.(1) leads to a value of the rock pressure p_n on the liner that exceeds the in situ stress p_0. This contradicts the physical evidence. According to the adopted simple scheme, if the liner is rigid no deformation should develop in the elastic medium surrounding the tunnel. Consequently, the load carried by the liner should coincide with the in situ hydrostatic stress p_0.

This contradiction depends on the fact that eq.(1) does not take into account the actual process of excavation. In fact, it was derived considering an initially unstressed elastic space, containing a circular inclusion, which is subsequently subjected to a uniform traction distribution p_0 at infinity. Quite obviously, the applied traction produces a stress concentration on the rigid inclusion represented by the circular liner.

Since the tunnel is not driven in an unstressed medium, but in a mass already subjected to a stress state, the elastic solution leading to eq.(1) does not consider the actual process of excavation and, hence, it cannot be adopted for the analysis of underground openings.

A way for reaching a more meaningful solution of this problem can be based on the so-called characteristic curve of the opening, which accounts for the process of excavation in a simplified manner.

The main assumptions on which this solution is based are similar to those previously mentioned, and can be listed as follows:

- The tunnel has a circular cross section of radius R, its length is unlimited, and the mass hosting it is homogeneous and isotropic.
- At the beginning of the excavation process the rock mass is subjected to a uniform hydro-

static stress distribution, defined by a normal stress p_0.
- The local effects of the rock own weight are neglected.
- The rock mass has an elastic-ideally plastic behaviour with Mohr-Coulomb yield condition (hence, it is characterised by its elastic constants, E and ν, and by cohesion c and friction angle ϕ). The non-associated plastic flow rule depends on the "dilatancy angle" ψ, which does not necessarily coincide with the friction angle ($\phi \geq \psi \geq 0$).

The process of excavation is simulated in plane strain and axisymmetric conditions (fig.1) by decreasing the pressure p, uniformly distributed on the tunnel perimeter ($p_0 \geq p \geq 0$). Due to the above assumptions, the only non-vanishing displacement component is the radial one u_r. The strain state within the rock mass is defined by its radial ε_r and circumferential ε_ϑ components, while the non vanishing normal stresses are σ_r and σ_ϑ, and the component σ_z in the direction of the tunnel axis.

The analytical solution of this simple elasto-plastic problem leads to the following expression of the "plastic radius" R_{pl}, defining the portion of the rock mass surrounding the opening within which the yield limit is attained and plastic strains develop,

$$R_{pl} = R \cdot \left(\frac{2}{1+k} \cdot \frac{A - p_0}{A - p} \right)^{\frac{1}{k-1}} , \tag{2}$$

where

$$A = \frac{c}{\tan\phi} \quad \text{and} \quad k = \frac{1+\sin\phi}{1-\sin\phi} . \tag{3a,b}$$

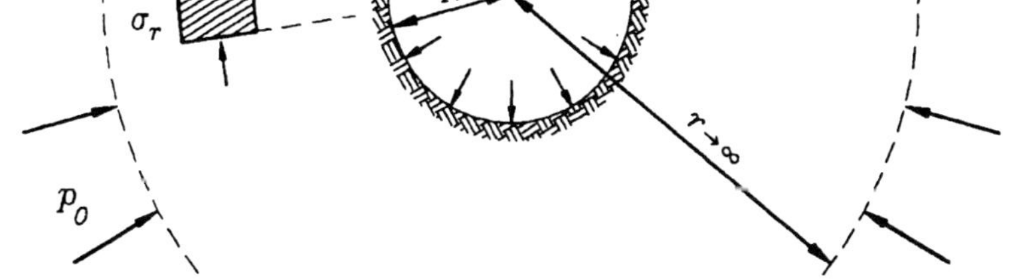

Figure 1. Plane strain and axisymmetric scheme adopted for deriving the characteristic curve of a deep tunnel.

Note that tensile stresses are assumed positive in the above equation. The normal pressure p_{pl} acting on the boundary between the plastic zone and the remaining (elastic) part of the rock mass is

$$p_{pl} = A-(A-p)\cdot\left(\frac{R_{pl}}{R}\right)^{k-1} \quad , \tag{4}$$

while the stress components within the plastic zone ($R_{pl}\geq r$) are

$$\sigma_r = A-(A-p)\cdot\left(\frac{r}{R}\right)^{k-1} \quad , \quad \sigma_\vartheta = A-k(A-p)\cdot\left(\frac{r}{R}\right)^{k-1} \quad , \tag{5a,b}$$

$$\sigma_z = \nu\left[2(A-p_0)-(1+k)\cdot(A-p)\cdot\left(\frac{r}{R}\right)^{k-1}\right]+p_0 \quad . \tag{5c}$$

The displacement δ of the tunnel wall, or convergence, can be easily evaluated on the basis of the following relationship between radial displacement and circumferential strain

$$u_r = r\cdot\varepsilon_\vartheta(r) \quad . \tag{6}$$

Hence, the following expression holds for δ,

$$\delta = R\left[\varepsilon_\vartheta^e(R)+\varepsilon_\vartheta^p(R)\right] \quad , \tag{7}$$

where the suffices e and p denote, respectively, elastic and plastic strains.

The strain components in eq.(7) are given by the following relationships

$$\varepsilon_\vartheta^e = \frac{\sigma_\vartheta}{R}-\frac{\nu}{E}(\sigma_r+\sigma_z) \quad , \quad \varepsilon_\vartheta^p = -\lambda\frac{1-\sin\psi}{2} \quad , \tag{8a,b}$$

and the plastic multiplier λ can be expressed as

$$\lambda = \frac{(1-\nu^2)\cdot(k^2-1)\cdot(A-p)}{E[2+(1-\sin\psi)\cdot(k-1)]}\cdot\left[\left(\frac{R_{pl}}{R}\right)^{\left(\frac{2}{1-\sin\psi}+k-1\right)}\cdot\left(\frac{r}{R}\right)^{-\frac{2}{1-\sin\psi}}-\left(\frac{r}{R}\right)^{k-1}\right] \quad . \tag{8c}$$

A schematic representation of the p vs. δ relationship (or characteristic curve), derived from the above equations, is provided by curve "a" in fig.2. In the same figure, curve "b" represents the p-δ relationship for the liner. The point of intersection between the two curves permits an approximated estimation of the load carried by the liner, after the completion of the tunnel, and of the corresponding convergence. Note that δ_0 represents the convergence that takes place before the installation of the liner, and that depends upon the technical details of the construction procedure.

Various modifications of this simple approach have been proposed in the literature, to broaden its range of applicability. However, the intrinsic limits of this solution make it necessary to adopt more sophisticated methods of analysis, like the finite element method (see e.g. Gioda and Swoboda, 1999), in a variety of cases met in engineering practice. Some examples concerning the application of this method to tunnelling problems are presented in the following.

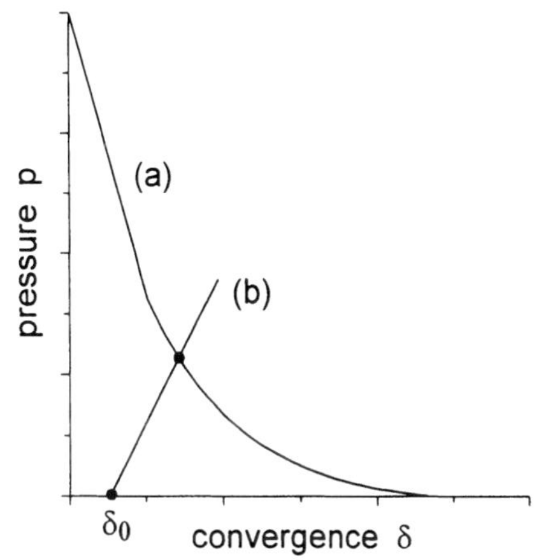

Figure 2. Representation of the characteristic curves of a circular tunnel (a) and of its lining (b).

3. Interaction Between Two Parallel Tunnels

A typical case in which the characteristic curve approach cannot be adopted concerns the interaction between adjacent tunnels. The example described here concerns the construction of two parallel tunnels as a part of a highway project (cf. fig.3). The tunnels have an internal diameter of about 13m and are located at different depths in a deposit of volcanic soil, having average friction angle and cohesion of about 30° and 0.02 MPa.

The lower tunnel was excavated first, without encountering major problems. In order to ensure its stability during construction, and to limit the surface settlements, jet grouted columns were set in place ahead of the excavation face, and a temporary lining was installed consisting of steel ribs and shotcrete.

To avoid possible instability problems or damages of the first tunnel during the excavation of the second one, caused by the relatively small distance between them, the contractor decided to construct its unreinforced concrete permanent lining before initiating the second excavation.

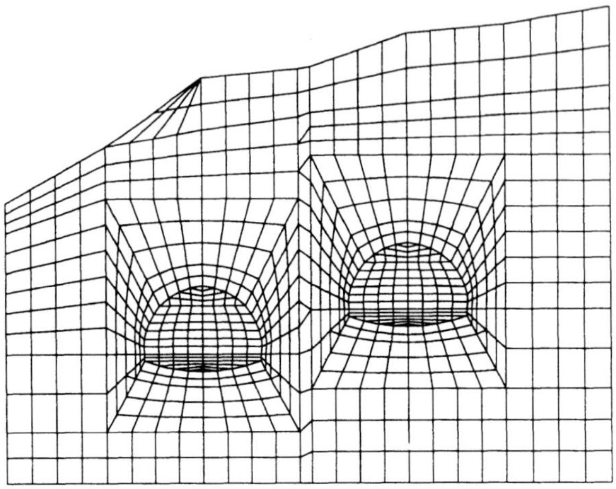

Figure 3. Detail of the finite element mesh for the two-tunnel problem.

Then, the excavation of the upper tunnel was initiated by adopting the already mentioned technique. During this process, a large fracture developed in the permanent lining of the first tunnel in the vicinity of its crown, on the side close to the second tunnel. This led to some questions about the overall stability of the lower tunnel.

To answer these questions some plane strain, elasto-plastic finite element analyses were carried out. Fig.3 shows a detail of the mesh in the vicinity of the openings. The analyses were subdivided into two stages. The first one, simulating the construction of the lower tunnel, concerns the excavation of the top portion of the tunnel, the installation of its temporary support, the completion of excavation and the installation of its permanent liner.

In the subsequent stage, the effects of the second excavation were simulated by gradually decreasing the nodal forces equivalent to the traction distribution on the perimeter of the upper tunnel.

The results of this second stage of analysis are summarised by the diagrams in fig.4, which shows the variation of bending moment and axial force in the permanent lining of the lower tunnel during the excavation of the upper one.

At the beginning of calculations no appreciable internal forces (dashed lines in fig.4) exist in the permanent liner, since most of the rock pressure is carried by the mentioned temporary supports.

The bending moment gradually increases during excavation, with a peak on the right hand side of the crown, which leads to appreciable tensile stresses at the inner surface of the liner. In the same zone, however, the compressive axial force has a very limited increase, while its increment is more pronounced on the left-hand side of the crown and in the vicinity of the invert arch.

These results provide a possible explanation for the formation of the crack and for the position in which the crack developed. They also indicate that no major stability problems should be expected, since the compressive stresses in the remaining portions of the liner are well below the compressive strength of concrete.

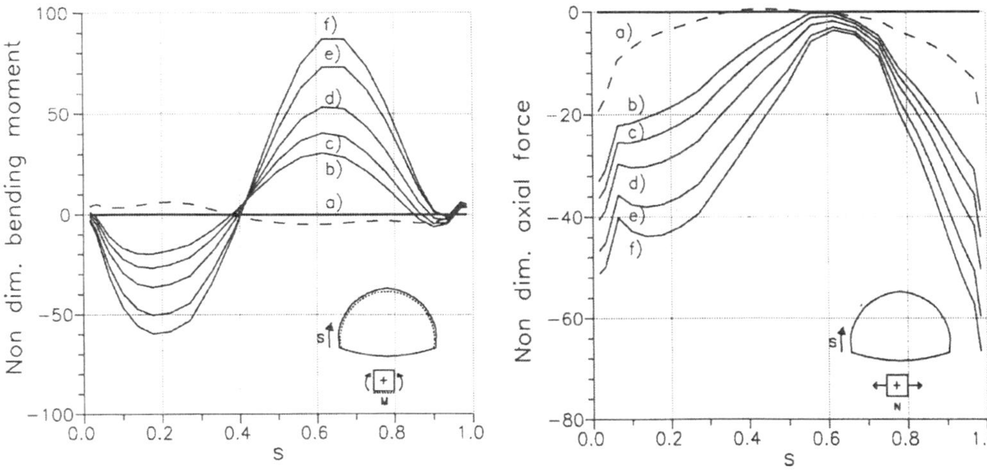

Figure 4. Bending moment (left) and axial force (right) vs. curvilinear co-ordinate s. The curves (a) to (f) refer to different excavation stages of the upper tunnel.

4. "Squeezing" Effects In Tunnels

A relevant time dependent effect in tunnels, often related to the presence of clay minerals in the surrounding rock, is referred to as "squeezing", or creep, behaviour (Sakurai, 1978; Gioda, 1981). In a squeezing rock the time dependent deformation is produced, basically at constant volume, by the concentration of stresses in the vicinity of the excavation. A slow movement of the rock towards the tunnel takes place, that in some instances is associated to a reduction of its shear strength and apparent viscosity, and that may lead to failure of unsupported openings.

In general terms, the time dependent behaviour of the rock can be subdivided into three parts: the so-called primary, secondary and tertiary creep stages. Under a stress state constant with time, primary creep is characterised by a strain rate decreasing with time and usually exhibits a reversible nature (see e.g. Ghaboussi and Gioda, 1977). During secondary creep, that appears when the stress level overcomes a given limit, the creep strain rate is approximately constant with time. Finally, tertiary creep might show up for higher stress levels, and is characterised by a strain rate increasing with time leading eventually to failure.

A rheological model able to describe the mentioned effects (Gioda, 1982) is depicted in fig.5. It consists of a visco-elastic Kelvin model, accounting for primary creep, connected in series with a visco-plastic Bingham model accounting for secondary creep.

The possible development of tertiary creep can be accounted for by providing a suitable law governing the reduction of the visco-plastic parameters (cohesion, friction angle and apparent viscosity) with increasing plastic strains.

This simple rheological model could provide a reasonable approximation of the time dependent behaviour of creeping rocks. Fig.6 shows the vertical creep strain vs. time diagrams obtained by the numerical simulation of triaxial creep tests under constant vertical load. In this figure, τ_F is the maximum shear stress carried by the material without undergoing secondary

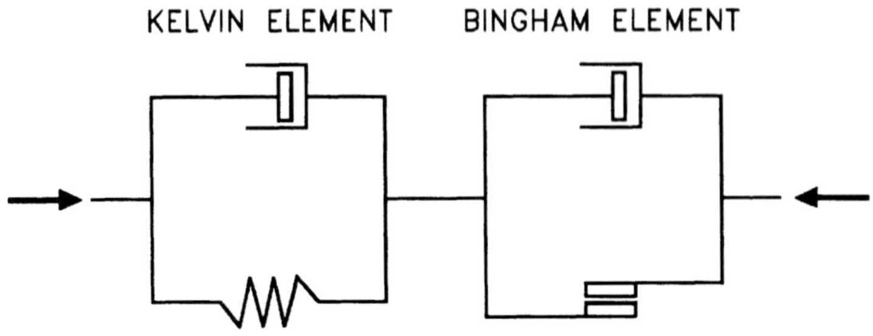

Figure 5. Non-linear rheological model.

creep; ε_{1L} is the maximum vertical strain reached when $\tau=\tau_F$; $t_L(90\%)$ is the time at which 90% of ε_{1L} develops when $\tau=\tau_F$.

This material model was applied to the analysis of a deep circular tunnel in plane strain and axisymmetric conditions. Consequently the only non-vanishing displacement component is the radial one.

At the beginning of calculations the stress state in the rock mass is constant and isotropic. The tunnel excavation is simulated by decreasing the uniform pressure on its perimeter, from the initial value to zero, at a constant rate.

Various situations have been considered after the end of excavation. Case I refers to the unlined situation. Cases II to V refer to lined situations in which different time spans separate

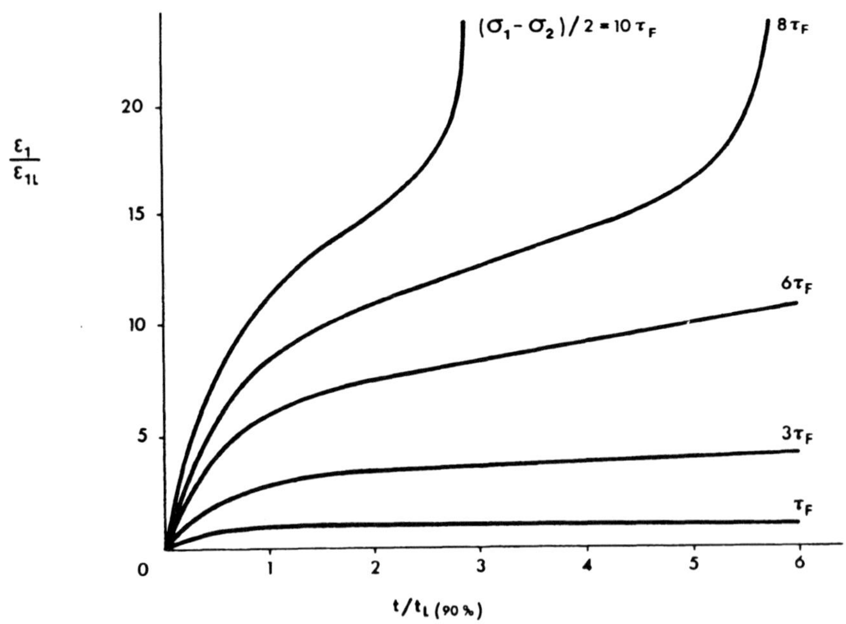

Figure 6. Strain vs. time diagrams from the numerical simulation of constant load triaxial creep tests.

the end of excavation (t_e) from the liner installation (t_i): case II) t_i=2.5t_e; III) t_i=2.0t_e; IV) t_i=1.5t_e; V) t_i=t_e.

Two sub-cases, denoted by subscripts (a) and (b), were considered for each lined analysis. The stiffness of the liner for case (b) is twice as large as the stiffness in case (a).

The results of calculations are shown in fig.7, in terms of radial displacement vs. time, and in fig.8, in terms of the variation of the yield function F along the radius for different values of time. Note that in fig.8 the yield function is evaluated considering the total stress state acting on both frictional and viscous components. Consequently, the yield function is greater than zero in the zone where the mechanical resistance of the frictional component is exceeded and where creep strains develop. In figs.7 and 8, δ_e denotes the radial displacement evaluated at time t_e, R is the radius of the tunnel and k is related to the cohesion of the rock mass.

The numerical results clearly show that squeezing would lead to failure of the unlined opening due to the large spreading of the plastic zone. A stable situation can be reached if a support is installed, of course at the cost of a large circumferential stress on the liner.

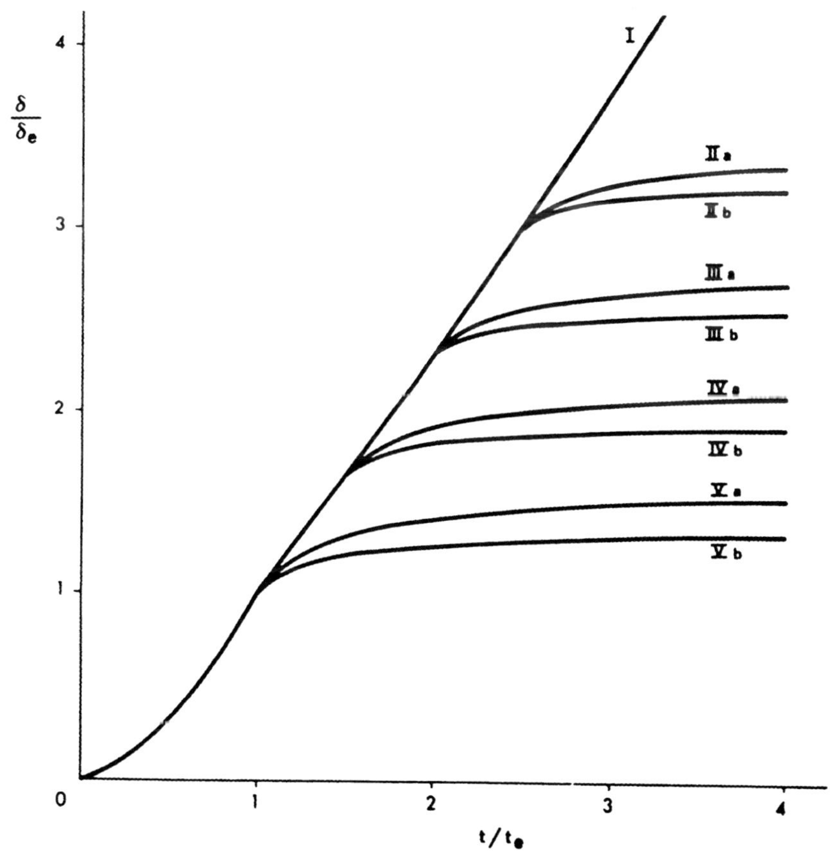

Figure 7. Radial displacement vs. time diagrams for the unlined (I) and lined (II to V) cases.

(a)

(b)

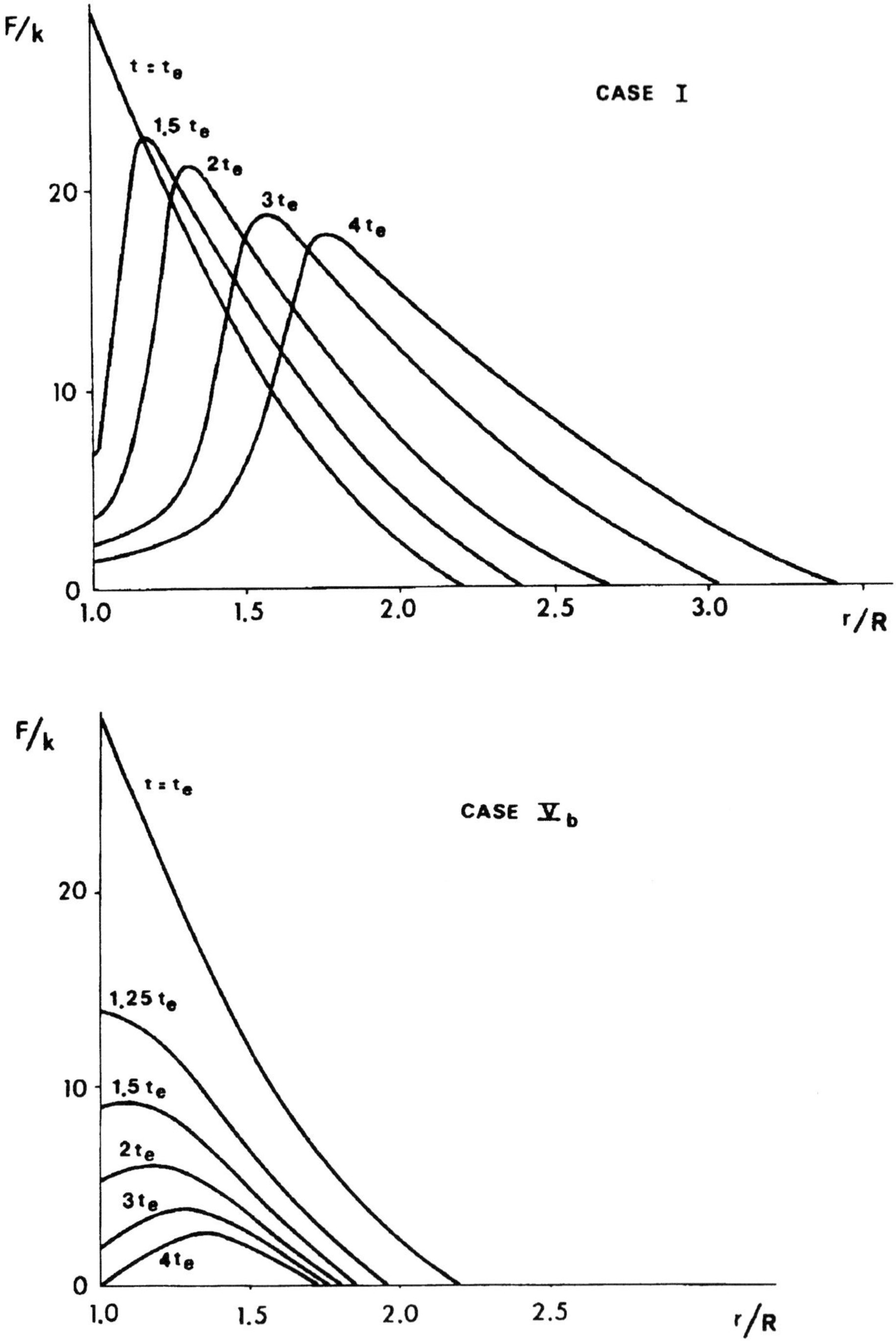

Figure 8. Variation of the yield function F along the radial co-ordinate for unlined (a) and lined (b) cases.

5. Tunnel in a Saturated Soil

The evaluation of the stress (neutral and effective) and strain distributions around an underground opening in a saturated soil deposit involves various aspects:

- the time dependent dissipation of the excess pore pressure induced by the excavation;
- the seepage flow towards the tunnel;
- the non-linear mechanical behaviour of soil;
- the possible viscous deformation of the skeleton (secondary compression).

A consistent coupled analysis of this non-linear, two-phase (consolidation) problem can be based on the finite element formulation of Biot consolidation equations (see e.g. Sandhu, 1968; Ghaboussi and Wilson, 1973). The visco-plastic material model previously mentioned permits to introduce in the analysis the time dependent behaviour of the skeleton, governed by the variation of the effective stress state.

A particular aspect of this analysis consists in the evaluation of the forces to be applied to the tunnel boundary for simulating the excavation process. In fact, three distinct contributions have to be separately determined (Cividini et al., 1985).

The first one depends on the total weight of the excavated soil and is calculated as the product of the transpose shape function matrix by the total body force vector, integrated over the volume of the excavated elements.

The second contribution is due to the removal of the total stress distribution on the perimeter of the opening. This, in turn, consists of two parts: a) the contribution of the effective stresses (evaluated as the integral over the excavated elements of the product between the matrix of the derivatives of the shape functions and the effective stress vector); b) the contribution of the pore pressure (calculated as the summation over the excavated elements of the product between the "coupling" matrix and the nodal pore pressure vector).

The third contribution does not represent a "physical" contribution to the excavation forces. It depends solely on the approximation introduced by the finite element discretization and, in particular, it is related to the size of the soil elements that have at least one node belonging to the perimeter of the excavation. The corresponding nodal forces are evaluated as the summation on these elements of the product between the "coupling" matrix and the difference between the pore pressures existing on the tunnel perimeter before and after the excavation.

Note that with decreasing size of the elements the first two contributions tend to a constant value, while the third one vanishes.

As a test example, a moderately deep unlined tunnel, excavated in a homogeneous and saturated soil deposit, was considered. It was assumed that the water table coincides with the ground surface and that its elevation does not change during time. The following consolidation analyses were performed (Cividini et al., 1985):

- ES linear elastic behaviour of the soil mass, with "short term" parameters;
- EL linear elastic behaviour with "long term" parameters;
- ESR as in case ES, but constraining the displacements of the tunnel perimeter;
- VE1 visco-elastic behaviour of the soil mass, with "low" viscosity;
- VE2 visco-elastic behaviour, with "high" viscosity;
- VP1 visco-plastic behaviour, with "low" viscosity.

The results of these calculations are summarised in figs.9 and 10. The first one refers to analysis ES and shows the variation with the non-dimensional time T of the pore pressure π along the vertical and horizontal axes through the tunnel centre. Fig.10 shows, for all analyses, the variation with time of the vertical displacements of three reference points.

The diagrams were obtained for the elastic material behaviour (ES and ESR in fig.10) and indicate that the surface settlements in the unlined case ES are only slightly smaller than those obtained when the displacements of the tunnel contour are constrained (ESR). This points out that, for deep tunnels, the deformation of the medium due to the reduction of pore pressure can be comparable to, or even larger than, that caused by the deformation of the opening.

A gradual variation of the displacements with time is observed in the visco-plastic case (VP1), until a rapid increase takes place, leading to the collapse of the opening. This "delayed" failure is due to the "dilatant" behaviour of the soil when its yield limit is reached. As a consequence, a volume increase of the medium surrounding the opening is necessary to reach collapse, and this effect is delayed by the two-phase nature of the soil and by its low permeability.

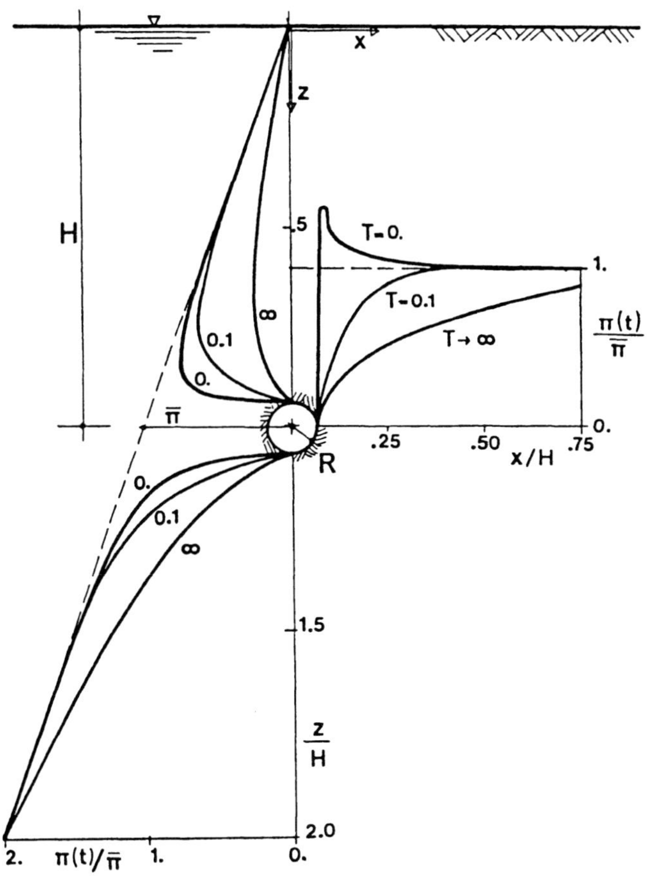

Figure 9. Variation with time of the pore pressure π for analysis ES.

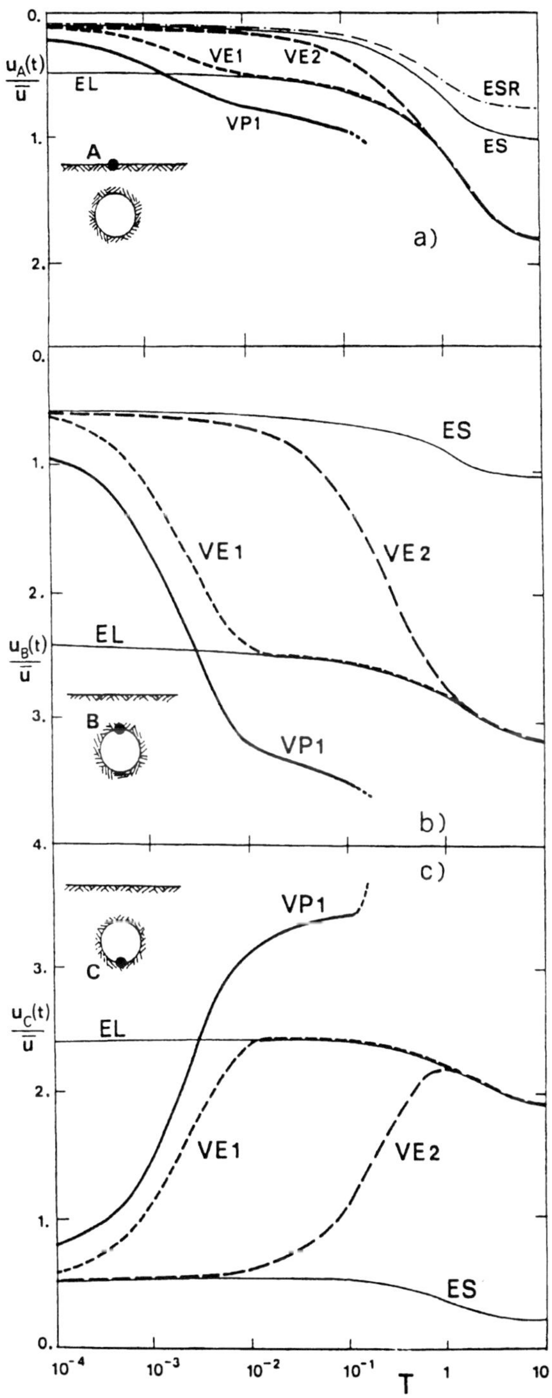

Figure 10. Variation with time of the vertical displacement of three reference points.

6. Three- vs. Two-Dimensional Analysis

The finite element solutions discussed so far have been based on axisymmetric or plane strain schemes, mainly because of the limited computational effort required.

It should be considered, however, that the results of these simplified analyses are less accurate than those obtained with more complex, and time consuming, three-dimensional calculations.

An attempt to quantitatively compare the accuracy of two-dimensional plane strain and three-dimensional calculations is based here on the finite element modelling of the excavation of a railroad tunnel.

The tunnel has a total length of 12.6 km, a maximum depth of 600m and its crown arch has an internal radius of 5.2m. The section considered in the calculations has a depth of cover of 43m and is excavated in a deposit of slightly cemented sand and gravel (Sterpi and Moro, 1995).

The excavation scheme is depicted in fig.11. First, 12m long jet grouted columns are set in place ahead of the tunnel face. Then the top part of the tunnel is excavated, for a length of 6m, and a temporary support consisting of steel ribs and shotcrete is installed.

Subsequently, the top part of the permanent concrete lining is constructed in the excavated zone and the central part of the bottom excavation is initiated at a distance of about 30m from the tunnel face. The foundations of the crown arch are constructed and the bottom excavation is completed at a distance of 60m from the face. Finally, the permanent invert arch is constructed.

The three-dimensional finite element analysis of the excavation/construction process was based on the mesh shown in fig.12. Eight node "brick" elements were used for discretizing both natural and grouted soils, while the temporary and permanent linings were introduced through 4 node shell elements.

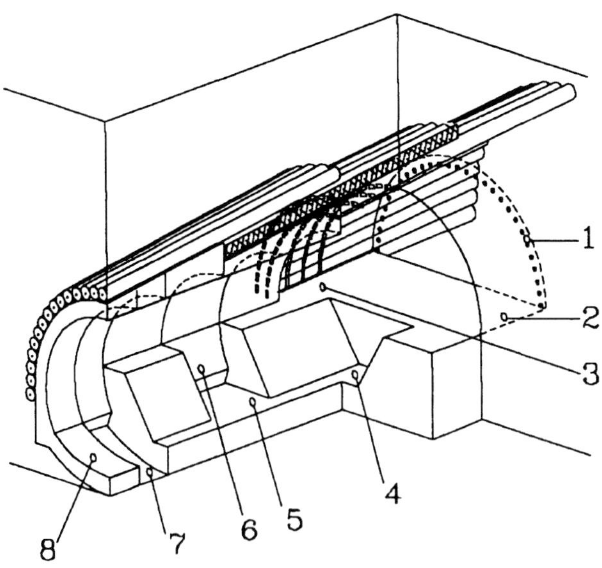

Figure 11. Excavation/construction stages of the railroad tunnel.

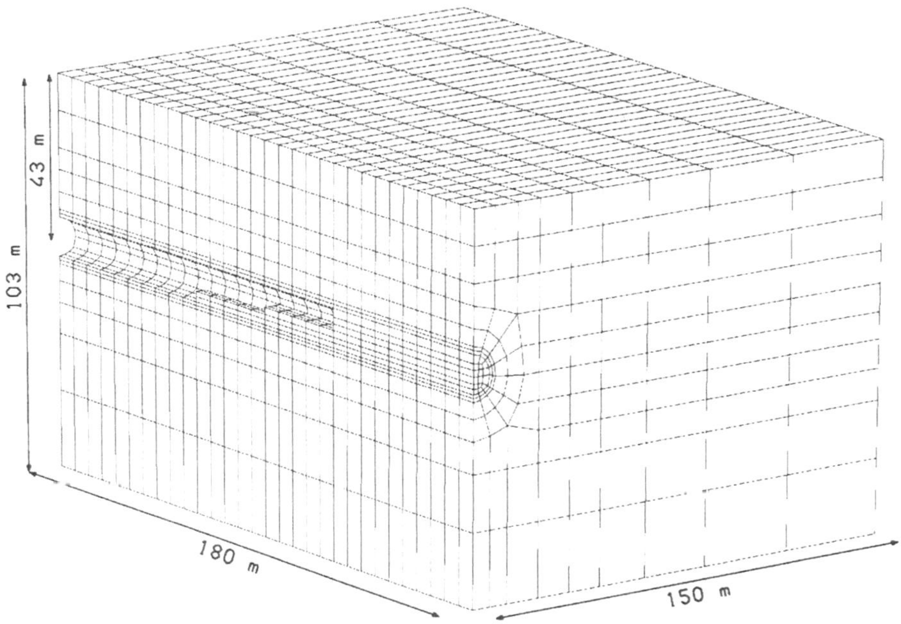

Figure 12. Three dimensional finite element mesh for the railroad tunnel.

The excavation process was reduced to three main steps: a) jet grouting, excavation of the crown and construction of its temporary and permanent linings; b) excavation of the central part of the tunnel and construction of the foundations of the crown arch; c) completion of excavation and construction of the invert arch.

Fig.13 shows the surface settlements obtained by the 3D analysis when a distance of 120m exists between the tunnel face and the beginning of excavation.

These results were used to "calibrate" a 2D plane strain analysis based on the construction steps previously mentioned. A key point of the 2D analysis consists in choosing the percentage of reduction of the excavation forces to be used at each step. It was assumed that this percentage coincides with the ratio between the surface settlement along the tunnel centreline, obtained at the corresponding stage of the 3D calculation, and the final surface settlement. It should be observed that this is an empirical criterion and that its effectiveness can be assessed only by a direct comparison of the results of 2D and 3D analyses.

Fig.14 shows such a comparison with reference to the surface settlements along the tunnel axis and in the direction normal to it. Fig.15 reports the distributions of the circumferential bending moment in the permanent lining at the end of each step of excavation.

These diagrams indicate that a 2D plane strain analysis, based on the mentioned equivalence criterion, is able to provide an acceptable approximation of the results of 3D calculations. This suggests the use of 2D analyses to control the tunnelling process, following a calibration based on a preliminary 3D calculation or on in situ measurements of the settlements due to excavation.

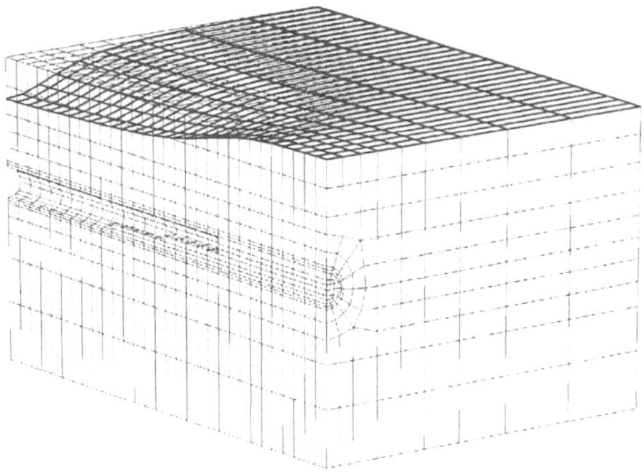

Figure 13. Surface settlement distribution from the 3D analysis.

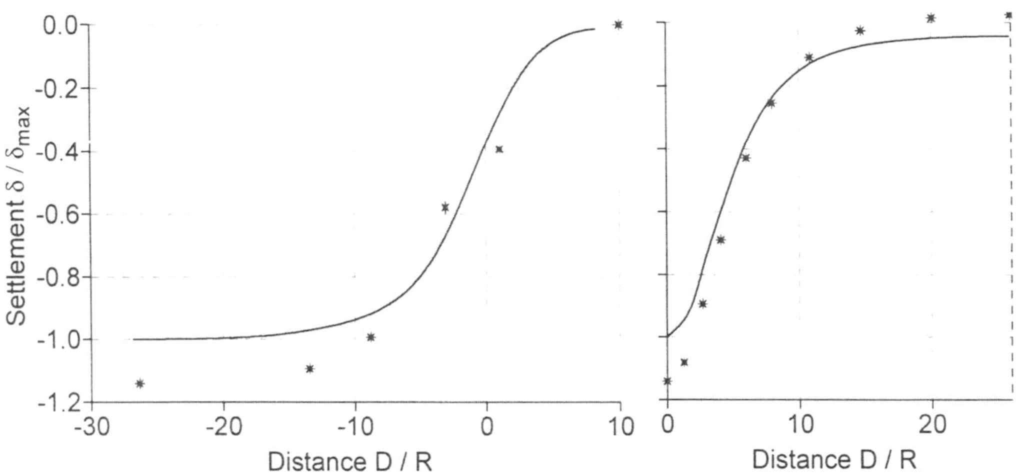

Figure 14. Surface settlements along the tunnel centreline (left) and in the direction normal to the tunnel axis (right) for 3D (solid lines) and 2D analyses (dots).

7. Influence of Strain Localisation and Softening Behaviour

The stability analysis of underground openings presents non negligible problems when dealing with rock masses and stiff soils, such as dense sands or overconsolidated clays, characterised by strain softening behaviour.

In geotechnical engineering the term "strain softening" customarily denotes materials that, when subjected for instance to a compression test, show a loss of their overall load carrying capacity with increasing deformation, after a peak load level has been reached. Quite often this behaviour is associated with the formation of shear bands within the sample (i.e. zones of limited thickness where irreversible strains localise).

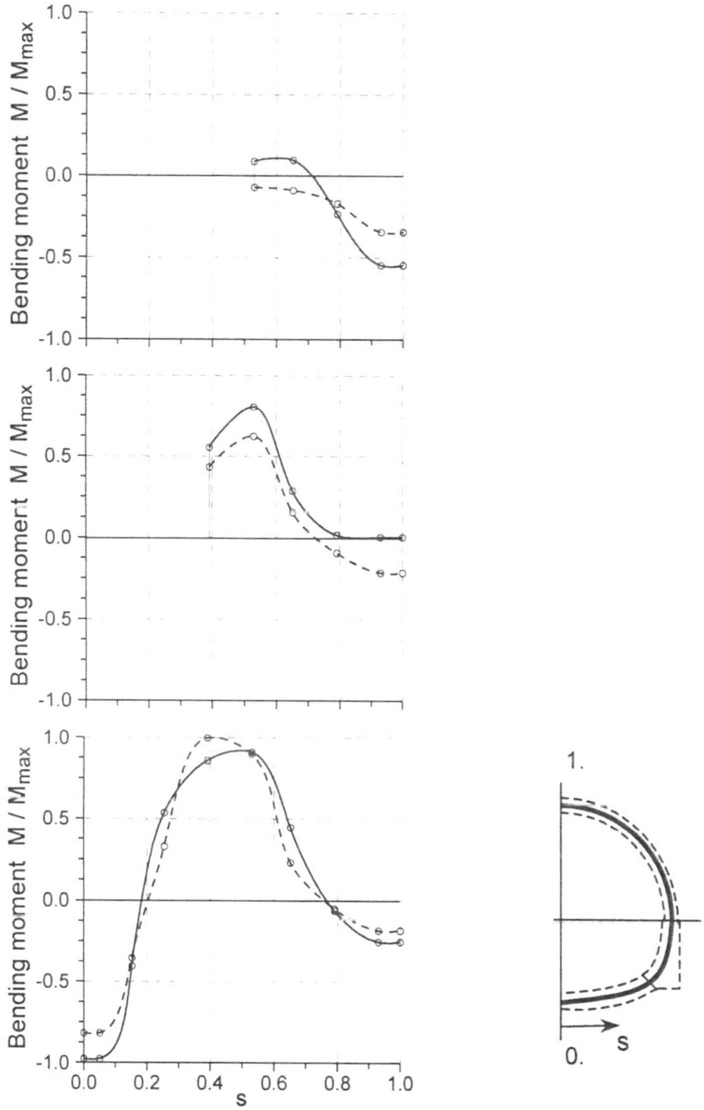

Figure 15. Bending moment distributions from 3D (solid lines) and 2D (dashed lines) analyses at the end of each excavation stage.

When dealing with large scale engineering problems, the shear bands represent potential sliding surfaces that could affect the overall stability by originating a mechanism of collapse. For this reason, the phenomenon of strain localisation, and the associated loss of the overall resistance of the soil/rock mass, is considered as one of the main causes of the so called "progressive failure".

When a tunnel is driven into a formation of stiff soil or weak rock, characterised by a low value of the coefficient of earth pressure at rest, zones of localisation of permanent strains may initiate at the tunnel springlines, if the opening is not properly supported. The extension of these zones might rapidly increase with the progress of excavation, developing upwards into arched shear bands that eventually join at a point above the tunnel crown, or that reach the ground surface in the case of shallow openings. In both cases a collapse mechanism may form with the consequent failure of the tunnel crown (Hansmire and Cording, 1985).

For geological media, the processes of initiation and of subsequent spreading of the localisation zones seem to depend on two aspects that are probably simultaneously present.

From a "constitutive" view point, the loss of the load carrying capacity can be seen as a consequence of an intrinsic property of the material, shown at a constitutive level as a loss of shear strength with increasing deformation (Bazant et al., 1984). The phenomenon is onset when a given condition on the strain state is attained. The subsequent shear band formation is caused by a progressive local damage of the material, in terms of its mechanical deformability and resistance, depending on the cumulated permanent strains or on the plastic strain energy.

From a "structural" point of view, the initiation of this phenomenon is due to a sort of structural instability of the material, which is not specifically related to a loss of its shear resistance introduced in the stress-strain relation. The local structural instability occurs when, for a given increment of the external actions, a bifurcation point is reached, i.e. the uniqueness in the solution of the equations governing the stress analysis problem is lost. This condition can be met even for perfectly plastic, or positive hardening, materials, if the plastic flow rule is non-associated (Hill, 1958; Rice, 1976).

The procedure for strain softening analysis here adopted considers separately two "phases" of the phenomenon, namely: the onset of localisation and the spread and coalescence of the shear bands. These two phases correspond to two stages of a non-linear finite element procedure, implemented in the code SoSIA-2, for Soil-Structure Interaction Analysis of 2-dimensional problems (Cividini and Gioda, 1992).

The first stage consists in checking a particular local condition, at each integration point and at the end of each loading step, the fulfilment of which indicates the local onset of strain localisation.

When the "constitutive" approach is adopted, this condition rests on the attainment of a given limit on the cumulated permanent strains. For instance, if the square root of the second invariant J_2 of the deviatoric plastic strains $\underline{\varepsilon}^p$ is assumed as a measure of the cumulated non reversible deformation, the condition reads:

$$\sqrt{J_2(\underline{\varepsilon}^p)} = \gamma^* \quad , \tag{9}$$

where γ^* is a characteristic parameter of the material, experimentally derived (Cividini and Gioda, 1992).

Alternatively, in the case of "structural" approach, the localisation condition depends on a shear band analysis of the bifurcation theory, which detects the occurrence of an alternative solution characterised by a planar discontinuity in the strain field. For a plane strain, ideally plastic, non-associated problem, this condition is expressed as:

$$\det\left[\underline{N}^T \, \underline{D}^{ep}(\underline{\sigma})\underline{N}\right] = 0 \quad , \tag{10}$$

where the entries of matrix $\underline{N}$ depend on the orientation of the planar discontinuity and $\underline{D}^{ep}$ is the tangent elasto-plastic constitutive matrix. Details on the derivation of eq.(10) were presented by Sterpi (1999).

The second stage of the analysis accounts for the actual coalescence of localisation zones into shear bands and, therefore, it is activated only for the integration points where the previous condition has been fulfilled.

Consider that the occurrence of a strain discontinuity causes a change in the local structure of the material. For instance, an increase of void ratio or a decrease of relative density, due to the dilatancy effects in the zone subjected to a shear mode of deformation. Consequently, a local loss of shear resistance and stiffness might occur.

This suggests to base the second stage of the analysis on a procedure in which the shear strength and stiffness parameters are gradually reduced, with increasing permanent strains, from their peak to their residual (or fully softened) values. In the calculations, this reduction is linearly related to the increment of the plastic shear strains (or $\sqrt{J_2}$) with respect to the corresponding value at the onset of localisation.

Once localisation has been detected at an integration point, the subsequent load increments lead in general to increments of the deviatoric plastic strains and, hence, to a reduction of the current values of the shear strength parameters. Consequently, the evaluated state of stress could be no longer admissible with respect to the modified yield surface. Following the modified Newton-Raphson method, the stress state is brought back to the surface and its non admissible portion is converted into equivalent unbalanced nodal forces, and applied as self equilibrated external loads. An iterative process is then initiated, which terminates when no further variation of the mechanical parameters is observed for the current load increment.

Some comments seem worthwhile, concerning the difficulties that may arise in a numerical strain softening analysis due to the dependence of the solution on the adopted discretisation. This may affect the thickness and the direction of the computed shear bands (Vardoulakis and Sulem, 1995). The mesh-dependence is induced by the very nature of the finite element method, since the loss of shear strength and stiffness is evenly distributed over a zone that depends on the size of the elements. As a consequence, the decrease of the material parameters adopted in the calculations should not only depend on the material properties, but it should be also related to the finite element size (Pietruszczak and Mroz, 1981). This provision has been adopted here, by keeping constant the product between the average element size and the rate of reduction of the mechanical parameters.

Particular attention should be also paid to the mechanical characterisation of the material from laboratory tests. In fact, when dealing with media showing loss of load-carrying capacity, a non-negligible dependence on the particular conditions imposed in the testing procedure might affect the test results, from both qualitative and quantitative points of view (Sterpi, 2000). In addition, the homogeneity of the stress/strain fields within the sample is in general lost when the material undergoes strain localisation. In this case, the derivation of the stress-strain relationship from the measured load-displacement diagram might lead to an overestimation of the loss of shear strength, due to the overestimation of the sample reacting area (Read and Hegemier, 1984).

The described approach has been applied to the numerical simulation of laboratory tests on 2D and 3D small-scale models of tunnels. This study concerned, in particular, the effects of strain localisation on the stability of the roof and of the excavation face of shallow tunnels driven into frictional materials.

A first series of 2D, plane strain tests was performed at the Rock Mechanics Laboratory of Kobe University (Japan), using an assemblage of aluminium bars as an "analogical" soil (Sterpi and Sakurai, 1997). The mechanical characterisation of this material was based on laterally constrained compression tests and direct shear tests performed, with non conventional devices, on bar assemblies having the same relative density adopted during the tunnel tests.

The tunnel tests were set up by lying the bars within a rigid frame, at the centre of which a steel cylinder is located that represents the cross section of a circular tunnel. The cylinder has a diameter of 15cm and contains a pressurised airbag. The excavation process was then simulated by removing the steel cylinder and by decreasing the air pressure in subsequent steps. The induced stresses and strains were obtained by embedded load cells and by digitised pictures of the aluminium bar assembly.

In all tests the collapse was originated by the development of two shear bands from the tunnel springlines towards the ground surface. Their shape and thickness depend on the tunnel depth and on the relative density of the medium. A negligible deformation was observed in the mass limited by the two bands, which underwent an almost rigid downward movement.

The numerical simulation of the laboratory tests showed a progressive spread of the localised zone, which eventually overlapped the experimentally observed shear bands (fig.16). The shape of the numerically evaluated collapse mechanism was rather similar to the one observed during the model tests and a fair quantitative agreement was also obtained between measured and calculated settlements. It should be finally observed that the numerical analyses correctly estimated the internal pressures corresponding to the collapse of tunnels located at various depths (fig.17).

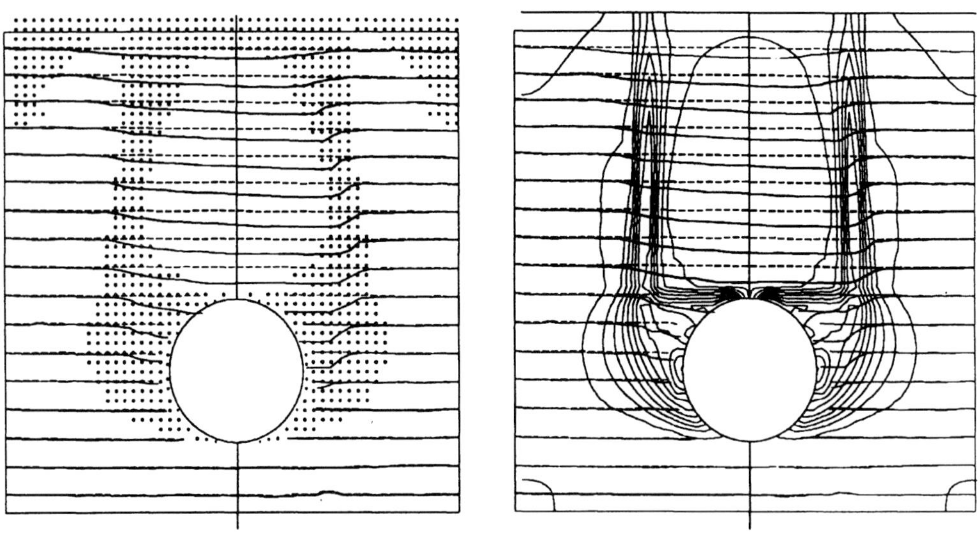

Figure 16. Measured settlements vs. numerically evaluated localisation zones (left) and contour lines of the square root of the 2nd invariant of deviatoric plastic strains (right), min=0.2%, Δ=0.4%.

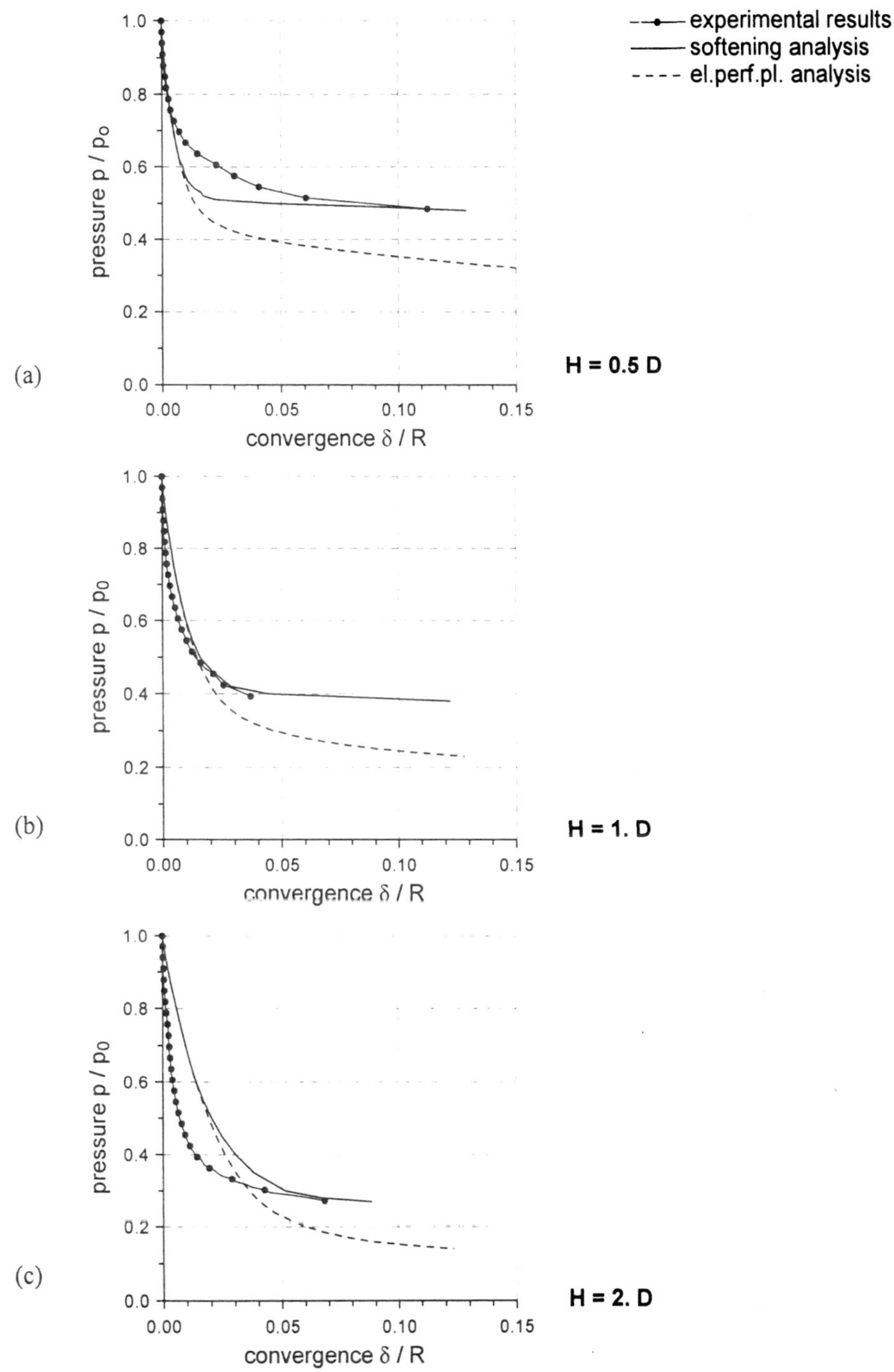

Figure 17. Comparison between measured and calculated pressure-convergence curves for various ratios between tunnel depth H and diameter D: H=0.5D (a), H=D (b), H=2D (c).

A second series of tests was performed at the Laboratories of Mitsubishi Heavy Industries in Takasago (Japan). They concern a 3D model of the portion close to the face of a horse shoe shaped, shallow tunnel. The tunnel model has a crown diameter of 1.2 m and the depth of the crown is 1.3 m.

The tunnel model is contained within a steel tank filled with a medium uniform sand, from an alluvial deposit of central Japan (Sterpi et al., 1996). Following a procedure similar to that used in the previous tests, the stability of the tunnel face was investigated measuring the displacements induced in the sand mass by the gradual reduction of the pressure of an airbag contained within the tunnel model.

Some 2D, plane strain calculations were then carried out with reference to the longitudinal section of the tunnel. Although these plane strain analyses do not allow a quantitative comparison between experimental and numerical results, they permit a qualitative estimation of the influence of softening on the overall behaviour of the tunnel. Fig. 18 shows that, for softening

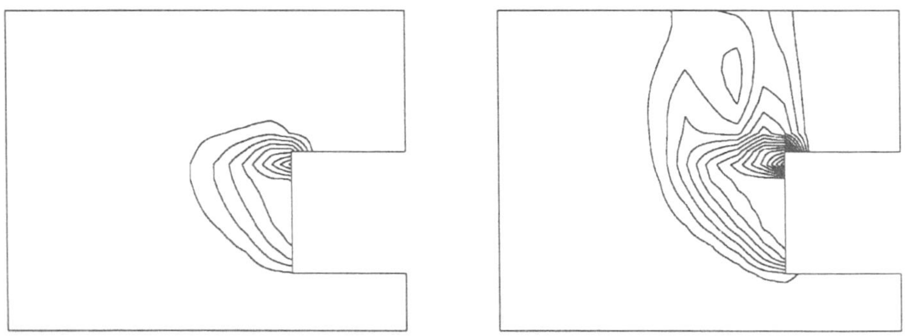

Figure 18. Contour lines of the square root of the 2nd invariant of the deviatoric plastic strains at failure (min=0.5%, Δ=0.5%), for elastic perfectly plastic analysis (left) and elasto-plastic softening analysis based on eq.10 (right), carried out on a coarse grid.

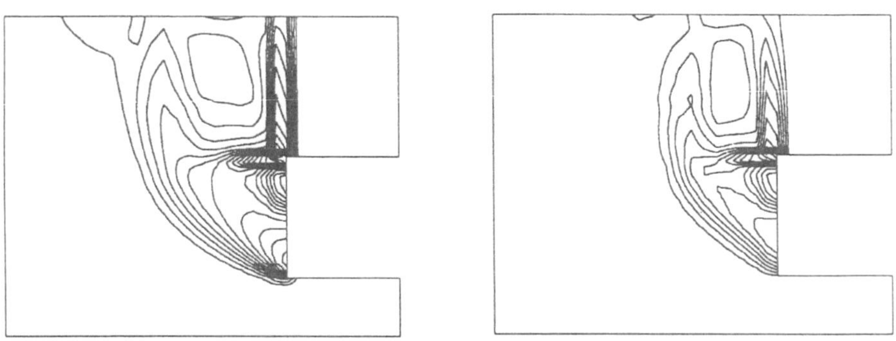

Figure 19. Contour lines of the square root of the 2nd invariant of the deviatoric plastic strains at failure (min=0.5%, Δ=0.5%), for elasto-plastic softening analyses based on eq.9 (left) and eq.10 (right), carried out on a refined grid.

analyses based on eq.(10), two shear bands develop from the tunnel crown and from its invert arch. They gradually reach the ground surface, in agreement with the experimental observation. Note that this effect is not present if the strains softening effects are neglected in the calculations. In this case, in fact, the shear strains concentrate only within a limited zone ahead of the tunnel face and the shape of the collapse mechanism is not properly predicted.

Finally, fig.19 shows the results obtained using a refined grid, and assuming eq. (9) or (10), alternatively, as criterion for the onset of localisation. It appears that the onset criterion does not appreciably modify the shape and the extension of the localised zone. The comparison between figs.18 and 19 indicates that the influence of the discretization on the numerical results can be limited if the softening parameters are modified according to the previously mentioned procedure.

8. Conclusions

Some applications have been discussed of the finite element method to the non-linear stress analysis of tunnels. The examples concern various problems of potential interest in tunnelling engineering, like the interaction between parallel tunnels, the effects of the time dependent or softening behaviour of the rock mass, the influence of its two-phase nature for tunnel driven in saturated media, etc.

In spite of its limits, the discussion shows that the numerical procedures for non-linear stress analysis can provide a valuable contribution to the solution of a variety of problems having a non negligible interest in tunnelling practice. In fact, they can be seen as an useful tool for the engineer, when dealing with relatively complex design problems that cannot be approached with standard methods of analysis or through simplified closed form solutions, like the characteristic line of deep openings.

Acknowledgements

The financial support of the Ministry of University and Research of the Italian Government (MURST) is gratefully acknowledged.

References

Bazant, Z., Belytschko, T., and Chang, T.P. (1984). Continuum theory for strain-softening. *J. of Eng. Mech. ASCE* 110:1666-1692.

Cividini, A., and Gioda, G. (1992). Finite element analysis of direct shear tests on stiff clays. *Int. J. Numer. Anal. Methods Geomech.* 16:869-886.

Cividini, A., Gioda, G., and Barla, G. (1985). A numerical analysis of tunnels driven in saturated two-phase media. *Proc. 11th Int. Conf. Soil Mech. Found. Eng.*, San Francisco, CA, 729-732.

Ghaboussi, J., and Gioda, G. (1977). On the time dependent effects in advancing tunnels. *Int. J. Numer. Anal. Methods Geomech.* 1:249-269.

Ghaboussi, J., and Wilson, E.L. (1973). Flow of compressible fluids in porous elastic media. *Int. J. Numer. Meth. Eng.* 1:75-100.

Gioda, G. (1981). A finite element solution of non-linear creep problems in rocks. *Int. J. Rock Mech. Min. Sci. & Geomech. Abst.* 18:35-46.

Gioda, G. (1982). On the non linear squeezing effects around circular tunnels. *Int. J. Numer. Anal. Methods Geomech.* 6:21-46.
Gioda, G., and Swoboda, G. (1999). Developments and applications of the numerical analysis of tunnels in continuous media. *Int. J. Numer. Anal. Methods Geomech.* 23:1393-1405.
Hansmire, W.H., and Cording, E.J. (1985). Soil tunnel test section: case history summary. *J. of Geotech. Eng., ASCE* 111(11):1301-1320.
Hill, R. (1958). A general theory of uniqueness and stability in elastic-plastic solids. *J. Mech. Phys. Solids* 6:236-249.
Pietruszczak, S., and Mroz, Z. (1981). Finite element analysis of deformation of strain-softening materials. *Int. J. Num. Meth. in Eng.* 17:327-334.
Poulos, H.G., and Davis, E.H. (1974). *Elastic solutions for soil and rock mechanics.* J.Wiley & Sons, New York.
Read, H.E., and Hegemier, G.A. (1984). Strain softening of rock, soil and concrete - A review article. *Mech. of Materials* 3:271-294.
Rice, J.R. (1976). The localization of plastic deformation. In Koiter, W.T., ed., *Proc. 14th Int. Conf. on Theoretical and Applied Mechanics*, North Holland, Amsterdam.
Sakurai, S. (1978). Approximated time-dependent analysis of tunnel support structure considering progress of the tunnel face. *Int. J. Numer. Anal. Methods Geomech.* 2:159-175.
Sandhu, R.S. (1968). *Fluid flow in saturated porous elastic media.* PhD Dissertation, University of California at Berkeley.
Sterpi, D. (1999). An analysis of geotechnical problems involving strain softening effects. *Int. J. Numer. Anal. Methods Geomech.* 23:1427-1454.
Sterpi, D. (2000). Influence of the kinematic testing conditions on the mechanical response of a sand. *Computers and Geotechnics* 26(1):23-41.
Sterpi, D., Cividini, A., Sakurai, S., and Nishitake, S. (1996). Laboratory model tests and numerical analysis of shallow tunnels. In Barla, G., ed., *Proc. Int. Symp. Prediction and Performance in Rock Mech. and Rock Eng., Eurock'96*, Balkema, 689-696.
Sterpi, D., and Moro, G. (1995). Finite element analyses of a shallow tunnelling process. *Gallerie e Grandi Opere Sotterranee* 46:58-67.
Sterpi, D., and Sakurai, S. (1997). Numerical analysis of laboratory tests on a model tunnel. In Asaoka at al., eds., *Proc. Int. Symp. Deformation and Progressive Failure in Geomechanics*, Elsevier Science, 757-762.
Vardoulakis, I., and Sulem, J. (1995). *Bifurcation analysis in geomechanics.* Chapman & Hall, Glasgow.

Nonlinear Problems During Consolidation Process

Janusz M. Dłużewski

Department of Environmental Engineering, Warsaw University of Technology
Poland

Abstract: The geometrical nonlinear problems oriented at geotechnical problems are presented herein. Lagrangian, Eulerian and updated Lagrangian procedures are discussed. The description of a two phase medium i.e. soil skeleton and water is formulated. Both geometrical and material nonlinearites for the fully coupled problem are considered. The updated Lagrange description is applied to model geometrically non-linear effects during consolidation process. The interaction problems between structure and soil foundation are studied. The description of the contact phenomenon which take place at interface soil-structure for a two-phase medium, i.e. skeleton and water, is formulated. Galerkin's and variational methods are used to derive the equations for the interface element in consolidation problems. The finite element equations for consolidation problems in large strains are formulated. The changes of the permeability in relation to the current porosity are discussed. The numerical tests for elastic and elasto-plastic skeleton models are shown and discussed.

1 Introduction, Review of Literature

Numerical analysis of the embankment rising on the soft subsoil requires definition and description of a few coupled nonlinear problems. The nonlinear effects, which play important role, from the engineering applications point of view, are listed bellow. First, the material nonlinearity, due to elasto-plastic behaviour of the soil skeleton should be modelled. Second, the large strain description is required. Next, the position of the phreatic line and saturation of the soil is often changed during the consolidation process. The numerical algorithms should allow for the changes of the volumetric weight due to the movement of the phreatic line and change of saturation. The embankment rising on the soft organic subsoil most often causes large changes in the permeability due to closing of the pores. The test results after Chaciński, Dłużewski (1995) shows that the changes of the permeability can reach even three orders in magnitude, in realistic scope of the peat deformation. This changes influence time of consolidation process.

The nonlinear effects listed above create coupled, strongly nonlinear system of equations. The finite element method is used to solve the initial boundary value problems. To solve the nonlinear coupled system the incremental iterative algorithms are used. During the iterations the nonlinear effects are superimposed on each other and they are solved simultaneously. If the strains are very large the mesh refreshment may be needed to avoid numerical difficulties.

Decoupling of the above nonlinear effects during consolidation is sometimes done. In such situations soil skeleton and fluid are considered sequentially in the time stepping process, but the control of the error generated during calculations is lost.

An analysis of soft soil behaviour during consolidation process plays a very important role in the geotechnical engineering. If we consider behaviour of soft soils as peat or mud both material and geometrical nonlinearities should be taken into account. For the geometrically non-linear problems a large variety of formulations and different strain and stress tensors can be encountered. The strain and stress tensors used in this paper can be found in monographs Truesdell and Toupin (1960), Truesdell and Noll (1965), Crisfield (1996) ,(1997). Lagrangian and Eulerian descriptions are often applied in computer procedure to solve the boundary value problems Kleiber (1975), Zienkiewicz and Taylor (1991), McMeeking and Rice (1975), Dłużewski (1988). The interest of the researchers is focused on the logarithmic stretches and their consistency with other strains and with stresses Hill (1968), (1978), Lee and McMeeking (1980). But the most common and widely used approach for both geometrical and material non-linear problems is the updated Lagrange description, Hibbit et al. (1970), Bathe (1982). The description constitutes a compromise between the theoretical formulation of the constitutive equations and numerical procedures. Most often the constitutive equations are defined in the current configuration, which is unknown at the beginning of the process. Thus the numerical algorithms have to begin the searching for the deformation field of the body, starting at the beginning from the previous, known, configuration.

In the paper the index notation as well as matrix notation are used to formulate the basic equations and numerically applicable finite element arrays.

Three approximation levels for consolidation problems (three theories or groups of theories) can be distinguished. The Terzaghi (1923), (1943) consolidation theory is the first one. It describes elastic consolidation for one-dimensional problems. The second is the Biot's (1941), (1956) theory, which describes a three-dimensional consolidation process. The assumption is made that, in spite of volumetric changes of the skeleton, the permeability is constant during the deformation process. In the original Biot's theory the isotropy and elasticity of the skeleton is assumed. The Darcy's flow rule is adopted. The small strains are assumed. In spite of such strict limitations, the Biot's theory extended to elasto-plastic models of the skeleton is the one which is most widely applied in numerical procedures. The number of the analytical solutions of practical problems is severely limited due to complexities of the formulation. A few of very simple cases involving only simple geometrical configurations were successfully solved analytically (see e.g., Gibson et al.(1970)), (Derski et al.(1988)).

The first numerical implementation of consolidation theory by means of the finite element method has been done by Sandhu (1968) and Sandhu and Wilson (1969) . The numerical formulations were enriched with new elements, e.g.: elasto-plastic models Norris and Lewis (1979). The fully saturated conditions were assumed by Zienkiewicz (1982), Zienkiewicz and Shiomi (1984), by Huyakorn and Pinder (1983).

The third level of approximation takes into consideration changes of the body configuration, porosity and permeability. At this level the research is spread in three main directions. The first one aims at one-dimensional consolidation theories, which take into account the configuration changes, Tan and Scott (1988), Mikasa(1965). This group uses the numerical applications. The second direction groups the theoretically advanced formulation,

Monte Kritzen(1976) , Szefer(1977) , but the coupling between various phenomena are so deep that the numerical applications are very limited. The third one takes into account chosen geometrically non-linear effects with numerical applications Meijer (1985), Asaoka *et al.* $(1997)_{1,2}$, (1998). Dłużewski (1997), (1998). A general model of saturated and partly saturated soil behaviour was developed by Zienkiewicz et al. $(1990)_{1,2}$. The interaction problems for a two-phase body during consolidation process were studied by Dłużewski (1991), (1993). Large strain effects are considered for fully and partly saturated model, static and dynamic behaviour by Meroi et al. (1995), Lewis & Schrefler (1998). The state of the art was summed up in the monograph by Lewis and Schrefler (1998).

The numerical modelling of soil-structure interactions requires models of the multiphase medium; soil skeleton and fluid as well as structure and definition of the behaviour in thin zones which are created at the soil-structure contact.

The interaction problems play a very important role in the soil mechanics and geotechnical engineering. Soil consists of three phases: solid skeleton, liquid and gas. The behaviour of such medium depends very strongly on the type of soil skeleton and water saturation of soil.

In practical engineering problems the great differences in material properties of structures and subsoil introduce an extra difficulty to the theoretical and numerical analysis of soil-structure interaction. The conjunction of the behaviour of concrete foundation blocks or steel impermeable sheet pile walls inserted into soil requires special theoretical and numerical treatment. In the thin zones at the contact between stiff structure and soil the assumption of continuum for the displacement of the skeleton and flow of fluid is no more fulfilled. Here the discontinuities appear and, in addition, they have very often an essential influence on the behaviour of structure on the subsoil.

In numerical modelling of soil-structure interactions the discontinuities of displacements and pore pressures produce an extra difficulty. The discontinuities tend to appear along stiff structures as geotextile reinforcements or sheet pile walls. They can also be found in the form of thin interlayers. In engineering problems the discontinuities occur very often, thus the appropriate description of this phenomenon is of practical importance. Introduction of the special interface elements into finite element mesh gives the possibility to simulate the slipping of the soil at the sheet pile wall during the decay of the excess pore pressure. Another type of problem where the proper description of soil-structure interaction produces an extra difficulty is the geotextile reinforced soil. The effect of pulling out of the geotextile reinforcement can be simulated with the use of interfaces allowing for slipping of soil at the geotextile.

A proper description of the interactions in the thin zones between structural and geological materials during consolidation process involves the coupled effects of the soil and fluid phases. Two following phenomena are observed in thin zones which usually appear at the contact between soil and structure. The first is the sliding of the skeleton at the stiff concrete or steel structure. The second one is the jump in the pore pressure. Several types of interfaces and joint elements has been developed to simulate soil-structure interactions in subsoil and rock masses. Three types of approaches can be distinguished. The first one treats the interfaces as true displacement discontinuities. The degrees of freedom are debounded (Ghaboussi et al.(1973)) and opening (or opening and closing) of the gap is allowed. Such formulations are used to solve the wedge problem or to simulate movement of the rock masses. The second approach stems from the zero-thickness interface element based on Goodman et al. (1968) conception.

The formulation is based on the introduction of the independent degrees of freedom which is considered to be a relative motion between surrounding solid elements. The third formulation takes its roots from the ordinary solid element, which is considered to be "thin". The "thin-layer" element can be defined as an isoparametric element (Desai et al.(1984). The proper choice of the element thickness is important here. If the thickness is too large (in comparison with the surrounding solid elements), the "thin-layer" element will behave as solid element. For too small thickness the numerical difficulty will arise. The ill-conditioning study for "thin-layer" interface elements can be found in the paper by Pande and Scharma (1978).

In the numerical modelling of flow problems the special types of elements can be inserted into finite element mesh to model barriers as steel and concrete structures or the flow through the thin layer close to such impermeable structures. Most often two types of such elements are placed close to the structures. The elements are considered "thin" or of zero thickness. The first one called fleece element is used to model barriers in the ground water flow (e.g. sheet pile wall). The fleece elements allow for the jump in the pore pressure but preserve the continuity equation. The second type of special elements are the drained elements. The elements allow for the large drainage along its length to simulate the thin drain layer (e.g. close to the retaining wall). The drained, one dimensional, elements are used to model vertical drains during consolidation of the soft organic subsoil.

The fourth chapter is aimed at studying the possibilities and limitations of the theoretical and numerical description of the phenomena which take place in the thin zones which appear at the contact between structures and multiphase medium. Special attention is paid to the finite element formulation of interaction problems (Dłużewski(1993)).

2 Displacements, Strains, Stresses -Definitions and Notation

In this chapter the definitions and convention of the kinematics and static quantities connected with the body are given. The chosen strains, stresses and evaluated quantities are defined. Since in literature different notations are used, the brief review of basic strains and stresses tensors is done. The motion of the body is considered in a stationary Cartesian coordinate system.

Two descriptions of the body kinematics are considered in classical mechanics. The first, Lagrangian (or material) description, combined with the initial undeformed configuration of the body and the second Eulerian (spatial) formulation defined in the final deformed body configuration. Both index and matrix notations are used in this section, Mróz (1985). The vectors and arrays are drown in bold.

In Lagrangian description a location of the material point is defined by means of vector **X** or coordinates X_K On the contrary in Eulerian description the location is defined by means of vector **x** or coordinates x_k. We can write in the material formulation :

$$\mathbf{x} = \mathbf{x}(\mathbf{X}, t) \quad \text{or} \quad x_k = x_k(\mathbf{X}, t) \tag{2.1}$$

on the contrary in the spatial coordinates

$$\mathbf{X} = \mathbf{X}(\mathbf{x}, t) \quad \text{or} \quad X_K = X_K(\mathbf{x}, t) \tag{2.2}$$

The capital letters are used in description combined with Lagrangian formulation and small letters combined with Eulerian description. The following deformations, strain and stress measurements are used herein.

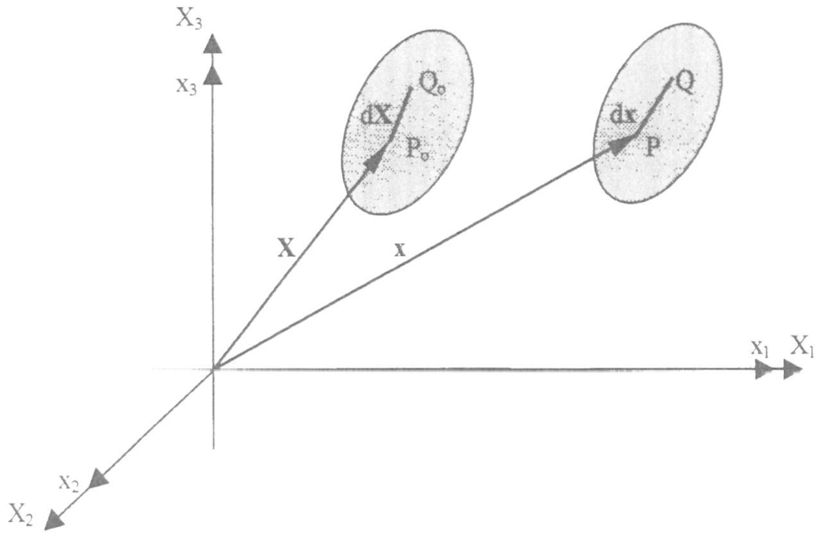

Figure 2.1. Motion and deformation of the body

2.1 Deformation Measures

Deformation gradient F_{IJ} Strain behaviour of the body is locally defined by the deformation gradient

$$F_{iJ}=\frac{\partial x_i}{\partial X_J} \qquad \mathbf{dx}=\mathbf{F}\ \mathbf{dX} \qquad dx_i=F_{iK}dX_K \tag{2.3}$$

Green deformation tensor C_{IJ} The square distance between points P and P_o, Fig. 2.1 after deformation, can be calculated as follows

$$dS^2=(\mathbf{dX})^2=\mathbf{dX}^T\cdot\mathbf{dX}=dX_k dX_k \tag{2.4}$$

$$ds^2=(\mathbf{dx})^2=\mathbf{dx}^T\cdot\mathbf{dx}=(\mathbf{F}\cdot\mathbf{dX})^T\cdot(\mathbf{F}\cdot\mathbf{dX})=\mathbf{dX}^T\left(\mathbf{F}^T\cdot\mathbf{F}\right)\mathbf{dX}=\mathbf{dX}^T\cdot\mathbf{C}\cdot\mathbf{dX} \tag{2.5}$$

where $\mathbf{C}$ is Green tensor expressed by deformation gradient $\mathbf{F}$ in index and matrix notation in the following form

$$\mathbf{C}=\mathbf{F}^T\cdot\mathbf{F} \qquad C_{IJ}=\frac{\partial x_k}{\partial X_I}\frac{\partial x_k}{\partial X_J} \tag{2.6}$$

$$ds^2=\frac{\partial x_k}{\partial X_I}\frac{\partial x_k}{\partial X_J}dX_I dX_J=C_{IJ}dX_I dX_J \tag{2.7}$$

Green strain tensor E_{IJ}. The difference in the square distance between points P and Q Fig. 2.1 after deformation is calculated as follows

$$ds^2 - dS^2 = \mathbf{dX}^T(\mathbf{C}-\mathbf{I})\,\mathbf{dX} = 2\mathbf{dX}^T\mathbf{E}\,\mathbf{dX} = (C_{IJ}-\delta_{IJ})\,dX_I dX_J = 2E_{IJ}\,dX_I dX_J \tag{2.8}$$

$$E_{IJ} = \frac{1}{2}(C_{IJ}-\delta_{IJ}) \qquad \mathbf{E} = \frac{1}{2}(\mathbf{C}-\mathbf{I}) \qquad E_{IJ} = \frac{1}{2}\left(\frac{\partial U_I}{\partial X_J} + \frac{\partial U_J}{\partial X_I} + \frac{\partial U_K}{\partial X_I}\frac{\partial U_K}{\partial X_J}\right) \tag{2.9}$$

where E_{IJ} is Green strain tensor (Lagrangian description), $\mathbf{I}$ is the zero unit array, δ_{ij} is Kronecker delta.

Cauchy deformation tensor c_{ij}. The square distance between points P and Q Fig. 2.1 after deformation can be calculated, in Eulerian description, as follows

$$dS^2 = dX_I dX_I = \left(\mathbf{F}^{-1}\mathbf{dx}\right)^T\left(\mathbf{F}^{-1}\mathbf{dx}\right) = \mathbf{dx}^T\left[\left(\mathbf{F}^{-1}\right)^T\mathbf{F}^{-1}\right]\mathbf{dx} = \mathbf{dx}^T\cdot\mathbf{c}\cdot\mathbf{dx} \tag{2.10}$$

Tensor Cauchy deformation tensor is defined as follows

$$ds^2 = \frac{\partial X_K}{\partial x_i}\frac{\partial X_K}{\partial x_j}dx_i dx_j = c_{ij}dx_i dx_j \qquad c_{ij} = \frac{\partial X_K}{\partial x_i}\cdot\frac{\partial X_K}{\partial x_j} \qquad \mathbf{c} = \left(\mathbf{F}^{-1}\right)^T\mathbf{F}^{-1} \tag{2.11}$$

Almansi strain tensor ε_{ij}. Considering the differences of square distance between points P i Q after deformation we have

$$ds^2 - dS^2 = d\mathbf{X}^T(\mathbf{I}-\mathbf{C})d\mathbf{X} = 2d\mathbf{X}^T\varepsilon\,d\mathbf{X} = \left(\delta_{ij}-c_{ij}\right)dx_i dx_j = 2\varepsilon_{ij}dx_i dx_j \tag{2.12}$$

where Almansi strain tensor is defined as follows

$$\varepsilon_{ij} = \frac{1}{2}\left(\delta_{ij}-c_{ij}\right) \qquad \boldsymbol{\varepsilon} = \frac{1}{2}(\mathbf{I}-\mathbf{C}) \qquad \varepsilon_{ij} = \frac{1}{2}\left(\frac{\partial u_i}{\partial x_j} + \frac{\partial u_j}{\partial x_i} - \frac{\partial u_k}{\partial x_i}\frac{\partial u_k}{\partial x_j}\right)$$

$$\varepsilon_{ij} = \frac{\partial X_K}{\partial x_i}\cdot\frac{\partial X_L}{\partial x_j}E_{KL} \qquad \varepsilon = \mathbf{F}^{-1}\mathbf{F}^{-1}\mathbf{E} \qquad \varepsilon_{ij} = e_{ij} - \eta_{ij} \tag{2.13}$$

$$e_{ij} = \frac{1}{2}\left(\frac{\partial u_i}{\partial x_j} + \frac{\partial u_j}{\partial x_i}\right) \qquad \eta_{ij} = \frac{1}{2}\left(\frac{\partial u_k}{\partial x_i}\frac{\partial u_k}{\partial x_j}\right)$$

Polar decomposition. The deformation gradient can be decomposed in two following ways

$$\mathbf{F} = \mathbf{R}\cdot\mathbf{U} = \mathbf{V}\cdot\mathbf{R} \qquad x_{i,k} = R_{il}U_{l,k} = V_{i,j}R_{jk} \tag{2.14}$$

where: **V** is the left-stretch (Euler) tensor, **U** is the right-stretch (Lagrange) tensor, **R** is the rotation tensor $\mathbf{R}^T \cdot \mathbf{R} = \mathbf{I}$.

Displacement velocities v_i, V_K. The following displacement velocities are used

$$\mathbf{x} = \mathbf{X} + \mathbf{U} \qquad \mathbf{v} = \dot{\mathbf{x}} \qquad \dot{\mathbf{X}} = \dot{\mathbf{U}} \tag{2.15}$$

The displacement velocity in the Lagrangian (material) formulation

$$V_K(X,t) = \dot{U}_K(X,t) = \frac{\partial U_K(X,t)}{\partial t} \tag{2.16}$$

The displacement velocity in the Eulerian (space) formulation

$$v_i(x,t) = \frac{\partial u_i(x(t),t)}{\partial t} \cdot \frac{\partial t}{\partial t} + \frac{\partial u_i}{\partial x_j} \frac{\partial x_j}{\partial t} \tag{2.17}$$

2.2 Stress Measures

Cauchy stress tensor σ_{ij} Cauchy stress tensor is defined in the current configuration of the body. The current acting force is divided by the current area.

Kirchhoff stress tensor τ_{ij}.

$$\tau_{ij} = \frac{\rho_o}{\rho} \sigma_{ij} \qquad \boldsymbol{\tau} = J\,\boldsymbol{\sigma} \tag{2.18}$$

I (First) Piola Kirchhoff stress tensor T_{Ji} To define I Piola Kirchhoff the current forces are related to the initial undeformed configuration of the body.

$$\mathbf{dP} = \mathbf{n}^o \mathbf{T} dS_o = \mathbf{n}\boldsymbol{\sigma} ds \tag{2.19}$$

where $\mathbf{n}^o$ is the normal vector to the initial surface dS_o. The moment equilibrium equation is not fulfilled what yields the nonsymmetry of the stress tensor $T_{Ji} \neq T_{iJ}$. The force equilibrium is confirmed. Using the Nanson's formula Crisfield (1996) we have

$$\mathbf{n}dS = \frac{\rho_o}{\rho} \mathbf{n}^o \mathbf{F}^{-1} dS_o \qquad n_m dS = \frac{\rho_o}{\rho} n_J^o \frac{\partial X_J}{\partial x_m} dS_o \tag{2.20}$$

The transformation formulas between Cauchy stresses and II Piola-Kirchhoff stresses can be written as follows

$$\mathbf{T} = \frac{\rho_o}{\rho} \mathbf{F}^{-1} \sigma \qquad T_{Ji} = \frac{\rho_o}{\rho} \frac{\partial X_J}{\partial x_m} \sigma_{mi} \tag{2.21}$$

II (Second) Piola-Kirchhoff stress tensor S_{IJ}. The transformation between initial and the current configuration takes the form $\mathbf{dx} = \mathbf{F}\ \mathbf{dX}$. I Piola-Kirchhoff stress tensor contains the area transformation without force transformation. II Piola-Kirchhoff stresses contains the force transformation the same as the area transformation. The transformation between II Piola-Kirchhoff stresses and Cauchy stresses can be written in the following form

$$S_{IJ} = \frac{\rho_o}{\rho} \frac{\partial X_J}{\partial x_m} \sigma_{mj} \frac{\partial X_I}{\partial x_j} \qquad \mathbf{S} = \frac{\rho_o}{\rho} \mathbf{F}^{-1} \sigma \left(\mathbf{F}^{-1}\right)^T \tag{2.22}$$

The transformation between I and II Piola-Kirchhoff stresses are defined as follows

$$S_{IJ} = T_{Jj} \frac{\partial X_I}{\partial x_j} \qquad \mathbf{S} = \mathbf{T}\left(\mathbf{F}^{-1}\right)^T \tag{2.23}$$

2.3 Stress Rates

Jaumann stress rate $\overset{\nabla}{{}_J\sigma_{ij}}$. Jaumann stress rate is defined in the co-rotational coordination system. Herein the Jaumann rate of Cauchy stresses is used in the following way

$$\overset{\nabla}{{}_J\sigma_{ij}} = \dot{\sigma}_{ij} - \sigma_{iq}\,\varpi_{jq} - \sigma_{pj}\,\varpi_{ip} \qquad \overset{\nabla}{{}_J\boldsymbol{\sigma}} = \dot{\boldsymbol{\sigma}} - \varpi\boldsymbol{\sigma} - \boldsymbol{\sigma}\varpi^T \tag{2.24}$$

where ϖ is a spin $\varpi = \frac{1}{2}(\mathbf{L} - \mathbf{L}^T)$, $\mathbf{L}$ is a velocity gradient $\mathbf{L} = \frac{\partial \mathbf{v}}{\partial \mathbf{x}}$

Oldroyd stress rate $\overset{\nabla}{{}_O\sigma_{ij}}$. Oldroyd stress rate is defined in the convected coordinate system. It is also sometimes known as the Cotter-Rivlin rate. The Oldroyd rate of the Cauchy stress is defined as follows

$$\overset{\nabla}{{}_O\sigma_{ij}} = \dot{\sigma}_{ij} + \sigma_{iq}\,L_{jq} + \sigma_{pj}\,L_{ip} \qquad \overset{\nabla}{{}_O\boldsymbol{\sigma}} = \dot{\boldsymbol{\sigma}} + \mathbf{L}^T\boldsymbol{\sigma} + \boldsymbol{\sigma}\mathbf{L} \tag{2.25}$$

where $\mathbf{L}$ is a velocity gradient $\mathbf{L} = \frac{\partial \mathbf{v}}{\partial \mathbf{x}}$ ϖ is a spin.

Truesdell stress rate $\overset{\nabla}{{}_T\tau_{ij}}$. Truesdell stress rate of the Kirchhoff stresses is defined in the following form

$$\overset{\nabla}{{}_T\tau_{ij}} = \dot{\tau}_{ij} + \tau_{iq}\,L_{jq} + \tau_{pj}\,L_{ip} \qquad \overset{\nabla}{{}_T\boldsymbol{\tau}} = \dot{\boldsymbol{\tau}} + \mathbf{L}^T\boldsymbol{\tau} + \boldsymbol{\tau}\mathbf{L} \tag{2.26}$$

$$\text{since} \quad \boldsymbol{\tau} = J\boldsymbol{\sigma} \quad \text{so} \quad \dot{\boldsymbol{\tau}} = \dot{J}\boldsymbol{\sigma} + J\dot{\boldsymbol{\sigma}} \quad \text{we have} \quad \dot{\boldsymbol{\sigma}} = \frac{1}{J}\dot{\boldsymbol{\tau}} - \boldsymbol{\sigma}\,\mathrm{tr}(\dot{\mathbf{e}}) \tag{2.27}$$

were $\mathrm{tr}(\dot{\mathbf{e}})$ is the trace of the strain rate tensor.

2.4 Equilibrium Equations

Equilibrium equations in Eulerian description

$$\frac{\partial \sigma_{ij}}{\partial x_j} + f_i = 0 \tag{2.28}$$

Equilibrium equation at the boundary condition.

$$\sigma_{ji} n_j = t_i \qquad \mathbf{n} \cdot \boldsymbol{\sigma} = \mathbf{t} \tag{2.29}$$

Equilibrium equation defined by means of I Piola-Kirchhoff stresses

$$\frac{\partial T_{Ji}}{\partial X_J} + \rho_o b_i = 0 \tag{2.30}$$

where $f_i = \rho \;\; b_i$

Equilibrium equation in terms of II Piola-Kirchhoff stresses

$$T_{Ji} = S_{Jk} \frac{\partial x_i}{\partial X_k} \qquad \frac{D}{\partial x_J}\left[S_{Jk} \frac{\partial x_i}{\partial X_k} \right] + \rho_o b_i = 0 \tag{2.31}$$

At the boundary

$$n_J \cdot T_{Ji} = t_i^o \qquad n_J S_{Jk} \frac{\partial x_i}{\partial X_k} = t_i^o \tag{2.32}$$

2.5 Virtual Work Principles

In the current configuration by means of Cauchy stress tensor

$$\int_V \sigma_{ij} \delta e_{ij} dV = \int_S t_i \delta u_i dS + \int_V f_i \delta u_i dV \tag{2.33}$$

where: e_{ij} - linear part of the Almansi strain tensor

$$e_{ij} = \frac{1}{2}\left(\frac{\partial u_i}{\partial x_j} + \frac{\partial u_j}{\partial x_i} \right) \tag{2.34}$$

In the initial configuration by means of Kirchhoff stresses

$$\int_{V_O} \tau_{ij} \delta e_{ij} \, dV_o = \int_{S_O} t_i^o \delta u_i dS_o + \int_{V_O} f_i^o \delta u_i dV_o \tag{2.35}$$

In the initial configuration by means of I Piola-Kirchhoff stress tensor

$$\int_{V_O} T_{Ji} \delta F_{iJ} dV_o = \int_{S_O} t_i^o \, \delta u_i dS_o + \int_{V_O} f_i^o \, \delta u_i dV_o \tag{2.36}$$

In the initial configuration by means of II Piola-Kirchhoff stress tensor

$$\int_{V_0} S_{IJ}\delta E_{IJ} dV_o = \int_{S_0} t_i^o \delta u_i dS_o + \int_{V_0} f_i^o \delta u_i dV_o \tag{2.37}$$

It can be easy shown that one virtual work principle is equivalent to any other listed above.

2.6 Conjugate Variables

The conjugate variables appear in the virtual work principles. The certain stresses are conjugated with the adequate strains in proper configuration of the body.

Table 2.1.. Conjugate variables

equation	stress measure	deformation measure	configuration
(2.33)	σ	**e**	current
(2.35)	τ	**e**	initial
(2.36)	**T**	**F**	initial
(2.37)	**S**	**E**	initial

2.7 Virtual Power Principles

The incremental form of the constitutive equations makes the power virtual principles often used as a starting point to formulate incremental iterative procedures to solve boundary value problems. The virtual power principals are defined below.
In the current configuration by means of Cauchy stress tensor

$$\int_V \sigma_{ij}\delta \dot{e}_{ij} dV = \int_S t_i \delta \upsilon_i dS + \int_V f_i \delta \upsilon_i dV \tag{2.38}$$

where $\dot{e}_{ij}$ is the rate strain tensor

$$\dot{e}_{ij} = \frac{1}{2}\left(\frac{\partial \upsilon_i}{\partial x_j} + \frac{\partial \upsilon_j}{\partial x_i}\right), \qquad \upsilon_i = \frac{du_i}{dt} \tag{2.39}$$

In the initial configuration by means of the Kirchhoff stresses

$$\int_{V_0} \tau_{ij}\delta \dot{e}_{ij} dV_o = \int_S t_i \delta \upsilon_i dS + \int_V f_i \delta \upsilon_i dV \tag{2.40}$$

In the initial configuration by means of the II Piola-Kirchhoff stresses

$$\int_{V_0} S_{IJ}\delta \dot{E}_{IJ} dV_o = \int_{S_0} t_i^o \delta \upsilon_i dS_o + \int_{V_0} f_i^o \delta \upsilon_i dV_o \tag{2.41}$$

2.8 Elasto-Plastic Deformations

In elasto-plastic analysis *the local reference configuration* is introduced, Lee (1969), Mandel (1972), Fig.2.2. In the literature it is also known as the intermediate configuration, Rice (1971), the local isoclinic relaxed configuration, Teodosiu (1970).

The reference configuration $\mathbf{x}^p$ corresponds to the configuration reached by the body when unloading. The deformation gradient has the representation in plastic and elastic parts as follows

$$F_{iJ} = \frac{\partial x_i}{\partial x_k^p} \frac{\partial x_k^p}{\partial X_J} = F_{ik}^e F_{kJ}^p \qquad \mathbf{F} = \mathbf{F}^e \mathbf{F}^p \qquad (2.42)$$

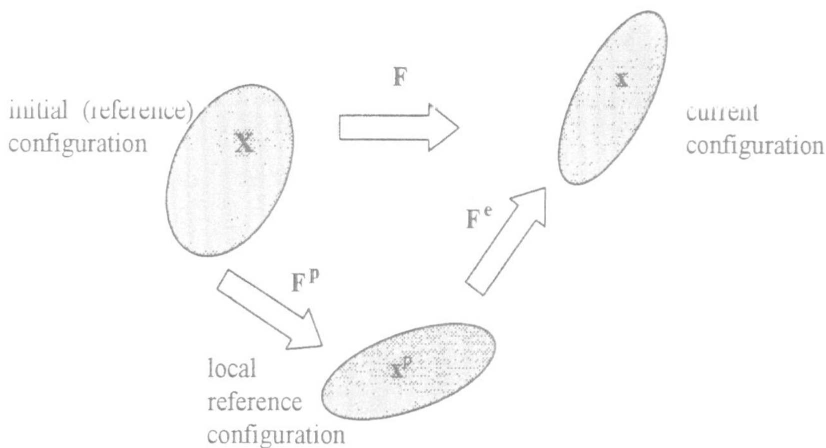

Figure 2.2. Local reference configuration in elasto-plastic analysis

In general the finite deformation measures are not linearly expressed in terms of displacements. Having assumed small elastic strain components, adding the elastic and plastic increments can be accepted.

$$d\varepsilon_{ij} \cong d\varepsilon_{ij}^e + d\varepsilon_{ij}^p \qquad \mathbf{d\varepsilon} \cong \mathbf{d\varepsilon}^e + \mathbf{d\varepsilon}^p \qquad (2.43)$$

2.9 Effective stresses

The general expression of the total and effective Cauchy stresses is assumed in the following form

$$\sigma_{ij} = \sigma_{ij}^{//} + \alpha\, \delta_{ij} p \qquad \boldsymbol{\sigma} = \boldsymbol{\sigma}^{//} + \alpha\, \mathbf{m}\, p \qquad (2.44)$$

where α is the Biot's constant, $\boldsymbol{\sigma}$ are the total Cauchy stresses, $\boldsymbol{\sigma}^{//}$ are the effective Cauchy stresses indicated by ($^{//}$) after Zienkiewicz et al. (1990). In the convention adopted herein, the stresses and pore pressure are defined positive for tension and negative for compression. The soil is assumed fully saturated. The Biot's constant is equal to

$$\alpha = 1 - \frac{K_T}{K_S} \tag{2.45}$$

where K_T is the bulk modulus of the overall skeleton, K_S is the bulk modulus of the solid grains. For deep layers (considered usually in mining applications) the Biot's constant α approaches zero. Then flow and mechanics are decoupled.

3. Numerical Formulations for Large Strains

Several numerical approaches are formulated for the geometrical nonlinear problems. The base numerical formulations take its roots from the virtual work or virtual power principles. The general classification of the large strain approaches is shown in the form of diagram on the Fig. 3.1.

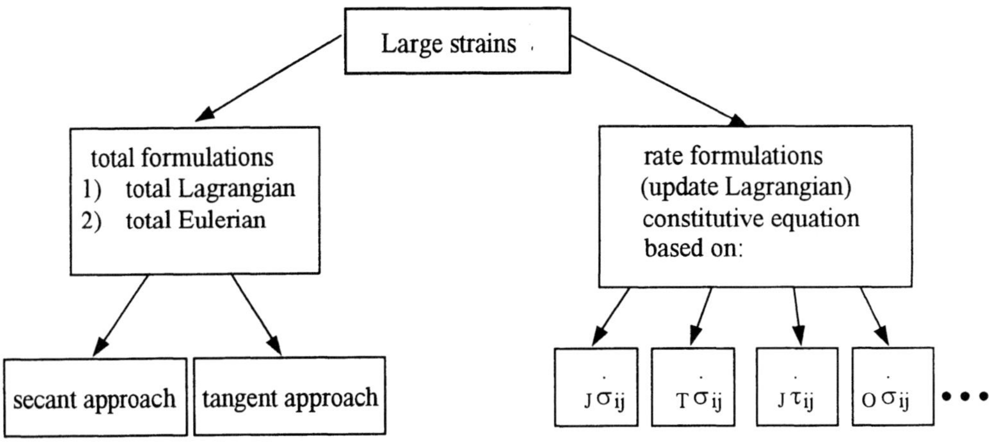

Figure 3.1. Large strain formulation

In general, the form of constitutive equations applied in the formulation generates the numerical procedure for the geometrical nonlinear problem. If the constitutive equation is defined in the total measures (e.g. between the Green strains and the II Piola-Kirchhoff stresses) the Lagrangian formulation is used.

During incremental iterative procedures various transformations are performed, so different errors are generated. It is good to remember that at any stage of the incremental iterative process the total equilibrium has to be kept through the whole numerical process. The virtual work principles (2.33), (2.35)-(2.37) with the conjugated variables Table 2.1 are the base to formulate the total large strain numerical procedures. The virtual work power principles (2.38), (2.40)÷(2.41) stands for the starting point to formulate the tangent stiffness matrix and incremental iterative algorithm.

3.1. Total Formulation of Large Strains

Total Lagrange formulation. Usually by a linear elasticity is meant the constitute equations defined in the initial undeformed configuration of the body i.e. in material configuration. The constitutive equations is defined in total form

$$\mathbf{S} = \mathbf{D}_{LG} \cdot \mathbf{E} \tag{3.1}$$

where S is the II Piola-Kirchhoff stress tensor, **E** is the Green strain tensor (2.9) and $\mathbf{D}_{LG}$ is the four order tensor of elastic constants. The virtual work principle (2.27) can be written in the following way

$$\int_{V_0} \delta\mathbf{E}^T \mathbf{S}\, dV_0 = \int_{S_0} \delta\mathbf{U}^T \mathbf{t}_0\, dS_0 + \int_{V_0} \delta\mathbf{U}^T \mathbf{f}_0\, dV_0 \tag{3.2}$$

where the Green strain tensor is composed of linear and nonlinear terms

$$\mathbf{E} = \mathbf{E}_0 + \mathbf{E}_L \tag{3.3}$$

The finite element approach with standard notation is used, Zienkiewicz and Taylor (1991)

$$\mathbf{E} = \mathbf{B}_{NL}\mathbf{U}$$

where
$$\mathbf{B}_{NL} = \mathbf{B}_0 + \mathbf{B}_L(\mathbf{U}) \tag{3.4}$$

Matrix $\mathbf{B}_L$ contains the nonlinear forms coming from the second order form of the Green strain tensor. Introducing equation (3.3) into virtual work principle (3.2) and eliminating kinematically admissible displacement field $\delta\mathbf{U}$ we arrive at

$$\int_{V_0} \mathbf{B}_{NL}(\mathbf{U})^T \mathbf{S}\, dV_0 = \int_{S_0} \mathbf{N}^T \mathbf{t}_0\, dS_0 + \int_{V_0} \mathbf{N}^T \mathbf{f}_0\, dV_0 \tag{3.5}$$

Incorporating the constitutive equation (3.1) into (3.5) the above matrix equation takes the form as follows

$$\int_{V_0} \mathbf{B}_{NL}^T(\mathbf{U}) \cdot \mathbf{D}_{LG} \cdot \mathbf{B}_{NL}(\mathbf{U})\, dV_0\, \mathbf{U} = \int_{S_0} \mathbf{N}^T \mathbf{t}_0\, dS_0 + \int_{V_0} \mathbf{N}^T \mathbf{f}_0\, dV_0 \tag{3.6}$$

The final matrix nonlinear equation of nodal equilibrium can be rewritten in the following form

$$\mathbf{K}_s \mathbf{U} = \mathbf{P} \tag{3.7}$$

In total approach the secant stiffness matrix $\mathbf{K}_s$ and the loading vector **P** are defined as follows

$$\mathbf{K}_S = \int_{V_0} \mathbf{B}_{NL}^T(\mathbf{U})\mathbf{D}_{LG}\mathbf{B}_{NL}(\mathbf{U})\, dV_0 , \tag{3.8a}$$

$$\mathbf{P} = \int_{S_0} \mathbf{N}^T \mathbf{t}_0 dS_0 + \int_{V_0} \mathbf{N}^T \mathbf{f}_0 dV_0 \tag{3.8b}$$

The matrix $\mathbf{K}_S$ can be split into linear and nonlinear parts as follows

$$\mathbf{K}_S = \mathbf{K}_0 + \mathbf{K}_L \tag{3.9}$$

where

$$\mathbf{K}_0 = \int_V \mathbf{B}_0^T \mathbf{D}_{LG} \mathbf{B}_0 dV_0 \tag{3.10}$$

$$\mathbf{K}_L = \int_{V_0} (\mathbf{B}_0^T \mathbf{D}_{LG} \mathbf{B}_L(\mathbf{U}) + \mathbf{B}_L^T(\mathbf{U}) \mathbf{D}_{LG} \mathbf{B}_0 + \mathbf{B}_L^T(\mathbf{U}) \mathbf{D}_{LG} \mathbf{B}_L(\mathbf{U})) dV_0 \tag{3.11}$$

The iterative process can be based on a direct secant stiffness array $\mathbf{K}_S$ as it is shown in the Fig 3.2.

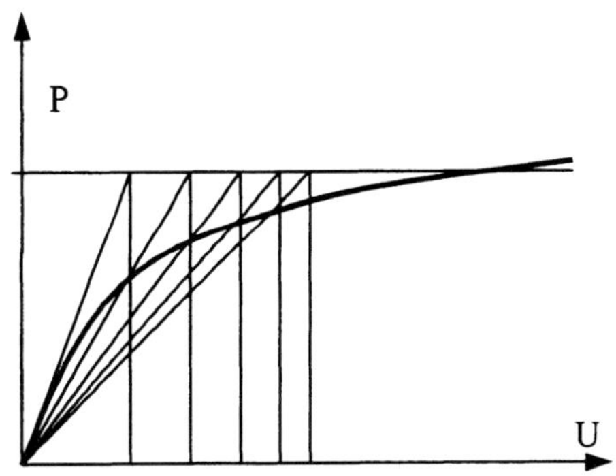

Fig.3.2 Direct secant iterative procedure

Total Eulerian formulation. An alternative formulation to total Lagrangian description is the Eulerian formulation. The constitutive equation can be defined in the spatial configuration as follows

$$\boldsymbol{\sigma} = \mathbf{D}_{EU} \cdot \boldsymbol{\varepsilon} \tag{3.12}$$

where $\boldsymbol{\sigma}$ is the Cauchy stress tensor, $\boldsymbol{\varepsilon}$ is Almansi strain tensor and $\mathbf{D}_{EU}$ is the tensor of elastic properties.

Although the constitutive equation is linear in one configuration (i.e.(3.12) in spatial reference system) it is nonlinear in other configuration (i.e. material reference system). Starting from equation (3.1) and transforming the II Piola-Kirchhoff stress tensor **S** to the Cauchy stress tensor $\boldsymbol{\sigma}$,equation (2.22), and Green stress tensor **E** to the Almansi strain tensor ε, equation (2.13), we get

$$\boldsymbol{\sigma} = \frac{\rho_0}{\rho}[\overline{\overline{\mathbf{FF}}}]^T \mathbf{D}_{LG} \overline{\overline{\mathbf{FF}}}\ \varepsilon \tag{3.13}$$

in index notation $\sigma_{ij} = \frac{\rho_0}{\rho} \frac{\partial x_i}{\partial X_I} \frac{\partial x_j}{\partial X_J} D^{LG}_{IJKL} \frac{\partial x_m}{\partial X_K} \frac{\partial x_n}{\partial X_L} \varepsilon_{mn}$ (3.14)

The virtual work principle defined in the current configuration of the body is the starting point for the numerical formulation in total Eulerian description

$$\int_v \delta \mathbf{e}^T \sigma dv = \int_s \delta \mathbf{u}^T \mathbf{t} ds + \int_v \delta \mathbf{u}^T \mathbf{f} dv \tag{3.15}$$

where **e** is the linear part of the Almansi strain tensor

$$\mathbf{e} = \mathbf{B} \cdot \mathbf{u} \qquad\qquad \varepsilon = (\mathbf{B} - \mathbf{B}_{NE}(\mathbf{u}))\mathbf{u} \tag{3.16}$$

Incorporating (3.16) and constitutive equation (3.12) into virtual work principle (3.15) and next eliminating the kinematicly admissible displacement field we get

$$\int_v \mathbf{B}^T \mathbf{D}_{EU} (\mathbf{B} - \mathbf{B}_{NE}(\mathbf{u})) dv \cdot \mathbf{u} = \int_s \mathbf{N}^T \mathbf{t} ds + \int_v \mathbf{N}^T \mathbf{f} dv \tag{3.17}$$

Finally it takes the following form

$$\mathbf{K}_S \cdot \mathbf{u} = \mathbf{P} \tag{3.18}$$

The stiffness array can be divided into two parts

$$\mathbf{K}_S = \int_v \mathbf{B}^T \mathbf{D}_{EU} \mathbf{B}^T dv - \int_v \mathbf{B}^T \mathbf{D}_{EU} \mathbf{B}_{NE}(\mathbf{u}) dv \tag{3.19}$$

$$\mathbf{P} = \int_s \mathbf{N}^T \mathbf{t} ds + \int_v \mathbf{N}^T \mathbf{f} dv \tag{3.20}$$

The difference between total Lagrangian and total Eulerian iterative process is that all arrays in Eulerian description are defined in the unknown configuration of the body, which is sought for via the iterative process. This approach has been applied for a simple sand model by Dłużewski (1988).

3.2. Rate Formulations of Large Strains

The rate formulation of the large strains requires proper definition of the constitutive equation. First, various stress measures are used to define constitutive law. Second, the stress rate have to be rotation objective. Two coordinate systems are used to define the objective stress measures.

The first is corotational coordinate system where the Jaumann stress rate and Truesdell stress rate are defined. The second is the convected coordinate system where the Oldroyd stress rate is defined.

Jaumann rate of Cauchy stress. The constitutive law is defined by means of the Eulerian strain rate $\mathbf{e}$ (2.39) (equivalent to ε in Crisfield (1997)) and Jaumann objective stress rate (2.24) of the Cauchy stress tensor

$$_J\dot{\sigma}^{//} = \mathbf{D}_T \cdot \dot{\mathbf{e}} \tag{3.21}$$

where $_J\dot{\boldsymbol{\sigma}}^{//}$ is the Jaumann stress rate of the effective Cauchy stress.

The updated Lagrange procedures confirms consistency with the constitutive law, provided the increments are small enough to allow neglecting the second order terms in the Green strain tensor. If the assumption of small elastic strain components is fulfilled the summing of the elastic and plastic parts of the strain components can be accepted. Chapter 2.8.

The proper definition of the volumetric changes is important. It requires three basic invariants of the Almansi or Green strain tensors and it is given in the Appendix B, after Eringer (1974).

In the updated Lagrangian procedure the calculation of the volumetric changes is simplified to the sum of the strains, $d\varepsilon_v = d\varepsilon_x + d\varepsilon_y + d\varepsilon_z$. Such summing up of the volume increase is acceptable provided small increments are used. The accumulated error can be checked by comparing sum of volumetric change increments with the exact definition given in the Appendix B.

The virtual power principle (2.38) is the starting point to formulate the numerical procedure for the rate formulation of the large strains. The rate form of the stress power π is given by Malvern (1969), Lewis and Schefler (1998)

$$\dot{\pi} = \frac{D}{Dt}\int_V \delta\dot{\mathbf{e}}^T \sigma dV = \int_V \left(\frac{d\delta\dot{\mathbf{e}}^T}{dt}\cdot\sigma + \delta\dot{\mathbf{e}}^T\dot{\sigma} + \sigma\delta\dot{\mathbf{e}}\frac{\dot{J}}{J}\right)dV \tag{3.22}$$

where $\frac{D}{Dt}$ denotes material time derivative.

Incorporating the constitute law (3.21) and the Jaumann rate of Cauchy stress (2.24) into the equation (3.18) and next eliminating the kinematically admissibly displacement rate field δv the tangent stiffness matrix is obtained.

$$\mathbf{K}_T \cdot \mathbf{v} = \dot{\mathbf{P}} \tag{3.23}$$

where the stiffness matrix $\mathbf{K}_T$ is the sum of the four terms

$$\mathbf{K}_T = \mathbf{K}_0 + \mathbf{K}_S + \mathbf{K}_{\sigma 1} + \mathbf{K}_{\sigma 2} \tag{3.24}$$

The first term $\mathbf{K}_0$ corresponds to small strain stiffness array

$$\mathbf{K}_0 = \int_V \mathbf{B}^T \cdot \mathbf{D}_T \cdot \mathbf{B} \cdot dv \tag{3.25}$$

where $\mathbf{B}$ is typical small strain geometrical array. The second matrix $\mathbf{K}_S$. is called initial stress stiffness array and takes the following form:

$$\mathbf{K}_S = \int_V \mathbf{B}_N^T \cdot \overline{\sigma} \cdot \mathbf{B}_N \cdot dv \tag{3.26}$$

where $\mathbf{B}_N$ and $\overline{\boldsymbol{\sigma}}$ are equal to

$$\mathbf{B}_N = \begin{bmatrix} \frac{\partial N_i}{\partial X_1} \cdot \mathbf{I} \\ \frac{\partial N_i}{\partial X_2} \cdot \mathbf{I} \\ \frac{\partial N_i}{\partial X_3} \cdot \mathbf{I} \end{bmatrix}, \qquad \overline{\boldsymbol{\sigma}} = \begin{bmatrix} \overline{\sigma}_{11} & \overline{\sigma}_{12} & \overline{\sigma}_{13} \\ \overline{\sigma}_{21} & \overline{\sigma}_{22} & \overline{\sigma}_{23} \\ \overline{\sigma}_{31} & \overline{\sigma}_{32} & \overline{\sigma}_{33} \end{bmatrix}, \qquad \overline{\sigma}_{ij} = \sigma_{ij}\mathbf{I} \tag{3.27}$$

where

$$\mathbf{I} = \begin{bmatrix} 1 & 0 & 0 \\ 0 & 1 & 0 \\ 0 & 0 & 1 \end{bmatrix} \tag{3.28}$$

The initial stress stiffness array $\mathbf{K}_S$ is symmetric and gives the contribution to $\mathbf{K}_T$ depending on the level of stresses in the certain step in the updated Lagrange procedure. The third and fourth arrays stem from Jaumann objective stress rate.

$$\mathbf{K}_{\sigma 1} = \int_v \mathbf{B}^T \boldsymbol{\sigma}_1 \mathbf{B} dV \tag{3.29}$$

where $\boldsymbol{\sigma}_1$ is equal to

$$\boldsymbol{\sigma}_1 = \begin{bmatrix} 2\sigma_{11}^{//} & 0 & 0 & 0 & \sigma_{13}^{//} & \sigma_{12}^{//} \\ 0 & 2\sigma_{22}^{//} & 0 & \sigma_{23}^{//} & 0 & \sigma_{21}^{//} \\ 0 & 0 & 2\sigma_{33}^{//} & \sigma_{32}^{//} & \sigma_{31}^{//} & 0 \\ 0 & \sigma_{32}^{//} & \sigma_{23}^{//} & \overset{=}{\sigma}{}_{23}^{//} & \sigma_{21}^{//} & \sigma_{31}^{//} \\ \sigma_{31}^{//} & 0 & \sigma_{13}^{//} & \sigma_{12}^{//} & \overset{=}{\sigma}{}_{13}^{//} & \sigma_{32}^{//} \\ \sigma_{21}^{//} & \sigma_{12}^{//} & 0 & \sigma_{13}^{//} & \sigma_{23}^{//} & \overset{=}{\sigma}{}_{12}^{//} \end{bmatrix} \tag{3.30}$$

$$\text{where } \overset{=}{\sigma}{}_{ij}^{//} = (\sigma_{ii}^{//} + \sigma_{jj}^{//})/2$$

The $\mathbf{K}_{\sigma 2}$ is equal to

$$\mathbf{K}_{\sigma 2} = \int_V \mathbf{B}^T \boldsymbol{\sigma}_2 \mathbf{B} dv \tag{3.31}$$

where σ_2 takes the following form

$$\sigma_2 = \begin{bmatrix} \sigma_{11}^{//} & \sigma_{11}^{//} & \sigma_{11}^{//} & 0 & 0 & 0 \\ \sigma_{22}^{//} & \sigma_{22}^{//} & \sigma_{22}^{//} & 0 & 0 & 0 \\ \sigma_{33}^{//} & \sigma_{33}^{//} & \sigma_{33}^{//} & 0 & 0 & 0 \\ \sigma_{23}^{//} & \sigma_{23}^{//} & \sigma_{23}^{//} & 0 & 0 & 0 \\ \sigma_{13}^{//} & \sigma_{13}^{//} & \sigma_{13}^{//} & 0 & 0 & 0 \\ \sigma_{12}^{//} & \sigma_{12}^{//} & \sigma_{12}^{//} & 0 & 0 & 0 \end{bmatrix} \tag{3.32}$$

The matrix $\mathbf{K}_{\sigma 2}$ is not symmetric what makes not symmetric the whole equation (3.32).

The form of the tangent matrix $\mathbf{K}_T$ depends in general on the strain and stress rates used in constitutive equation and the virtual power principle. Depending on the form of the above, different tangent stiffness matrix can be used. Crisfield (1997) derives also tangent stiffness matrix using: the Jaumann rate of Kirchhoff stress, Truesdell rate of Kirchhoff stress and others. The detail form of tangent stiffness arrays for several formulations can be found in his book, Crisfield (1997). The updated Lagrangian procedure for two phase medium based on notation proposed by Bathe (1982) is discussed in the Chapter 5.

4 Interfaces in Consolidation

In this section the basic equations for the interface element are derived. The virtual work principle and the continuity equation for the fluid are the starting points for the analysis. The variational approach is applied to check the consistency of the formulation. The formulation ensures full coupling between fluid and the skeleton during the consolidation process.

4.1. Description of Discontinuities

Numerical modelling of soil structure interaction requires special interface elements to model shearing in thin zones, vertical drains etc., Fig.4.1. Experimental data on the behaviour of joints in shear and normal stresses can be divided (Goodman et al. (1968), (1972)) into two classes, which can be linked to the two classes of natural soils: sand and clay. In clay-type soil the shear joints behaviour reflects peaked or nonpeaked behaviour depending on the filling material which is usually present in this kind of discontinuities. The compaction is observed in compression, during opening of the joint the peak tensile strength is reached and then the stresses drop down to zero. Dense sand displays a peaked stress deformation curve reaching a residual strength of even one-half of the peak strength after 0.02 m to 0.03 m (or rarely more) of difference in displacements Δu_s. During compression the compaction is observed, and no tension stresses will appear in the opening of the joint.

a) b)

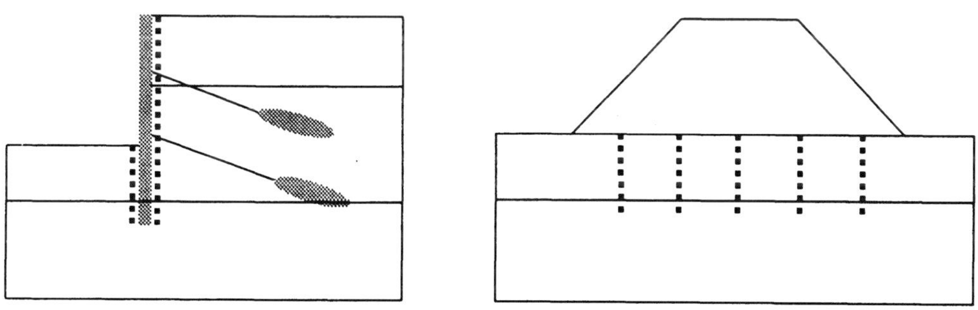

Figure 4.1. a) Interaction between a sheet pile wall and soil. Interface elements should be placed along the contact zones on both sides of the wall, b) Vertical drains modelled by the interfaces.

The attention is focused on the clay-type soil, where the two-phase coupled problems are important. Two directions of flow in the thin zone are considered, the first along the thin zone and the second normal to the thin zone. If normal flow is not observed ($v_n = 0$) due to impermeability of the solid blocks, then the distribution of v_s is found to be close to parabolic in the thin zone (Huyakorn and Pinder (1983)), Fig.4.2. Here v_s is understood as the mean value of the $\vec{v}_s$ distribution. From the phenomenological point of view the flow in the normal direction is obvious, provided the surrounding blocks are permeable. The proper description of the v_n is very important from the viewpoint of solving the boundary value problems. After the introduction of impermeable material (e.g. membrane) whose mechanical part is modelled by a membrane element (no coupling with water), the flow part of the interface gives the possibility to block the flow through the membrane. Thus, the creation of barriers for water flow can successfully be done by blocking $v_n \cong 0$. It can easily be obtained by introducing very low value of the permeability coefficient in the normal direction.

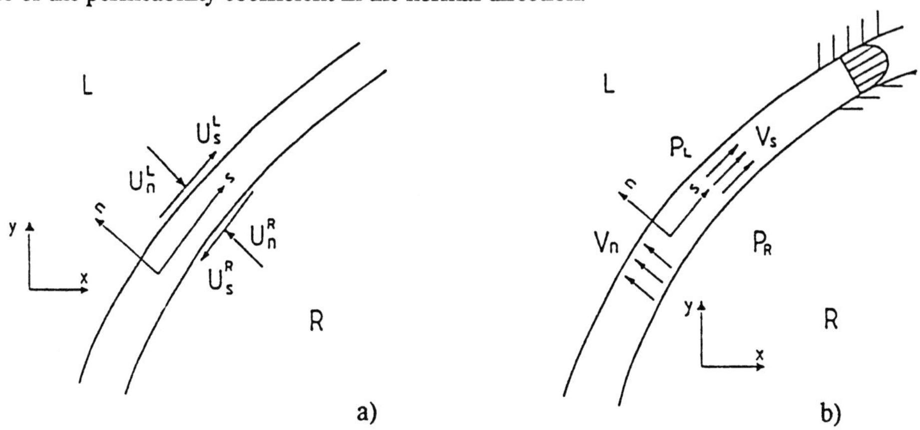

Figure 4.2. Model of the contact zone. Discontinuities in displacements and flow in thin zones.

In this chapter no debounding of the solid blocks is considered. No gap can be created and neither opening nor closing are admissible. In the mechanics of soils the initial stresses due to gravity forces ensure in most cases the compression at the interfaces.

4.2. Discontinuities in the Solid Phase

The virtual work principle for the continuum with the displacement discontinuity surface can be written in the following form

$$\int_V \delta\varepsilon^T \boldsymbol{\sigma} dV + \int_{S_c} \delta(\Delta\mathbf{u}_{s-n})^T \mathbf{t}_c dS_c = \int_V \mathbf{f}^T \delta\mathbf{u} dV + \int_S \mathbf{s}^T \delta\mathbf{u}\, dS, \tag{4.1}$$

where σ is the stress tensor, $\delta\mathbf{u}$, $\delta\varepsilon$ are the virtual displacements and strains, $\mathbf{t}_c$ is the traction transmitted at the interface, **f**, **s**, are the body forces and the surface tractions, $\delta(\Delta\mathbf{u})$ is the virtual displacement discontinuity. Large deformation effects are not considered here. The second term in the above equation corresponds to the work at the discontinuity, thus the attention will be focused on this term. The following description does not include other terms, but it is stated that the second term produces only a contribution to the final total equilibrium equations.

The stresses at the discontinuity for a two-phase medium are decomposed after Terzaghi

$$\mathbf{t}_c = \mathbf{t}'_c + \mathbf{n}p_t, \tag{4.2}$$

where

$$\mathbf{t}_c = \begin{bmatrix} \tau \\ \sigma_n \end{bmatrix}, \quad \mathbf{t}'_c = \begin{bmatrix} \tau \\ \sigma'_n \end{bmatrix}, \quad \mathbf{n} = \begin{bmatrix} 0 \\ 1 \end{bmatrix}, \quad \Delta\mathbf{u}_{s-n} = \begin{bmatrix} \Delta\mathbf{u}_s \\ \Delta\mathbf{u}_n \end{bmatrix}, \tag{4.3}$$

$\mathbf{t}_c'$ is the effective traction at the interface and **n** takes the value one for the normal stress component and zero for the shear stress component. The constitutive equation for the interface can written in the two forms. The first defined by the viscoplastic flow rule tangent material array

$$\dot{\mathbf{t}}_c' = \mathbf{D}\ (\Delta\dot{\mathbf{u}}_{s-n} - \Delta\dot{\mathbf{u}}_{s-n}^{vp}) \tag{4.4a}$$

where **D** is the elastic material array, $\Delta\mathbf{u}_{s-n}^{vp}$ is the increment of viscoplastic relative displacements. The second constitutuve equation is defined by the general tangent material array

$$\dot{\mathbf{t}}_c' = \mathbf{D}_T \Delta\dot{\mathbf{u}}_{s-n} \tag{4.4b}$$

The above equations are defined in the local coordinate system under the assumption that the vectors of normal and shearing stresses coincide with the s- and n-axis. The array **D** for the elastic behaviour is taken without coupling terms between shearing and normal stresses

$$\mathbf{D} = \begin{bmatrix} d_s & 0 \\ 0 & d_n \end{bmatrix}, \tag{4.5}$$

where in the elastic case d_s and d_n are he shear and bulk moduli, respectively.

In the general case the array **D** contains the non-zero off-diagonal terms. The sliding inside the interfaces occurs in the plastic range of the material behaviour. Since the zero thickness interface is considered, hence the elastic displacements are undesired inside the element. If a small value of elastic modulus is taken for the interfaces, the nearby solid elements will be superimposed on each other. To minimise the elastic displacements, the values of the shear and bulk moduli are multiplied by a coefficient κ. To keep the consistency of the units, the bulk and shear moduli are divided by the "virtual thickness" of the element. From numerical tests the thickness d is found to be 1/10 of the element length, thus the bulk and shear moduli for the interface take the following form

$$d_s = \chi\frac{G}{l};\ d_n = \chi\frac{G}{(1-2\nu)l}, \tag{4.6}$$

where χ includes 1/10 of the element length l and the multiplier κ. In the calculations χ is taken from the interval $100 > \chi > 10$. The proper choice of χ is important. Too large value will effectively block the elastic behaviour of the interface but it may produce the ill-conditioning of the global stiffness array.

The pseudo-viscoplastic iteration method is used to model elastoplastic behaviour. Since the pseudo-viscous approach is applied, the value of the elastic properties has the influence on the pseudo-time step (Cormeau (1975), (1976)). Too large value can lead to very short pseudo-time step and slow down very much the iterative process.

The simple Coulomb-Mohr yield criterion is accepted for the interface element

$$f(\mathbf{t}') = |\tau| + \sigma_n' \tan\phi - c, \tag{4.7}$$

where ϕ and c are the friction angle and cohesion, respectively. The rate of plastic slip inside the interface depends on the plastic potential function

$$g(\mathbf{t}')=|\tau|+\sigma_n' \tan\psi - c, \tag{4.8}$$

where ψ is the dilatation angle. For $\psi = \phi$ we have the associated flow rule, acceptance of which produces the overprediction of the plastic dilatation. This effect is not in accordance with that observed in experiments on sliding. The numerical results for $\psi = \phi$ show the increase of volume in the interface what leads to the artificial over-prediction of volume increase between the surrounding solid elements and undesired swelling of interface. Thus the non-associated flow rule is used.

The formulation of constitutive equations used in the numerical applications stems from the theory of viscoplasticity proposed by Perzyna (1966) and developed in the finite element method by (Cormeau (1975), (1976)), Zienkiewicz and Cormeau (1973), (1974). The formulation was successfully applied to elastoplastic problems by Winnicki, et al. (1988) and Dłużewski (1997), and developed in the finite element code HYDRO-GEO.

The relative velocity of the viscoplastic displacements is defined as follows

$$\Delta\dot{\mathbf{u}}_{s-n}^{vp} = \gamma < \phi(f) > \frac{\partial g}{\partial \mathbf{t}'}, \tag{4.9}$$

where g is understood as the viscoplastic potential, f is the yield function, γ is a viscosity coefficient and $< \phi(f) >$ is a special function. The brackets $< >$ stand for an on/off switch. If $f > 0$, then the switch is on, if $f < 0$, the switch is off. We assume that the function $\phi(f)$ takes one of the following forms

$$\phi(f)=f^n, \; n>1, \tag{4.10}$$

$$\phi(f)=kf, \; k>0. \tag{4.11}$$

The choice of the function $\phi(f)$ exerts an influence on the rate of the viscoplastic displacements.

To calculate the relative increment of the displacement used in the constitutive equation (4.4), the following integral should be determined

$$\Delta\mathbf{u}_{s-n}^{vp} = \int_0^t \Delta\dot{\mathbf{u}}_{s-n}^{vp} dt = \int_0^t \gamma < \phi(f) > \frac{\partial g}{\partial \mathbf{t}'} dt. \tag{4.12}$$

The analytical calculation of the integral is possible in special cases, herein the numerical calculation is applied. The Eulerian scheme with step by step calculation of the increment of relative displacements is used

$$d\Delta\mathbf{u}_{s-n}^{vp} = \int_{t_i}^{t_{i+1}} \Delta\dot{\mathbf{u}}_{s-n}^{vp} dt \cong \Delta\dot{\mathbf{u}}_{s-n}^{vp}\left(t_{i+1}-t_i\right) = \Delta\dot{\mathbf{u}}_{s-n}^{vp} \cdot dt_i . \tag{4.13a}$$

The accumulated viscoplastic relative increments are calculated as a following sum

$$\Delta \mathbf{u}^{vp}_{s-n} = \sum_{i=1}^{n} d\Delta \mathbf{u}^{vp}_{s-n}. \tag{4.13b}$$

The time which appeared in the relaxation process can be considered as the real time or as the numerical parameter only without physical interpretation. In such situation the pseudo-time is a step in numerical come back at the plastic surface, Fig.4.3. In the paper the time plays the role of the numerical parameter and should not be mixed with the real time of the consolidation process.

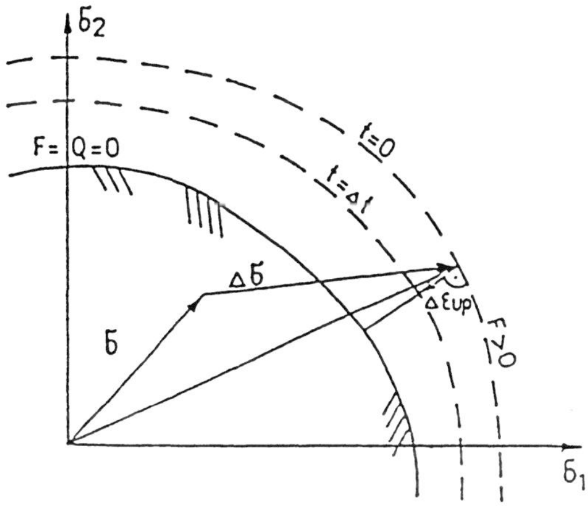

Figure 4.3. Visualisation of the pseudo-viscoplastic procedure. The return to the yield surface after elastic overshooting.

Since in the contact zones, inside the interface element, no changes of volume are observed and therefore a non-associated flow rule has to be applied to describe the behaviour of the material properly. One of the main advantages of the pseudo-viscoplastic approach is that the non-associated flow rule does not lead to the nonsymmetrical arrays.

The definition of the viscoplastic flow rule is proposed in the local coordinate system. To obtain the constitutive equations in the x, y coordinate system, the equations should suitably be transformed.

If the tangent formulation is used, for the Coulomb-Mohr yield criterion (4.7) and plastic potential (4.8) the $\mathbf{D}_T$ array takes the following form Van Langen (1991)

$$\mathbf{D}_T = \frac{d_s d_n}{d_s + d_n \operatorname{tg}\phi} \begin{bmatrix} \operatorname{tg}\phi \operatorname{tg}\psi & -\operatorname{tg}\phi \\ -\operatorname{tg}\psi & 1 \end{bmatrix} \tag{4.14}$$

In the virtual work principle, defined by equation (4.1), the second term describes the work at the discontinuity. We will focus our attention on this term bearing in mind that it gives an extra contribution to the global system of equilibrium in the finite element method. Rewriting the second term from the virtual work principle (4.1) in the velocity form and incorporating equations (4.2) - (4.5), we have

$$\int_{S_c} \delta(\Delta \mathbf{u}_{s-n})^T \dot{\mathbf{t}}_c dS = \int_{S_c} \delta(\Delta \mathbf{u}_{s-n})^T (\mathbf{D}(\Delta \dot{\mathbf{u}}_{s-n} - \Delta \dot{\mathbf{u}}^{vp}_{s-n}) + \mathbf{n}\dot{\mathbf{p}}) dS_c. \tag{4.15a}$$

$$\int_{S_c} \delta(\Delta \mathbf{u}_{s-n})^T \dot{\mathbf{t}}_c dS = \int_{S_c} \delta(\Delta \mathbf{u}_{s-n})^T (\mathbf{D}_T \Delta \dot{\mathbf{u}}_{s-n} + \mathbf{n}\dot{\mathbf{p}}) dS_c. \tag{4.15b}$$

The transformation between the local and the global systems of coordinates has the following form

$$\Delta \mathbf{u}_{s-n} = \mathbf{c}\, \Delta \mathbf{u}_{xy}, \tag{4.16}$$

where **c** is the rotation array in the form

$$\mathbf{c} = \begin{bmatrix} \cos\alpha & -\sin\alpha \\ \sin\alpha & \cos\alpha \end{bmatrix} \tag{4.17}$$

To calculate the pore pressure p and jumps in the displacements $\Delta \mathbf{u}_{s\text{-}n}$ inside the interface from the nodal values, the arrays **R** and **T** are introduced and defined in the Appendix A

$$p = \mathbf{N}_p \mathbf{R} \mathbf{p}_e, \tag{4.18}$$

$$\Delta \mathbf{u}_{s-n} = \mathbf{c} \mathbf{N}_u \mathbf{T} \mathbf{u}_e \tag{4.19}$$

Substituting the above equations into (4.15), we arrive at

$$\begin{aligned} \int_{S_c} \delta(\Delta \mathbf{u}_{s-n})(\mathbf{D}(\Delta \dot{\mathbf{u}}_{s-n} - \Delta \dot{\mathbf{u}}^{vp}_{s-n}) + \mathbf{n}\dot{\mathbf{p}}) dS_c &= \delta \mathbf{u}_e \int_{S_c} \mathbf{T}^T \mathbf{N}_u^T \mathbf{c}^T \mathbf{D}\ \mathbf{c} \mathbf{N}_u \mathbf{T} dS_c \dot{\mathbf{u}}_e - \\ \delta \mathbf{u}_e \int_{S_c} \mathbf{T}^T \mathbf{N}_u^T \mathbf{c}^T \mathbf{D} \Delta \dot{\mathbf{u}}^{vp}_{s-n}\, dS_c &+ \delta \mathbf{u}_e \int_{S_c} \mathbf{T}^T \mathbf{N}_u^T \mathbf{c}^T \mathbf{n} \mathbf{N}_p \mathbf{R} dS_c \dot{\mathbf{p}}_e \end{aligned} \tag{4.20a}$$

$$\begin{aligned} &\int_{S_c} \delta(\Delta \mathbf{u}_{s-n})(\mathbf{D}_T \Delta \dot{\mathbf{u}}_{s-n} + \mathbf{n}\dot{\mathbf{p}}) dS_c = \\ &= \delta \mathbf{u}_e \int_{S_c} \mathbf{T}^T \mathbf{N}_u^T \mathbf{c}^T \mathbf{D}_T \mathbf{c} \mathbf{N}_u \mathbf{T} dS_c \dot{\mathbf{u}}_e + \delta \mathbf{u}_e \int_{S_c} \mathbf{T}^T \mathbf{N}_u^T \mathbf{c}^T \mathbf{n} \mathbf{N}_p \mathbf{R} dS_c \dot{\mathbf{p}}_e \end{aligned} \tag{4.20b}$$

The kinematically admissible displacement field $\delta \mathbf{u}_e$ can be eliminated from the above equation. Summing up the equations (4.20) element by element, we get

$$\mathbf{K}_{in} \dot{\mathbf{u}} + \mathbf{L}_{inl} \dot{\mathbf{p}} = \dot{\mathbf{F}}, \tag{4.21}$$

where $\mathbf{K}_{in}$ is the stiffness matrix for the interface, $\mathbf{L}_{in}$ is the coupling matrix for the interface, $\dot{\mathbf{F}}$ contains the load velocity $\dot{\mathbf{P}}$ and the nodal forces due to viscoplastic behaviour, $\dot{\mathbf{u}}$ and $\dot{\mathbf{p}}$

are the velocities of nodal displacements and pore pressures. $\mathbf{K}_{in}$ has the following form for a single element

$$\mathbf{K}_{in}^{e}=\int_{S_c}\mathbf{T}^{T}\,\mathbf{N}_{u}^{T}\mathbf{c}^{T}\mathbf{D}\,\mathbf{c}\mathbf{N}_{u}\,\mathbf{T}dS_{c} \tag{4.22a}$$

$$\mathbf{K}_{in}^{e}=\int_{S_c}\mathbf{T}^{T}\,\mathbf{N}_{u}^{T}\mathbf{c}^{T}\mathbf{D}_{T}\mathbf{c}\mathbf{N}_{u}\,\mathbf{T}dS_{c} \tag{4.22b}$$

The coupling array and $\dot{F}_{in}^{e}$ vector for the element takes the form

$$\mathbf{L}_{in}^{e}=\int_{S_c}\mathbf{T}^{T}\mathbf{N}_{u}^{T}\mathbf{c}^{T}\mathbf{n}\mathbf{N}_{p}\,\mathbf{R}dS_{c} \tag{4.23}$$

$$\dot{\mathbf{F}}_{in}^{e}=\dot{\mathbf{P}}+\int_{S_c}\mathbf{T}^{T}\mathbf{N}^{T}\mathbf{c}^{T}\mathbf{D}\Delta\dot{\mathbf{u}}_{s-n}^{vp}dS_{c}$$

The right - hand side vector $\dot{\mathbf{F}}_{in}^{e}$ contains the load velocity as well as the term due to viscoplasticity. If the pseudo-viscoplasticity is analysed, the time must be split into real time and the parameter of the pseudo-visco iterative process.

The equation (4.21) contains arrays for interfaces only and it supplies the contribution to the system of global equilibrium equations. The above arrays are derived for a zero thickness element, thus the left-hand side nodes and the right-hand side nodes of the interface element occupy exactly the same position. On the other hand there exists the "virtual thickness" d which enters directly into constitutive array **D** and ensures the consistency of units. The proper choice of d (4.6) is important due to numerical difficulties which may arise in calculations.

4.3. Flow in Thin Zone

Considering the flow problem, the continuity equation is the starting point to formulate an interface element for the consolidation problems. The formulation discussed here allows two directions of water flow. The first direction is assumed to be along the discontinuity line. The second one has the normal direction to the discontinuity line, Fig.4.2. The continuity equation for a plane problem is defined in the s-n system of coordinates as

$$\nabla_{s-n}^{T}\mathbf{v}_{s-n}+\frac{\partial\varepsilon_{v}}{\partial t}-\frac{n}{K_{w}}\frac{\partial p}{\partial t}=0\,, \tag{4.24}$$

where ε_v are the volume changes, φ porosity, K_w bulk modulus of water. Operator ∇_{s-n}^{T} has the following form

$$\nabla_{s-n}^{T}=\left[\frac{\partial}{\partial s};\frac{\partial}{\partial n}\right]. \tag{4.25}$$

The flow velocity $\mathbf{v}_{s-n}$ is defined as

$$\mathbf{v}_{s-n}^{T}=\left[v_{s},v_{n}\right]. \tag{4.26}$$

The flow is defined according to the Darcy's rule

$$\mathbf{v}_{s-n}^{T} = \frac{\mathbf{k}_{s-n}}{\gamma_w} \nabla^T p_t \tag{4.27}$$

where the flow array takes the following form

$$\mathbf{k}_{s-n} = \begin{bmatrix} k_s & 0 \\ 0 & k_n \end{bmatrix}, \tag{4.28}$$

k_s and k_n denote the coefficient of permeabilities in the s and n directions.

4.4. Coupling

The coupling between fluid and skeleton is described by the volumetric changes (the second term in the equation (4.24)). The definition of the volumetric changes has an essential influence on the coupling between two phases. The volumetric changes ε_v are composed of two parts

$$\varepsilon_v = \varepsilon_v^n + \varepsilon_v^s, \tag{4.29}$$

where ε_v^n is associated with the n-direction and ε_v^s is, associated with the s-direction. To define the first one, the thickness d of the flow zone is introduced. The ε_v^n is defined as

$$\varepsilon_v^n = \frac{\Delta u_n}{d}, \tag{4.30}$$

where Δu_n is the jump in the normal displacements, $\Delta u_n = u_n^L - u_n^R$ (Fig. 4.2., L - Left, R - Right). The second part ε_v^s is defined as follows

$$\varepsilon_v^s = \frac{\partial u_s}{\partial s}, \tag{4.31}$$

where u_s is the mean value of the tangent displacements at the discontinuity

$$u_s = \frac{1}{2}\left(u_s^L + u_s^R\right) \tag{4.32}$$

so we have

$$\varepsilon_v^s = \frac{1}{2}\frac{\partial\left(u_s^L + u_s^R\right)}{\partial s} \tag{4.33}$$

The transformation between the global and local normalised coordinates is defined in the Appendix A (A.22), so we arrive at

$$\varepsilon_v^s = \frac{1}{2D}\frac{\partial\left(u_s^L + u_s^R\right)}{\partial \xi}. \tag{4.34}$$

Introducing the arrays: $\mathbf{l}$, $\mathbf{B}_\partial$, defined in the Appendix A (A.15), (A.16), ε_v^s can be formulated as follows

$$\varepsilon_v^s = \frac{1}{D}\mathbf{l}^T \mathbf{c}\mathbf{B}_\partial \mathbf{u}_e \tag{4.35}$$

The volumetric changes due to the jump in the displacements in the directions normal to the discontinuity, (4.31), can be rewritten in the following form

$$\varepsilon_v^n = \frac{1}{d}\mathbf{n}^T \mathbf{c}\mathbf{B}_u \mathbf{u}_e \tag{4.36}$$

The $\mathbf{B}_u$ array is defined in the Appendix A (A.18). The differential $\frac{\partial p}{\partial t}$, which appeared in the third term of (4.24), responsible for the water compressibility inside the interface, takes the form

$$\frac{\partial p}{\partial t} = \mathbf{N}_p \mathbf{R} \cdot \dot{\mathbf{p}}_e, \tag{4.37}$$

where $\mathbf{N}_p, \mathbf{R}$ are defined in the Appendix A.

Applying the Galerkin's formulation to the equation (4.20), introducing the boundary conditions for the outflow of fluid

$$\dot{\mathbf{q}} = \int \mathbf{N}\dot{\mathbf{q}}_n \, dS \tag{4.38}$$

applying the Green's theorem and multiplying the whole equation by d, we arrive at

$$\int_{S_c} \overset{*}{\mathbf{B}}{}_f^T \mathbf{k} \overset{*}{\mathbf{B}}_f \, dS_c \, \mathbf{p} + \left(\int_{S_c} (\mathbf{N}_p \mathbf{R})^T \mathbf{n}^T \mathbf{c} \mathbf{N}_u \mathbf{T} dS_c + \int_{S_c} (\mathbf{N}_p \mathbf{R})^T \mathbf{l}^T \mathbf{c} \mathbf{B}_\partial \frac{d}{D} dS_c \right) \dot{\mathbf{u}} -$$

$$- \int_{S_c} (\mathbf{N}_p \mathbf{R})^T \frac{n}{K_w} \mathbf{N}_p \mathbf{R} dS_c \, \dot{\mathbf{p}} = \dot{\mathbf{q}}. \tag{4.39}$$

The above equation can be rewritten as

$$\mathbf{H}_{in} \mathbf{p} + \left(\mathbf{L}_{inI}^T + \mathbf{L}_{inII}^T \right) \dot{\mathbf{u}} - \mathbf{S}_{in} \dot{\mathbf{p}} = \dot{\mathbf{q}}. \tag{4.40}$$

where the relevant arrays take the following forms:

- the flow array

$$\mathbf{H}_{in}^e - \int_{S_c} \overset{*}{\mathbf{B}}{}_f^T \overset{*}{\mathbf{k}} \overset{*}{\mathbf{B}}_f \, dS_c, \tag{4.41}$$

- the coupling arrays

$$\mathbf{L}_{inI}^{eT} = \int_{S_c} \mathbf{R}^T \mathbf{N}_p^T \mathbf{n}^T \mathbf{c} \mathbf{N}_u \mathbf{T} dS_c, \tag{4.42}$$

$$\mathbf{L}_{\mathrm{inII}}^{\mathrm{eT}} = \int_{S_c} \mathbf{R}^{\mathrm{T}} \mathbf{N}_{\mathrm{p}}^{\mathrm{T}} \mathbf{1}^{\mathrm{T}} \mathbf{c} \mathbf{B}_{\partial} \frac{\mathrm{d}}{\mathrm{D}} \mathrm{d}S_c\,, \tag{4.43}$$

- the compressibility array

$$\mathbf{S}_{\mathrm{in}}^{\mathrm{e}} = \frac{n}{K_w} \int_{S_c} \mathbf{R}^{\mathrm{T}} \mathbf{N}_{\mathrm{p}}^{\mathrm{T}} \mathbf{N}_{\mathrm{p}} \mathbf{R} \mathrm{d}S_c\,. \tag{4.44}$$

Definitions of the arrays which appear under the integrals can be found in the Appendix A. Herein two coupling arrays appeared. The first one stems from the compressibility of a thin zone in the normal direction. The second one takes its roots from the compressibility along the thin zone. The array $\mathbf{L}_{\mathrm{inI}}$ is equal to $\mathbf{L}_{\mathrm{in}}$ defined by the equation (4.23).

The matrix equation (4.21), describing the equilibrium of the solid phase, can be rewritten with the continuity equation (4.40) in the coupled form

$$\begin{bmatrix} \mathbf{K}_{\mathrm{in}} & \mathbf{L}_{\mathrm{in}} \\ \mathbf{L}_{\mathrm{in}}^{\mathrm{T}} + \mathbf{L}_{\mathrm{inII}}^{\mathrm{T}} & -\mathbf{S}_{\mathrm{in}} \end{bmatrix} \begin{bmatrix} \dot{\mathbf{u}} \\ \dot{\mathbf{p}} \end{bmatrix} = \begin{bmatrix} 0 & 0 \\ 0 & \mathbf{H}_{\mathrm{in}} \end{bmatrix} \begin{bmatrix} \mathbf{u} \\ \mathbf{p} \end{bmatrix} + \begin{bmatrix} \dot{\mathbf{F}} \\ \dot{\mathbf{q}} \end{bmatrix}. \tag{4.45}$$

The above system describes the behaviour of the two-phase medium in the thin zone in terms of the finite element method.

The coupling array $\mathbf{L}_{\mathrm{inII}}^{\mathrm{T}}$ introduced nonsymmetry to the global system (4.45). The physical interpretation of this effect can be explained in the following way. We allow for the flow to proceed along an element. At the same time there is no mechanical resistance which can take the stresses due to the outflow of water from the interface.

The mechanical part of the interface offers the resistance in two cases: first in the normal direction to the discontinuity (what produces the coupling term $\mathbf{L}_{\mathrm{inI}}^{\mathrm{T}}$) and second, the shearing resistance. The interface has no stiffness along the element which would take the stresses resulting from the decay of the excess pore pressure inside the interface. In the next section the variational approach is employed to shed the new light on the coupled effects.

The system of equations (4.45) can be solved with the help of the step by step integrating rule, as used to be in consolidation problems. Most often the interfaces are considered together with the solid elements, thus the equation (4.45) should be added to the global system for the two-phase medium.

4.5. Variational Approach

The finite element equations for the consolidation problems are often derived by means of the Galerkin's method. The variational approach is equivalent to the solving of differential equations with certain boundary conditions. Since for the interface element the flow part introduced nonsymmetry to the global equation set (4.45), the variational approach is used to check this effect. The variational methods lead to the direct way of obtaining the finite element equations, but often at the cost of losing the interpretations of the quantities involved. The equations obtained from the virtual work principle plus the Galerkin's method are compared with the variational approach. The functional for the two-phase medium is accepted after Verruijt (1977). The value of displacements and pore pressures are sought for when the functional has a stationary value. For elastic problems the functional is defined as follows

$$U=\frac{1}{2}\int_{v}\Big((K-\frac{2}{3}G)v_{i,i}v_{j,j}+G(v_{i,j}+v_{j,i})v_{i,j}-2f_iv_i+$$
$$+2p(v_{i,i}-v_{i,i}^{o})-(1-\theta)\Delta t\frac{k}{\mu_w}p_{,i}p_{,i}-n\beta p^2+2n\beta pp^{o}\Big)dv- \tag{4.46}$$
$$-\int_{S_1}t_iv_idS_1+\int_{S_4}(1-\theta)\Delta thdS_4$$

where u_i and p are arbitrary functions throughout the volume v with the constraints, (Verruijt (1977))

on the boundary $S_1:\sigma_{ij}n_j=t_i$, on the boundary $S_2:u_i=a_i$,

on the boundary $S_3:p=g$, on the boundary $S_4:q_in_i=h$.

The proof that a stationary value of U occurs if $u_i=\bar{u}_i$ and $p=\bar{p}$ can be given in the usual way. The mean values $\bar{p}$ and $\bar{u}_i$ are:

$$\bar{p}=\frac{1}{\Delta t}\int_{t_o}^{t_+}pdt=\theta p^{o}+(1-\theta)p^{+},\quad \bar{u}_i=\frac{1}{\Delta t}\int_{t_o}^{t_+}u_i\,dt=\theta u_i^{o}+(1-\theta)u_i^{+}, \tag{4.47}$$

where $p^{o}=p(t+t_o)$, $p^{+}=p(t+t_+)$, $u_i^{o}=u_i(t=t_o)$, $u_i^{+}=u(t=t_+)$. The q parameter is a constant and depends on the time integration scheme.

Considering the continuous domain with a very thin zone, where high variations of pore pressures and flow take place and the jump in displacements is admitted, the functional U_{TOT} is composed of U and U_c

$$U_{TOT}=U+U_c, \tag{4.48}$$

where U_c is responsible for the coupled problem in the thin zone

$$U_c=\frac{1}{2}\Bigg[\int_{S_c}\Delta\mathbf{u}_{s-n}^{T}\mathbf{D}^{e}\Delta\mathbf{u}_{s-n}dS_c+\int_{S_c}\Big[2p(\varepsilon_v-\varepsilon_v^{o})-$$
$$-(1-\theta)\Delta tp\nabla_{s-n}^{T}\mathbf{k}_{s-n}\nabla_{s-n}p-\frac{n}{K_w}p^2+\frac{2n}{K_w}pp^{o}\Big]\Bigg]dS_c- \tag{4.49}$$
$$-\int_{S_1}\mathbf{s}^{T}\mathbf{u}dS_1+\int_{S_4}(1-\theta)\Delta thdS_4$$

By analogy to the derivations in the previous section we will focus our attention on the functional U_c, bearing in mind that in the former derivation it gives only a certain contribution due to thin zone to the global set of equations.

To simplify analysis, the functional U_c will be divided into five parts

$$U_c=I+II+III+IV+V, \tag{4.50}$$

where

$$I=\frac{1}{2}\int_{S_C}\Delta\mathbf{u}_{s-n}^T\,\mathbf{D}^e\,\Delta\mathbf{u}_{s-n}\,dS_c\ ,\tag{4.51}$$

$$II=\int_{S_C}p(\varepsilon_v-\varepsilon_v^o)dS_c\ ,\tag{4.52}$$

$$III=-\frac{1}{2}(1-\theta)\Delta t\int_{S_C}p\nabla_{s-n}^T\mathbf{k}_{s-n}\nabla_{s-n}\,dS_c\ ,\tag{4.53}$$

$$IV=-\frac{1}{2}\frac{n}{K_w}\int_{S_C}(p^2+2pp_o)dS_c\ ,\tag{4.54}$$

$$V=-\int_{S_1}\mathbf{s}^T\mathbf{u}dS_1+\int_{S_4}(1-\theta)\Delta t\,h\,dS_4\ .\tag{4.55}$$

The detailed description of variational approach to the finite element modelling can be found in Huyakorn and Pinder (1983).

In the first step the function **u**, which minimises the functional U_c, will be sought for. Introducing the finite element approximation, incorporating equations (A.8) into (4.51) and limiting the analysis to the first part, the functional I takes the form

$$I=\frac{1}{2}\int_{S_C}\mathbf{u}_e^T\mathbf{T}^T\mathbf{N}_u^T\mathbf{c}^T\mathbf{D}\,\mathbf{c}\mathbf{N}_u\,\mathbf{T}\mathbf{u}_e\,dS_c\ .\tag{4.56}$$

If the stationary value of the above integral is satisfied, then

$$\delta I=\int_{S_C}\mathbf{T}^T\mathbf{N}_u^T\mathbf{c}^T\mathbf{D}\mathbf{c}\mathbf{N}_u\mathbf{T}dS_c\mathbf{u}_e\ .\tag{4.57}$$

Having in mind the initial state (when t = 0) and incorporating equations (A.1), (4.29), (A.14), (A.17) into (4.52), we obtain the second term in the form

$$II=\int_{S_C}p(\varepsilon_v^n+\varepsilon_v^s)dS_c\ =\int_{S_C}\mathbf{u}_e^T\mathbf{T}^T\mathbf{N}_u^T\mathbf{n}\mathbf{N}_p\,\mathbf{R}\mathbf{p}_e\,dS_c+\int_{S_C}\mathbf{u}_e^T\mathbf{B}_\partial^T\mathbf{c}^T\mathbf{l}\mathbf{N}_p\,\mathbf{R}\mathbf{p}_e\,dS_c\tag{4.58}$$

The first variation with respect to **u** yields

$$\delta II=\int_{S_C}\mathbf{T}^T\mathbf{N}_u^T\mathbf{n}\mathbf{N}_p\mathbf{R}dS_c\mathbf{p}_e+\int_{S_C}\mathbf{B}_\partial^T\mathbf{c}^T\mathbf{l}\mathbf{N}_p\mathbf{R}\frac{1}{D}dS_c\mathbf{p}_e\ .\tag{4.59}$$

Incorporating equations (A.1), (A.6), (4.25), we can rewrite the third integral responsible for flow in the following form

$$\begin{aligned}III&=-\frac{1}{2}(1-\theta)\Delta t\int_{S_C}p\nabla_{s-n}^T\mathbf{k}_{s-n}\nabla_{s-n}\,p\,dS_c=\\&=-\frac{1}{2}(1-\theta)\Delta t\int_{S_C}\mathbf{p}_e^T\overset{*}{\mathbf{B}}\overset{*}{\mathbf{k}}\overset{*}{\mathbf{B}}_f\,\mathbf{p}_e\,dS_c\end{aligned}\ ,\tag{4.60}$$

where $\mathbf{B}_{f,}\ \mathbf{k}$ are defined in the Appendix A. The variation of part III with respect to **u** is zero

$$\delta III=0 \tag{4.61}$$

The fourth integral, after incorporating equations (A.1), takes the form

$$IV=-\frac{1}{2}\frac{n}{K_w}\int_{S_c} p^2 dS_c=-\frac{1}{2}\frac{n}{K_w}\int_{S_c}\mathbf{p}_e^T\mathbf{R}^T\mathbf{N}_p^T\mathbf{N}_p\mathbf{R}\mathbf{p}_e dS_c\,. \tag{4.62}$$

The variation vanishes

$$\delta IV=0\,. \tag{4.63}$$

The last integral can be rewritten as

$$V=-\int_{S_1}\mathbf{u}_e^T\mathbf{N}^T\mathbf{s}dS_1+(1-\theta)\Delta t\int_{S_4}hdS_4\,. \tag{4.64}$$

The variation takes the form

$$\delta V=-\int_{S_1}\mathbf{N}^T\mathbf{s}dS_1\,. \tag{4.65}$$

To sum up, the function $\dot{\mathbf{u}}$, which minimises the functional U_c, is determined from $\delta U_c=0$. The integration over time is defined by equation (4.47). Multiplying the whole equation by d, we obtain

$$\mathbf{K}_{in}\dot{\mathbf{u}}+\mathbf{L}_{in}\dot{\mathbf{p}}=\dot{\mathbf{Q}}, \tag{4.66}$$

where: $\mathbf{K}_{in}$ is defined by the equation (4.22), $\mathbf{L}_{in}$ is composed of two terms

$$\mathbf{L}_{in}=\mathbf{L}_{inI}+\mathbf{L}_{inII}, \tag{4.67}$$

in which $\mathbf{L}_{inI}$ is defined by (4.23) (transposed by (4.42)) and $\mathbf{L}_{inII}$, transposed is defined by the equation (4.43). $\dot{\mathbf{Q}}$ is the load yielding from equation (4.65). The equation (4.66) is different from equation (4.21). The difference is caused by the array $\mathbf{L}_{inII}$, what will be discussed in the next section in more detail.

In the next step the function **p**, which minimises the functional U_c, will be sought for. The following variations will be calculated with respect to **p**

$$\delta I=0, \tag{4.68}$$

the variation of the second integral

$$\delta II=\int_{S_c}\mathbf{R}^T\mathbf{N}_p^T\mathbf{n}^T\mathbf{c}\mathbf{N}_u\mathbf{T}dS_c\mathbf{u}^e+\int_{S_c}\mathbf{R}^T\mathbf{N}_p^T\mathbf{1}^T\mathbf{c}\mathbf{B}_\partial\frac{1}{2}dS_c\mathbf{u}_e\,, \tag{4.69}$$

the variation of the third integral

$$\delta III=-(1-\theta)\Delta t\int_{S_c}\overset{*}{\mathbf{B}}{}_f^T\mathbf{k}\overset{*}{\mathbf{B}}_f\,dS_c\mathbf{p}_e\,, \tag{4.70}$$

the variation of the fourth integral

$$\delta IV = -\frac{n}{K_w}\int_{S_c} \mathbf{R}^T \mathbf{N}_p^T \mathbf{N}_p \mathbf{R} dS_c \mathbf{p}_e \,. \tag{4.71}$$

The variation of the fifth integral yields the term due to outflow of fluid

$$\delta V = (1-\theta)\Delta t \int_{S_c} \mathbf{N} \mathbf{q}_n \, dS_c \,. \tag{4.72}$$

The function $\dot{\mathbf{p}}$, which minimises the functional U_c, is determined from $\delta U_c = 0$. The integration over time is defined by the equation (4.47), thus after multiplying by d we have

$$\mathbf{H}_{in}\mathbf{p} + \left(\mathbf{L}_{inI}^T + \mathbf{L}_{inII}^T\right)\dot{\mathbf{u}} - \mathbf{S}_{in}\dot{\mathbf{p}} = \dot{\mathbf{q}} \,. \tag{4.73}$$

The above equation is identical with (4.40) obtained by means of the of Galerkin's method.

4.6. Discussion of Discontinuities

The variational method, when minimising U_c with respect to **p**, is equivalent to the Galerkin's approach. At the same time the minimising of U_c, to find **u**, yields the results different to those starting from the virtual work principle (4.1) after Koiter (1960). One part of the interaction between two phases is lost in this way, L_{inII}. Mathematics proves that the solving of differential equations is equivalent to the minimising of the functional. Herein the starting point to solve the coupled problem is the virtual work principle for continuum with the surface discontinuity (after Koiter), and not the equilibrium equations. Coming back to the virtual work principle (4.1) and reconsidering the discontinuity allowed for in the skeleton, we can draw the conclusion that the coupling with liquid phase via volumetric changes is defined worse than in the continuity equation (4.24). For skeleton we allow for two kinds of discontinuities Δu_s and Δu_n. At the same time, we do not consider the possibilities of extension or contraction of the discontinuity line. When we introduce the strains in the s direction, defined in the standard way,

$$\varepsilon_s = \frac{\partial u_s}{\partial s} \,, \tag{4.74}$$

where u_s is defined by (4.32) as a mean value, then the consistent derivation of the finite element equations is enriched in terms due to ε_s. The jumps in displacements can be extended by ε_s in the following way

$$\Delta \mathbf{u}_{s-n-s} = \begin{bmatrix} \Delta u_s \\ \Delta u_n \\ \dfrac{\partial u_s}{\partial s} \end{bmatrix} . \tag{4.75}$$

Subsequently, the constitutive array should be enriched

$$\mathbf{D}^{e}=\begin{bmatrix} d_s & 0 & 0 \\ 0 & d_n & 0 \\ 0 & 0 & d_{ss} \end{bmatrix}. \tag{4.76}$$

For elastic isotropic material d_{ss} depends on d_s and d_n. Let us assume for simplicity that $d_{ss}=0$. Then the constitutive equation takes the form

$$\mathbf{t}_{cs}=\begin{bmatrix} \tau \\ \sigma_{nn} \\ \sigma_{ss} \end{bmatrix}=\begin{bmatrix} d_s & 0 & 0 \\ 0 & d_n & 0 \\ 0 & 0 & 0 \end{bmatrix}=\begin{bmatrix} \Delta u_s \\ \Delta u_n \\ \dfrac{\partial u_s}{\partial s} \end{bmatrix}+p\begin{bmatrix} 0 \\ 1 \\ 1 \end{bmatrix}. \tag{4.77}$$

Reconsidering the second term of the equation (4.1), which leads to (4.20), now we have

$$\begin{aligned}\int_{S_C} \delta(\Delta \mathbf{u}_{s-n-s})^{T}\dot{\mathbf{t}}_{cs}\, dS_c &=\delta\mathbf{u}_e \int_{S_C} \mathbf{T}^{T}\mathbf{N}^{T}{}_{u}\mathbf{c}^{T}\mathbf{D}\mathbf{c}\mathbf{N}_u\mathbf{T}dS_c\dot{\mathbf{u}}_e+ \\ &+\delta\mathbf{u}_e \int_{S_C} \mathbf{T}^{T}\mathbf{N}_u^{T}\mathbf{c}^{T}\mathbf{n}\mathbf{N}_p\mathbf{R}dS_c\dot{\mathbf{p}}_e+\delta\mathbf{u}_e \int_{S_C} \mathbf{B}_{\partial}^{T}\mathbf{c}^{T}\mathbf{1}\frac{1}{D}\mathbf{N}_p\mathbf{R}dS_c\dot{\mathbf{p}}\end{aligned} \tag{4.78}$$

The last integral describes the coupling due to the contraction of discontinuity line, what leads to the equation

$$\mathbf{K}_{in}\dot{\mathbf{u}}+\mathbf{L}_{in}\dot{\mathbf{p}}=\dot{\mathbf{F}}, \tag{4.79}$$

where $\mathbf{L}_{in}$ is composed of two arrays according to the equation (4.67). Now the symmetry of the total equation set (4.45) is satisfied. The described formulation of interface elements has some disadvantages. The discontinuity in the n-direction gives consistent symmetrical coupling between the skeleton and the liquid. The jump in displacements at the s-direction can fully be independent of ε_v^s. The introduction of $\varepsilon_s \neq 0$ leads to symmetrical coupling between the skeleton and the liquid. On the other hand, it introduces some inconsistency to the constitutive equation and its physical interpretation. In the development of software and further calculations the assumption is made that $\varepsilon_v^s=0$, thus $\varepsilon_v=\varepsilon_v^n$, which leads to symmetrical formulation, where $L_{in}=L_{inI}$. Such an *ad hoc* formulation for the flow in the s-direction (without coupling for the consolidation) was used by Russel (1992) and compared with the analytical solution.

4.7. Time Stepping Process

Most often the contact zone is analysed with the surrounding bodies. In such situation the equations for the interfaces are assembled with others in the allocation process. Considering the plain problem, the derived herein arrays are added to the 2D elements. The equation of nodal equilibrium coupled with the nodal continuity equation is written in the following form

$$\begin{bmatrix} \mathbf{K} & \mathbf{L} \\ \mathbf{L}^{T} & -\mathbf{S} \end{bmatrix}\begin{bmatrix} \dot{\mathbf{u}} \\ \dot{\mathbf{p}} \end{bmatrix}=\begin{bmatrix} \mathbf{0} & \mathbf{0} \\ \mathbf{0} & -\mathbf{H} \end{bmatrix}\begin{bmatrix} \mathbf{u} \\ \mathbf{p} \end{bmatrix}+\begin{bmatrix} \dot{\mathbf{F}} \\ \mathbf{q} \end{bmatrix} \tag{4.80}$$

The above system of equations describes the behaviour of a two-phase medium during the consolidation process. To solve the set of equations (4.80), the time integration scheme has to be engaged Lewis and Schrefler (1998). The time domain is divided into a number of time steps and the integration over one time step is carried out to obtain the velocity of the excess pore pressures and displacements. The step by step integration procedure allows to trace the changes in displacements and pore pressure and to sum up all the incremental variables. Applying the time integration scheme to the set of equations (4.80), we have

$$\int_{t_k}^{t_k+\Delta t_k}\begin{bmatrix}\mathbf{K} & \mathbf{L}\\ \mathbf{L}^T & -\mathbf{S}\end{bmatrix}\begin{bmatrix}\dfrac{d\mathbf{u}}{dt}\\ \dfrac{d\mathbf{p}}{dt}\end{bmatrix}dt=\int_{t_k}^{t_k+\Delta t_k}\begin{bmatrix}\mathbf{0} & \mathbf{0}\\ \mathbf{0} & -\mathbf{H}\end{bmatrix}\begin{bmatrix}\mathbf{u}\\ \mathbf{p}\end{bmatrix}dt+\int_{t_k}^{t_k+\Delta t_k}\begin{bmatrix}\dfrac{d\mathbf{F}}{dt}\\ \mathbf{q}\end{bmatrix}dt, \tag{4.81}$$

where Δt_k is the length of the k-th time step between the time instants t_k and $t_k + \Delta t_k$. The linear variation for the first order of the time derivatives of the displacements and pore pressures is assumed

$$[\mathbf{u},\ \mathbf{p}]=\left[w_1^t,\ w_2^t\right]\begin{bmatrix}\mathbf{u}^t & \mathbf{p}^t\\ \mathbf{u}^{t+\Delta t} & \mathbf{p}^{t+\Delta t}\end{bmatrix}, \tag{4.82}$$

where w_1^t, w_2^t are defined as follows

$$w_1^t=1-\theta,\ w_2^t=\theta,\ \theta=\frac{t-t_k}{\Delta t_k}. \tag{4.83}$$

The derivatives of w_1^t, w_2^t take the form

$$\frac{d}{dt}\left[w_1^t, w_2^t\right]=\left[-\frac{1}{\Delta t_k},\frac{1}{\Delta t_k}\right]. \tag{4.84}$$

The introduction of the equations (4.82) - (4.83) into (4.81) leads to

$$\begin{aligned}&\int_{t_k}^{t_k+\Delta t_k}\begin{bmatrix}\mathbf{K}_T & \mathbf{L}\\ \mathbf{L}^T & -\mathbf{S}\end{bmatrix}\left\{-\frac{1}{\Delta t_k}\begin{bmatrix}\mathbf{u}\\ \mathbf{p}\end{bmatrix}_{t_k}+\frac{1}{\Delta t_k}\begin{bmatrix}\mathbf{u}\\ \mathbf{p}\end{bmatrix}_{t_k+\Delta t_k}\right\}dt=\\ &=\int_{t_k}^{t_k+\Delta t_k}\begin{bmatrix}\mathbf{0} & \mathbf{0}\\ \mathbf{0} & -\mathbf{H}\end{bmatrix}\left\{(1-\theta)\begin{bmatrix}\mathbf{u}\\ \mathbf{p}\end{bmatrix}_{t_k}+\theta\begin{bmatrix}\mathbf{u}\\ \mathbf{p}\end{bmatrix}_{t_k+\Delta t_k}\right\}dt+\int_{t_k}^{t_k+\Delta t_k}\begin{bmatrix}\dfrac{d\mathbf{F}}{dt}\\ \mathbf{q}\end{bmatrix}dt.\end{aligned} \tag{4.85}$$

After integration over time and writing the value of the nodal parameters at time $t_k + \Delta t_k$ in the incremental form

$$\begin{bmatrix}\mathbf{u}\\ \mathbf{p}\end{bmatrix}_{t+\Delta t_k}=\begin{bmatrix}\mathbf{u}\\ \mathbf{p}\end{bmatrix}_t+\begin{bmatrix}\Delta\mathbf{u}\\ \Delta\mathbf{p}\end{bmatrix}_t, \tag{4.86}$$

and after rearrangement, the equation (4.85), takes the following form as follows

$$\begin{bmatrix}\mathbf{K} & \mathbf{L}\\ \mathbf{L}^T & -(\mathbf{S}+\theta\Delta t_k\mathbf{H})\end{bmatrix}\begin{bmatrix}\Delta\mathbf{u}\\ \Delta\mathbf{p}\end{bmatrix}_t=\begin{bmatrix}\mathbf{0} & \mathbf{0}\\ \mathbf{0} & -\Delta t\mathbf{H})\end{bmatrix}\begin{bmatrix}\mathbf{u}\\ \mathbf{p}\end{bmatrix}_t+\begin{bmatrix}\dfrac{d\mathbf{F}}{dt}\\ \dot{\mathbf{q}}\end{bmatrix}\Delta t_k. \tag{4.87}$$

The above equation system constitutes the basic formulation for the time stepping procedure. The iteration parameter θ takes values from 0 to 1. In particular, it can be kept constant during the time stepping process, or can vary in relation to time t_k and time increment Δt_k. Taking constant values of $\theta = 0, \theta = \frac{1}{2}, \theta = 1$, from the finite difference method we have: forward differences (Euler), mid-differences (Crank-Nicholson), backward differences, respectively. The time-stepping process is unconditionally stable if $\theta \geq 0.5$ (Booker and Small (1975)).

The detailed study of the equation (4.87) leads to the conclusion that the left-handside matrix is often ill-conditioned. Considering clay-type soil the differences in values of the terms of stiffness array **K** and **W** = (**S** + θ Δt**H**) may reach several orders of magnitude. The ill-conditioning degree of this matrix can be very high. At the diagonal mixed positive and negative values are located what constrains the types of solving procedures. The numerical experiences show that starting the arrangement of global array from displacements degrees of freedom followed by pore pressure, both the Gauss elimination and the triangle decomposition solvers preserve the correct solution even for considerable differences in terms of the arrays **K** and **W**.

To improve the property of the global array, the preconditioning can be, applied Song (1990). The columns and rows corresponding to **L** array are multiplied by a scalar factor s. The corresponding terms of the right-handside are multiplied by s and the unknown pore pressures by $1/\lambda$. The equation (4.87) is rewritten in the form as follows

$$\begin{bmatrix} \mathbf{K}_T & \lambda \mathbf{L} \\ \lambda \mathbf{L}^T & -\lambda^2 \mathbf{W}_t \end{bmatrix} \begin{bmatrix} \Delta \mathbf{u} \\ \frac{1}{\lambda} \Delta \mathbf{p} \end{bmatrix}_t = \begin{bmatrix} \Delta \mathbf{F} \\ \lambda \Delta \mathbf{Q} \end{bmatrix}_t . \tag{4.88}$$

The scaling factor s can be determined as

$$\lambda = \sqrt{(K + 4G/3)/(\Delta t k/\mu)}. \tag{4.89}$$

Several procedures stemming from elastoplastic modelling can be coupled with the time stepping algorithm during the consolidation process. The finite element formulation of the elastoplastic consolidation which takes its roots from pseudo-viscous iteration process can be applied. The interface elements are included in the formulation.

The pseudo-viscoplastic algorithm for numerical modelling of elastoplastic behaviour was developed by Zienkiewicz and Cormeau et al. (1973), (1974). The stability of the time marching scheme was proved by Cormeau et al. (1975), (1976). The pseudo-viscous algorithm developed in finite element computer code HYDRO-GEO was successfully applied to solve a number of boundary value problems by Winnicki et al. (1988). The procedure was extended to cover the geometrically nonlinear problems by Kanchi et al. (1978) and also developed for large strains in consolidation by Dłużewski (1991). The pseudo-viscous procedure is adopted herein for modelling elastoplastic behaviour in consolidation process with interface elements. In the procedure two times appear, the first is the real time of consolidation and the second time is only a parameter of the pseudo- relaxation process. The global set of equations is rewritten in the form below where ΔR^1 is the vector caused by pseudo-viscous relaxation process. The contributor due to the interface element is calculated from the equation (4.23).

For each time step the material nonlinearity is solved via iterative method. The steps for the i-th iteration are listed below

$$\begin{bmatrix} \mathbf{K} & \mathbf{L} \\ \mathbf{L}^T & -(\mathbf{S}+\theta\Delta t^i\mathbf{H}) \end{bmatrix}\begin{bmatrix} \Delta\mathbf{u}^i \\ \Delta\mathbf{p}^i \end{bmatrix}=\begin{bmatrix} \mathbf{0} & \mathbf{0} \\ \mathbf{0} & -\Delta t\mathbf{H} \end{bmatrix}\begin{bmatrix} \mathbf{u} \\ \mathbf{p} \end{bmatrix}+\begin{bmatrix} \Delta\mathbf{F}+ & \Delta\mathbf{R}^i \\ \Delta\mathbf{q} & \end{bmatrix} \tag{4.90}$$

where $\Delta\mathbf{R}^i$ takes now the form

$$\Delta\mathbf{R}^i=\int_v \mathbf{B}^T\mathbf{D}\Delta\varepsilon_i^{vp}dv \tag{4.91}$$

where for the interface element we have from the equation (4.23)

$$\Delta\mathbf{R}_{in}^i = \int_{S_c} \mathbf{T}^T\mathbf{N}^T\mathbf{c}^T\mathbf{D}\Delta\mathbf{u}_i^{vp}dS_c \tag{4.92}$$

$\Delta\mathbf{R}^i$ stands for the nodal vector which results from the relaxation of stresses.

The local criterion for terminating the iterative process is adopted. The iterations are continued until the calculated stresses are acceptable close to the yield surface, Fig. 5.2.

$$F < \text{Tolerance} \tag{4.93}$$

at all checked points. No global criterion for this procedure is used, the global equilibrium is fulfilled in every iterative pseudo-time step.

4.8. Numerical Integration

To calculate the arrays and vectors for interface elements, the integration scheme should be applied. The analytical integration can be made for the simplest cases (i.e. not curved elements, elastic material, small strains, linear flow rule). For more complicated cases the numerical integration scheme has to be involved. What is more, the numerical results show that the Gauss integration leads to the undesirable effects of oscillation in the computed stresses. To improve the results most often Newton-Cotes integration scheme is applied. The oscillations of stresses has been reported by several researchers, Hohberg (1990), Langen (1991), Langen and Vermeer (1991), especially in the regions where the stress concentrations are observed. On the contrary to others, Day and Potts (1994) show results where the Newton-Cotes integration scheme has no benefit over the Gaussian integration.

The interface elements are inserted into continuum very often side by side with the solid elements, which are usually integrated by the Gauss quadratures. Thus, two types of integration are used, the Gauss scheme for solid elements and the Newton-Cotes one for interfaces. To study the effects of the stress oscillation and the influence of the order of shape functions on these effects in consolidation process, the following problems are approached, Figs. 4.4, 4.5, 4.6. The oscillations usually appear when the stress concentrations exist, thus the compressed subsoil is assumed to be glued via interfaces to the vertical walls. The loading was applied during 30 days. For simplicity the elastic behaviour of material is assumed. The material parameters are as follows: E = 2000 kPa, ν = 0.3, $k_x = k_y = 0.0001$m/d,

$\gamma_{wat} = 10 kN/m^3$. Two types of integration schemes are applied; Gauss and Newton-Cotes. The distribution of stresses σ_n inside interfaces for different types of elements are shown in Figs. 4.4, 4.5, 4.6.

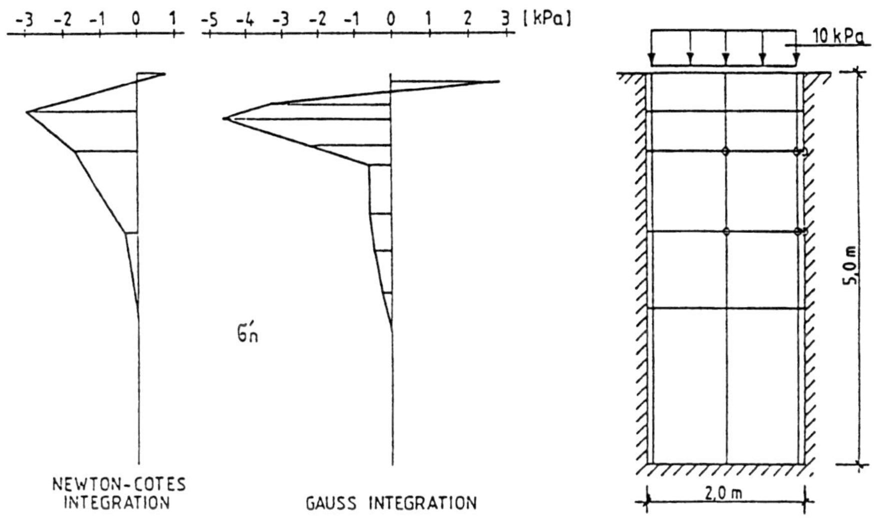

Figure 4.4. Comparison of effective normal stresses at interfaces for two types of integration rules for 4-noded elements.

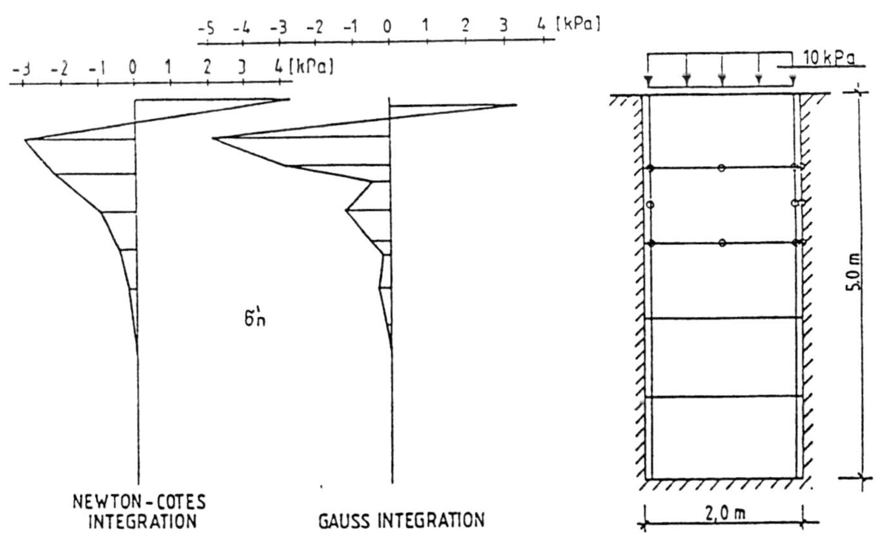

Figure 4.5. Comparison of effective normal stresses at interfaces for two types of integration rules for 8-noded elements.

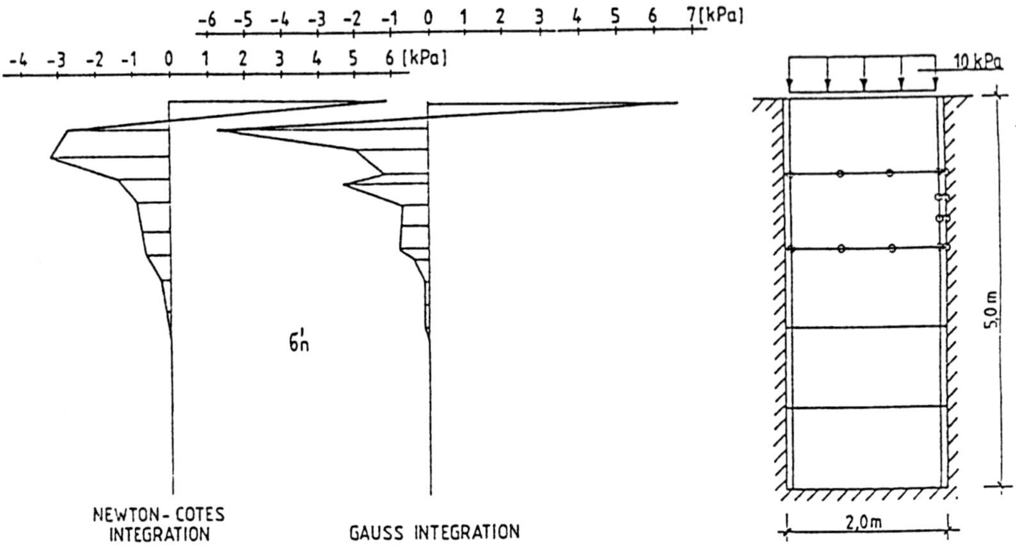

Figure 4.6. Comparison of effective normal stresses at interfaces for two types of integration rules for 12-noded elements.

The concentration of the stresses takes place at the top of the vertical boundary lines. Since the sliding is restricted by the interfaces, the extension in the x-direction appeared at the top of the subsoil. The horizontal displacements (which are by two orders smaller than vertical ones) change their sign close to the top of the analysed area. The distribution of σ_n for four noded interfaces does not show the oscillation of the normal stresses inside interfaces for the Gauss integration scheme. The situation worsens when the six-noded interface is applied and this effect is amplified when the eight-noded interface is used. The increase in the order of the shape function causes the undesired effects of stress oscillation. These effects can be reduced when the Newton-Cotes integration scheme is applied. In none of these tests the oscillation was observed.

The oscillations tend to appear at the beginning of the consolidation process when the time step is short (close to the critical time step). During the consolidation process the effect is smeared out and oscillations vanish. This phenomenon is especially undesired when the elastoplastic material behaviour is considered. The artificial amplification of stresses can cause the plastic yielding which would not appear when we damp the oscillations. The oscillations can also appear in the excesses of pore pressure, especially when the order of the shape function is high and two interfaces are side by side. Such situation can occasionally appear when the internal nodes of the interfaces cause a local weak conditioning of the stiffness array. The oscillation can also appear when the triangular solid elements are used side by side with the interface. In some problems when the 15-noded triangles are used with 10-noded interfaces the oscillation in pore pressure also takes place when the Newton-Cotes integration scheme is applied. The effect is caused by the triangles which introduce some disturbances at the corners and which are transferred to the interfaces.

In the second problem the investigation of the influence of the time step length for the special oscillation is done. The stiff foundation is placed on the subsoil, Fig. 4.7. The material data for the soil are given in the table below.

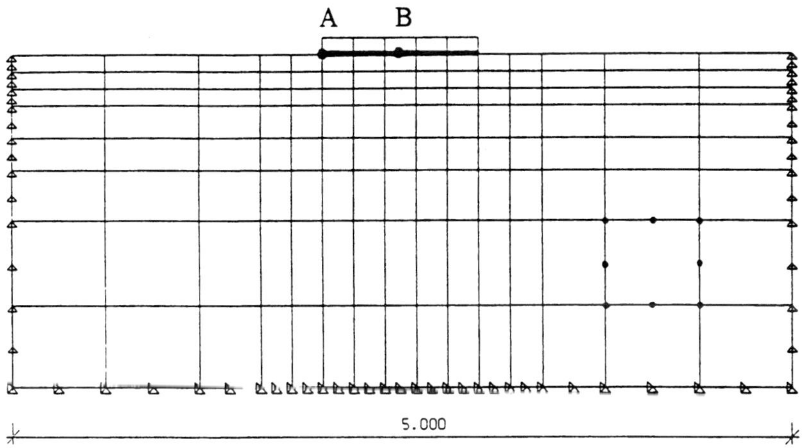

Figure 4.7. Mesh with dimensions and boundary conditions. The loading is applied in the form of prescribed displacements. Displacements of value 0.1 m are applied to the stiff block.

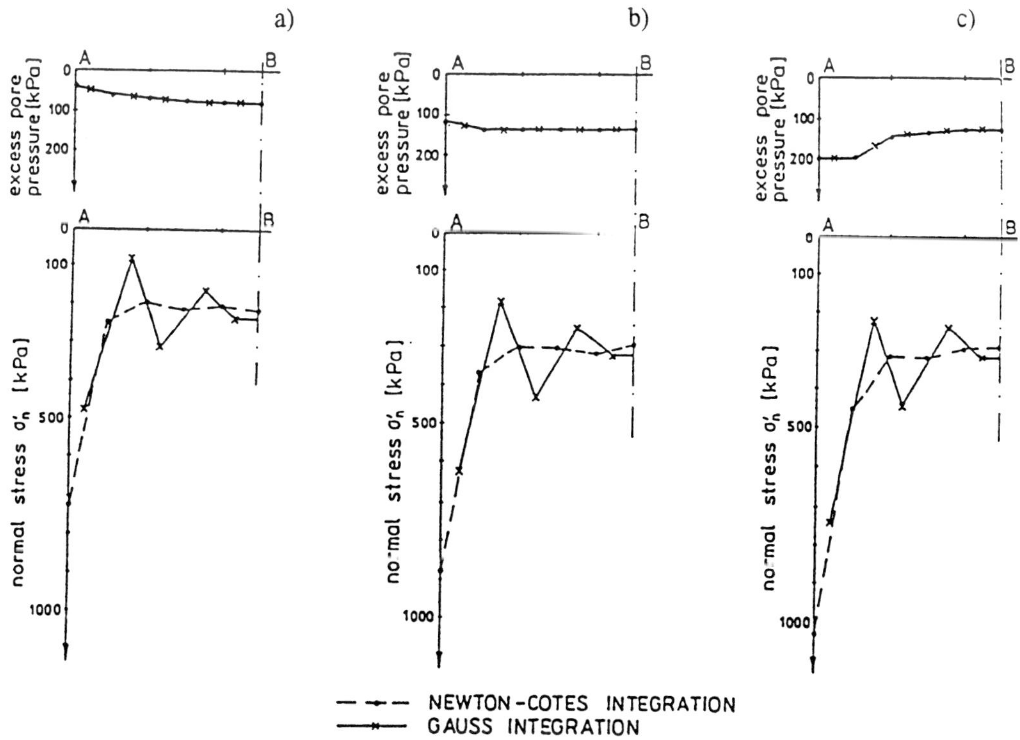

Figure 4.8. Comparison of excess pore pressure and effective stresses for three different time steps a) 200 days, b) 20 days, c) 2 days.

Table 4.1. Material parameters.

Young modulus E [kPa]	Poisson ratio ν	Permeability $k_x = k_y$ [m/d]	Water weight γ_w [kN/m^3]
1000	0.3	0.00002	10

The loading is applied in the prescribed displacements form. The displacements of value 0.1 m is applied to the stiff impermeable block. The calculations are performed for the three different time steps. For the first case the taken time step is 200 days, for the second - 20 days and for the third - 2 days. In each case only one time step is applied. The calculations are performed for the Gauss and Newton-Cotes integration schemes.

The numerical results are shown in Figs. 4.8-4.15. The comparison for two integration schemes is depicted in Fig. 4.8. Due to the symmetry of the problem the figures are drawn for the left-hand side half. The oscillations of the pore pressures are not observed for both integration methods even for a short time step. At the same time the oscillations exists in the effective normal stresses at the interfaces when the Gauss integration scheme is used. This undesired effects are removed by the use of the Newton-Cotes integration. The normal stresses are much smoother, although slight oscillation is still observed.

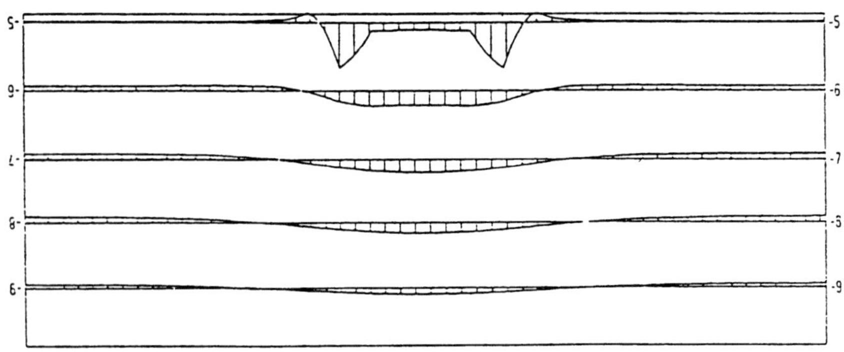

Figure 4.9. Distributions of vertical effective stresses at the horizontal cross-section. No oscillation of stresses is observed in the vicinity of the foundation.

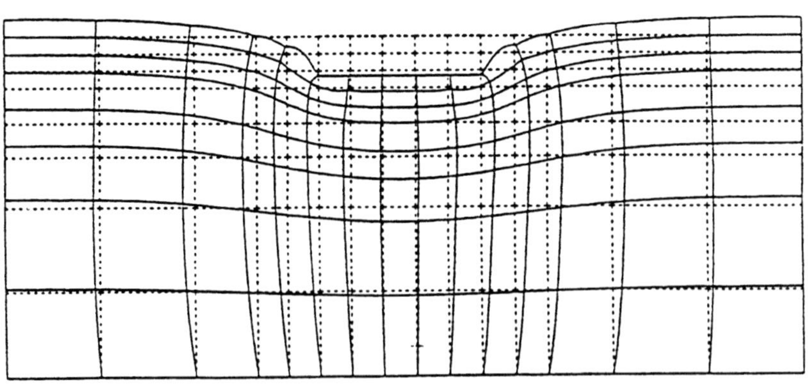

Figure 4.10. The form of deformation below the foundation.

Neither oscillation of stresses, nor the excess of pore pressure, is observed in the foundation below the contact zone. The smooth distribution of the stresses and pore pressure are shown in Figs. 4.9, 4.11, 4.14, 4.15 and Figs. 4.12, 4.13, respectively. No oscillation of pore pressure appeared in the interfaces, neither for the Gauss nor the Newton-Cotes integration rule in both examples.

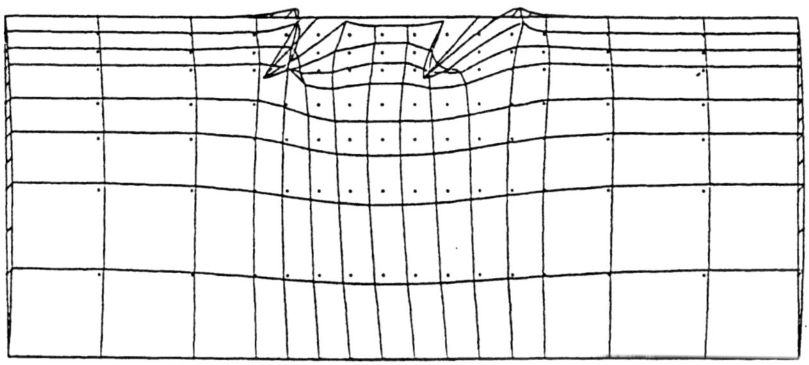

Figure 4.11. Vertical effective stresses in the form of the axonometric view. Concentration of stresses just below the corners of the foundation.

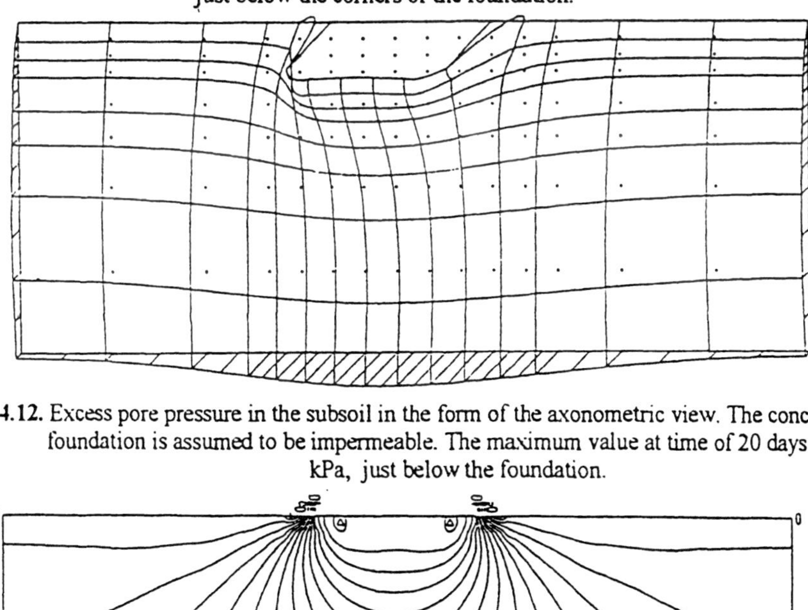

Figure 4.12. Excess pore pressure in the subsoil in the form of the axonometric view. The concrete foundation is assumed to be impermeable. The maximum value at time of 20 days is 140 kPa, just below the foundation.

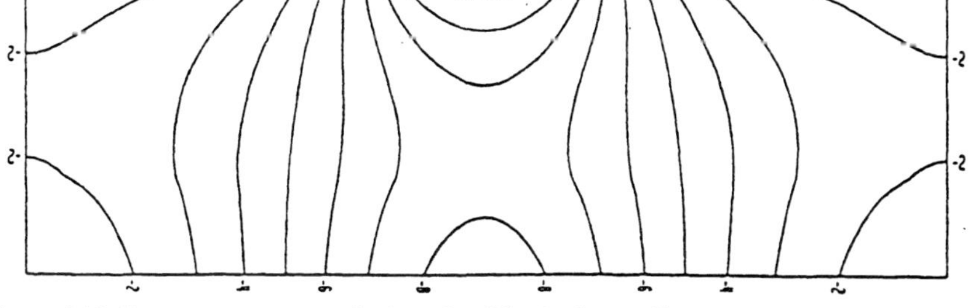

Figure 4.13. Excess pore pressure in the subsoil in the form of isoline drawn every 10 kPa.

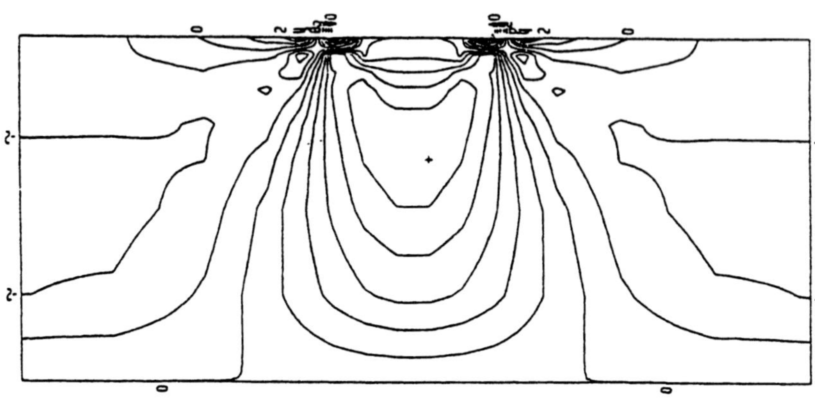

Figure 4.14. Horizontal stresses in the form of isolines drawn every 10 kPa.

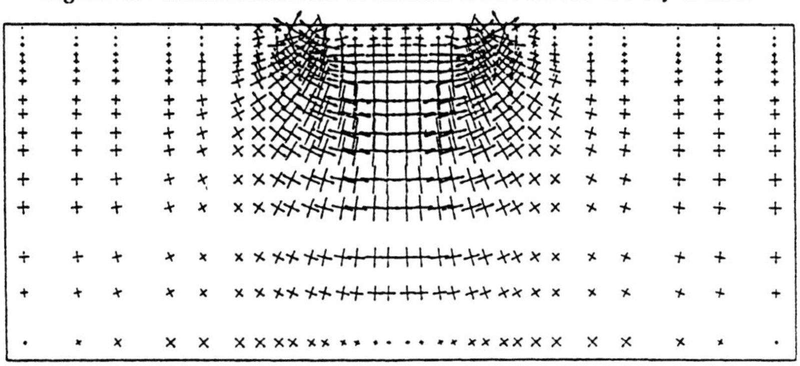

Figure 4.15. Main stresses in the form of crosses. Extensions are depicted in the form of arrows

The Newton-Cotes integration scheme ensures the continuity in stresses and pore pressures from one element to another, which is in an excellent agreement with the theoretical formulation. The oscillations for elastoplastic formulations reported by other researchers appeared also in the consolidation calculations. The introduction of the Newton-Cotes integration scheme brought about a considerable improvement of the results and that is why the scheme.

The finite element solutions of consolidation problems often exhibit spatial oscillation of the pore pressure when using very short time steps. The oscillation tends to increase when the time steps are reduced. The critical time step, i.e. the shortest one which still gives a smooth distribution of the pore pressure, depends on the material parameters as well as the finite element mesh and the type of elements being used. For the simplest case (assuming linear approximation of the pore pressures and the same distance from node to node) the critical time step was found to be (Vermeer and Verruijt (1981)). The introduction of the interface-fleece elements into subsoil will also have an influence on the critical step. Of course the results will strongly be dependent on the position of the interface-fleece elements in the mesh and the material data for these elements. The most sensitive place which determines the critical time step is the drained boundary of the saturated porous medium. The tests of the influence of the interface-fleece element on the critical time step for the one-dimensional consolidation of homogeneous elastic porous medium has been done by Dłużewski (1993). The introduction of the interface elements at the outflow boundary improves the situation, i.e. the critical time step can be even shorter than without the interface and oscillations are not observed.

The discussed interface offers the possibility to model various engineering situations such as sliding of the soil at the structure or another layer of the soil, interaction between soil and a sheet pile wall or concrete foundation. Sliding of soil very often takes place where geotextile reinforcement is introduced into subsoil. The covering of a water reservoir with a membrane causes very often sliding of the membrane over soil. Other types of problems are connected with weak interlayers and lens in the subsoil. Even very thin weak interlayer can considerably reduce the strength of the subsoil and completely change the mechanism of its damage. The interface element is composed of two phases: skeleton and liquid. Special types of elements are also used in the pure flow problems. In flow problems two types of special elements are of great value. The first one, often called a drained element, allows for flow along thin, highly permeable zones. The second one, called a fleece element, offers the possibilities to block the flow perpendicular to the element. Herein, the above possibilities are coupled with the mechanical part of the two-phase medium. The formulation can successfully be applied to block the flow through a membrane or a sheet pile wall. At the same time the sliding of soil can be allowed or not. If e.g. the geotextile reinforcement is permeable, the element can model only sliding of soil when the flow part is engaged to allow for flow through the geotextile. The element enables to model flow along a weak interlayer or a thin vertical drain (geotextile drain).

The integration according to the Gauss quadratures scheme leads to the oscillation of stresses. For the Gauss integration scheme the oscillations of stresses and excess pore pressure appeared in the consolidation calculations. For consolidation the same philosophy as for solid interfaces is applied to suppress the oscillations. All arrays should take the lumped form. The most elegant way leading to such a form is to apply the Newton-Cotes integration rule, where (according to the degree of the polynomial of the shape function) the integration points coincide with the positions of the nodes. Such situations lead to the lumped form of the stiffness, flow and coupling arrays but it does not necessarily mean that the Newton-Cotes rule improved the integration results.

The next important effect is the introduction of an interface element into the finite element mesh, what causes the increase of the critical time step. The detailed study shows that the introduction of the interface at the outflow boundary improves the situation, i.e. the critical time step can be even shorter than without the interface. Most often the interfaces are not situated on the free outflow boundary, so the criterion of the critical time integration step for the solid element should be used.

When the pseudo-viscous approach is applied the longest pseudo-time step, which still ensures the convergence of the iterative process, depends on a few material parameters, among others, on the Young's modulus. The pseudo-time step is inversely proportional to the Young's modulus. Applying the interface element, the high value for the Young's modulus is often accepted to suppress the elastic effects and focus the attention on the plastic yielding. Such approach may lead to the short pseudo-time steps which still ensure the convergence of the iterations but, as a consequence, a great number of iteration steps is required. Fortunately, the most stiff material determines this step. In the engineering soil-structure interaction problems it is not very often that the yielding is observed in the strongest material. When the plastic flow is observed in the stiffer material, the compromise between elastic effects and the pseudo-time step should be reached.

5. Geometrically Nonlinear Description of Deformation for a Two-Phase Medium

5.1 Soil Skeleton

Let us consider the field description of strain and stress combined with large deformations. During the deformation process the particle is moved. The whole analysis is related to one repeatable stage of the deformation which takes place between time t and t + Δt. The following notation is adopted after Bathe (1982), Fig.5.1. The subscripts (as t + Δt) denote, that the considered magnitude is defined in relation to the configuration at time t + Δt. The superscripts define time when the magnitude is analysed: e.g. ${}^{t+\Delta t}_{\ t}\varepsilon_{ij}$ means the strain tensor at time t + Δt in relation to the configuration at time t.

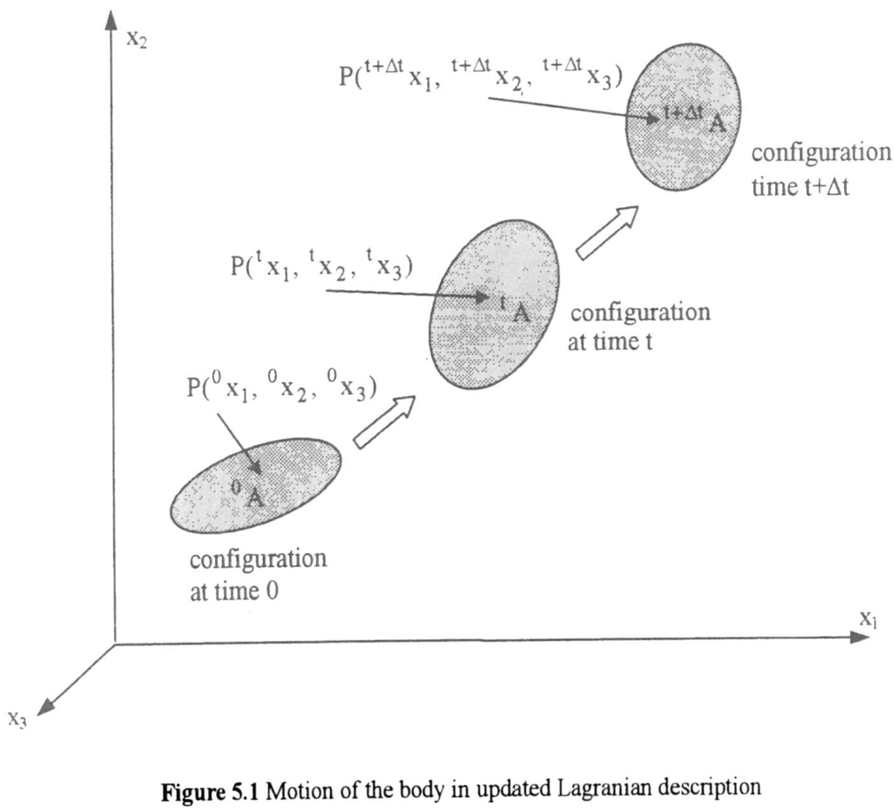

Figure 5.1 Motion of the body in updated Lagranian description

The energy balance in the form of the virtual work principle in the Eulerian description (current configuration of the body) is defined at time t + Δt. The Cauchy stress tensor is conjugated with the linear part of the Almansi strain tensor at the current configuration of the body, (2.33)

$$\int_{^{t+\Delta t}V} {}^{t+\Delta t}_{t+\Delta t}\sigma_{ij}\,\delta^{t+\Delta t}_{t+\Delta t}e_{ij}\;{}^{t+\Delta t}dV = {}^{t+\Delta t}R \tag{5.1}$$

where ${}^{t+\Delta t}_{t+\Delta t}\sigma_{ij}$ is the Cauchy stress tensor, ${}^{t+\Delta t}_{t+\Delta t}e_{ij}$ is the linear part of the Almansi strain tensor. The final state after loading and deformation is defined at time t + Δt. The external virtual work is equal to

$${}^{t+\Delta t}R = \int_{^{t+\Delta t}S} {}^{t+\Delta t}t_i\,\delta\, u_i\,{}^{t+\Delta t}dS + \int_{^{t+\Delta t}V} {}^{t+\Delta t}f_i\,\delta\, u_i\,{}^{t+\Delta t}dV \tag{5.2}$$

where: ${}^{t+\Delta t}t_i$ - surface traction, ${}^{t+\Delta t}f_i$ - body forces, δu_i - virtual displacements.

Let us consider a porous medium fully saturated with water. The total stress, which appeared in equation (5.1), is the sum of effective stress and the pore-water pressure, (2,44)

$${}^{t+\Delta t}_{t+\Delta t}\sigma_{ij} = {}^{t+\Delta t}_{t+\Delta t}\sigma''_{ij} + \delta_{ij}\,\alpha\,{}^{t+\Delta t}_{t+\Delta t}p \tag{5.3}$$

where ${}^{t+\Delta t}_{t+\Delta t}\sigma''_{ij}$ is the effective stress tensor, δ_{ij} is the Kronecker delta, p is the pore-water pressure, α is a Biot's constant (2.45).

The fundamental difficulty in solving equation (5.1) is the fact that the configuration of the body is unknown at time t +Δt. Thus the updated Lagrange procedure is applied to solve equation (5.1).

Let us assume that the current configuration of the body and tensor fields (i.e. displacements, strains and stresses) at the beginning of each repeatable step (at time t) are known. In the beginning (at time t = 0) the initial state of the undeformed body is the starting configuration. The virtual work principle defined in the Lagrange description at time t is the starting point to formulate the incremental iterative procedure

$$\int_{^{t}V} {}^{t+\Delta t}_{t}S_{ij}\,\delta\,{}^{t+\Delta t}_{t}E_{ij}\;{}^{t}dV = {}^{t+\Delta t}_{t}R \tag{5.4}$$

where: ${}^{t+\Delta t}_{t}S_{ij}$ - II Piola-Kirchhoff stress tensor,

${}^{t+\Delta t}_{t}E_{ij}$ - Green strain tensor.

The right hand side of the above equation takes the following form

$${}^{t+\Delta t}_{t}R = \int_{^{t}S} {}^{t+\Delta t}_{t}t_i\,\delta^{t+\Delta t}u_i\,{}^{t}dS + \int_{^{t}V} {}^{t+\Delta t}_{t}f_i\,\delta^{t+\Delta t}u_i\,{}^{t}dV \tag{5.5}$$

The strain tensor can be written as follows

$$ {}^{t+\Delta t}_{t}E_{ij} = \frac{1}{2}\left[{}^{t+\Delta t}_{t}u_{i,j} + {}^{t+\Delta t}_{t}u_{j,i} + {}^{t+\Delta t}_{t}u_{k,i} \cdot {}^{t+\Delta t}_{t}u_{k,j}\right] \tag{5.6} $$

and subsequently it can be divided into linear and non-linear parts. To make the notation much clearer, the superscript t + Δt will be omitted, bearing in mind that the tensor is defined at the end of increment

$$ {}^{t+\Delta t}_{t}E_{ij} = {}_{t}E_{ij} \tag{5.7} $$

$$ {}_{t}E_{ij} = {}_{t}e_{ij} + {}_{t}\eta_{ij} \tag{5.8} $$

$$ {}_{t}e_{ij} = \frac{1}{2}\left[{}_{t}u_{i,j} + {}_{t}u_{j,i}\right] \qquad {}_{t}\eta_{ij} = \frac{1}{2}\,{}_{t}u_{k,i} \cdot {}_{t}u_{k,j} \tag{5.9} $$

where: ${}_{t}e_{ij}$ contains the linear part of the Green strain tensor, while ${}_{t}\eta_{ij}$ contains the non-linear one. At the reference configuration it is valid that

$$ {}^{t+\Delta t}_{t}S_{ij} = {}^{t}_{t}S_{ij} + \Delta_{t}S_{ij} \tag{5.10} $$

where ${}^{t}_{t}S_{ij}$ is the II Piola-Kirchhoff stress tensor, which at the beginning of load increment is equal to the Cauchy stress tensor ${}^{t}_{t}\sigma_{ij}$. ${}_{t}\Delta_{t}S_{ij}$ is the increment of the II Piola-Kirchhoff stress tensor. In other terms the equation (5.10) can be rewritten as follows

$$ {}^{t+\Delta t}_{t}S_{ij} = {}^{t}_{t}\sigma_{ij} + \Delta_{t}S_{ij} \tag{5.11} $$

The total stress is the sum of effective stresses and pore-water pressure at the current configuration of the body

$$ {}^{t}_{t}\sigma_{ij} = {}^{t}_{t}\sigma_{ij}^{//} + \delta_{ij}\alpha\,{}^{t}_{t}p \tag{5.12} $$

where δ_{ij} is the Kronecker symbol. Expressing the increment of the II Piola-Kirchhoff stress tensor in terms of equations (5.11), (5.12), we arrive at

$$ {}^{t+\Delta t}_{t}S_{ij} = {}^{t}_{t}\sigma_{ij}^{//} + \delta_{ij}\alpha\,{}^{t}_{t}p + \Delta_{t}S_{ij} \tag{5.13} $$

The attention should be given to the fact that ${}^{t}_{t}S_{ij} = {}^{t}_{t}\sigma_{ij}$ and ${}^{t}_{t}S_{ij}^{//} = {}^{t}_{t}\sigma_{ij}^{//}$. Introducing equations (5.10) and (5.7) into (5.4), we obtain

$$ \int_{{}^{t}V} {}^{t}_{t}S_{ij}\,\delta_{t}E_{ij}\,{}^{t}dV + \int_{{}^{t}V} \Delta_{t}S_{ij}\,\delta_{t}E_{ij}\,{}^{t}dV = {}^{t+\Delta t}_{t}R \tag{5.14} $$

Having decomposed the strain tensor into linear and non-linear parts (5.8), we can rewrite the above equation in the following form

$$\int_{{}^{t}V} {}^{t}_{t}S_{ij}\,\delta_{t}e_{ij}\,{}^{t}dV + \int_{{}^{t}V} {}^{t}_{t}S_{ij}\,\delta_{t}\eta_{ij}\,{}^{t}dV + \int_{{}^{t}V} \Delta_{t}S_{ij}\,\delta_{t}E_{ij}\,{}^{t}dV = {}^{t+\Delta t}_{\quad t}R \tag{5.15}$$

Let us move the first term of equation (5.15) from the left to the right hand side. After elimination of the kinematically admissible displacement field δu_i all variables in this term will be known. The tensor ${}^{t}_{t}S_{ij}$ is at the same time the Cauchy stress tensor ${}^{t}_{t}\sigma_{ij}$ at the reference configuration t at time t.

$$\int_{{}^{t}V} {}^{t}_{t}S_{ij}\,\delta_{t}\eta_{ij}\,{}^{t}dV + \int_{{}^{t}V} \Delta_{t}S_{ij}\,\delta_{t}E_{ij}\,{}^{t}dV = {}^{t+\Delta t}_{\quad t}R - \int_{{}_{t}V} {}^{t}_{t}\sigma_{ij}\,\delta_{t}e_{ij}\,{}^{t}dV \tag{5.16}$$

The right hand side of equation (5.16) stands for the increment of the external force work caused by load increment. Decomposing the total stress into effective and pore-water pressure (5.11), (5.12), we arrive at

$$\int_{{}^{t}V} ({}^{t}_{t}S^{//}_{ij} + \delta_{ij}\alpha\,{}^{t}_{t}p)\,\delta_{t}\eta_{ij}\,{}^{t}dV + \int_{{}^{t}V} \Delta_{t}S_{ij}\,\delta_{t}E_{ij}\,{}^{t}dV = \Delta R \tag{5.17}$$

where

$$\Delta R = {}^{t+\Delta t}_{\quad t}R - \int_{{}^{t}V} ({}^{t}_{t}\sigma^{//}_{ij} + \delta_{ij}\alpha\,{}^{t}_{t}p)\,\delta_{t}e_{ij}\,{}^{t}dV \tag{5.18}$$

Equations (5.17) and (5.18) are derived with neither simplification nor linearization and form the starting point for the further analysis.

According to the updated Lagrange limitations all starting equations have to be formulated in Lagrangian updated configuration. Let us assume that the load increments are small what allows for the following simplifications:

1) the total increment of the II Piola-Kirchhoff stress is composed of the increments of effective stress tensor and pore-water pressure defined at the updated configuration of the body for the previous load increment,
2) the constitutive equation between increments of the Green strain tensor and the II Piola-Kirchhoff stress tensor is accepted,
3) at the end of increment the transformations between increments of the II Piola-Kirchhoff stress tensor and the Cauchy stress tensor should be performed

According to the first assumption we can write

$$\Delta_{t}S_{ij} = \Delta_{t}S^{//}_{ij} + \delta_{ij}\alpha\,\Delta_{t}p \tag{5.19}$$

In the above equation the increments of the stress and pore-water pressure are defined at the updated configuration of the body. The configuration is known for the previous load increment.

All the time we are conscious of the error we make due to increment of pore-water pressure defined at the previous configuration of the body. We will minimize this error in the iterative process of the equilibrium equations, searching for deformed configuration of the body, which is unknown at the beginning of the increment.

Introducing decomposition (5.19) into (5.17), we get

$$\int_{{}^tV} ({}^t_tS^{//}_{ij} + \delta_{ij}\alpha^t_t p)\delta_t\eta_{ij}\,{}^t dV + \int_{{}^tV} {}_t\Delta S^{//}_{ij}\delta_t E_{ij}\,{}^t dV + \int_{{}^tV} \delta_{ij}\alpha\,\Delta_t p\delta_t E_{ij}\,{}^t dV = \Delta R \qquad (5.20)$$

In the Chapter 3 the hypoelastic constitutive equations are used (3.21). The alternative formulation can be based on pseudo-viscoplastic approach. To model the elasto-plastic behaviour of soil the theory of viscoplasticity, developed by Perzyna (1966), is used. The pseudo visco-plastic iterative procedure, proposed by Zienkiewicz and Cormeau (1975, 1976), is applied in the numerical algorithm. The Perzyna's theory has been applied successfully to characterise elasto-plastic behaviour in terms of the generalised hierarchical rate-independent models proposed by Desai and Zang (1987). The viscous-plastic strain rate is defined in the current configuration of the body

$$\dot{e}^{vp}_{ij} = \gamma <\phi(F)> \frac{\partial Q}{\partial \sigma_{ij}} \qquad (5.21)$$

where Q - potential, F - yield function, γ - viscosity coefficient, $<\phi(F)>$ - special function. The brackets $< >$ stand for an on/off switch. If $F > 0$ then the switch is on, if $F < 0$ the switch is off. We assume that the function $\phi(F)$ takes the following form

$$\phi(F) = \alpha F^{\beta} \qquad (5.22)$$

where α and β are the material parameters.

The strain rate is defined in the current configuration of the body as

$$\dot{e}_{ij} = \frac{1}{2}\left[\, v_{i,j} + v_{j,i} \right] \qquad (5.23)$$

where v_i - displacement rate, (2.13), (2.15).

We will apply the visco-plastic formulation to model the elasto-plastic behaviour of soil. The time which appears in equation (5.21) is assumed to be a parameter of the process and is called pseudo-time. Applying the time integrating scheme, we have

$$\Delta_t e^{vp}_{kl} = \Delta\tau \left[(1-\theta)\, {}^{\tau}_{\tau}\dot{e}^{vp}_{kl} + \theta\, {}^{\tau+\Delta\tau}_{\tau}\dot{e}^{vp}_{kl} \right] \qquad (5.24)$$

where τ is understood here as a time parameter, not a real time, $\Delta_t e_{kl}^{vp}$ is the pseudo visco-plastic strain increment during a psudo time interval $\Delta\tau_n = \tau_{n+1} - \tau_n$. Assuming $\theta = 0$ the simple Euler integration scheme is obtained

$$\Delta_t e_{kl}^{vp} = \Delta\tau \; {}^{\tau}_{\tau}\dot{e}_{kl}^{vp} \tag{5.25}$$

The t configuration is the only known at the beginning of the iterative process.

Let us introduce the following linearization at the increment Bathe (1982)

$$\Delta_t E_{kl} = \Delta_t e_{kl} \tag{5.26}$$

where e_{kl} contains the linear terms of the strain tensor equation (5.9)

According to the second assumption the constitutive equation is written in the incremental form as follows

$$\Delta_t S_{ij}^{//} = {}_t D_{ijkl}(\Delta_t e_{kl} - \Delta_t e_{kl}^{vp} - \Delta_t e_{kl}^{c} - \Delta_t e_{kl}^{0}) \tag{5.27}$$

where: D_{ijkl} - elastic material tensor and, $\Delta_t e_{kl}$ - total strain increment, $\Delta_t e_{kl}^{vp}$ - pseudo visco-plastic strain increment, $\Delta_t e_{kl}^{c}$ - strain increment caused by creep, $\Delta_t e_{kl}^{0}$ - strain increment caused by other effects (thermal, chemical, etc.).

The strain increment which yields, $\Delta_t e_{kl}^{c}$ is defined as follows

$$\Delta_t e_{kl}^{c} = c_{kl}\Delta t \tag{5.28}$$

where: c_{kl} – thermal expansion tensor. Herein we assume

$$\Delta_t e_{kl}^{c} = 0 \tag{5.29}$$

$$\Delta_t e_{kl}^{0} = 0 \tag{5.30}$$

Finally it leads to

$$\int_{{}^tV} ({}_t^t S_{ij}^{//} + \delta_{ij}\alpha_t^t p)\delta_t \eta_{ij}\, {}^t dV + \int_{{}^tV} {}_t D_{ijkl}\Delta_t e_{kl}\delta_t e_{ij}\, {}^t dV + \int_{{}^tV} \delta_{ij}\,\alpha\,\Delta_t p \delta_t e_{ij}\, {}^t dV =$$

$$= \Delta R + \int_{{}^tV} D_{ijkl}\Delta_t e_{kl}^{vp}\delta_t e_{ij}\, {}^t dV \tag{5.31}$$

The above equation serves to be the starting point to formulate the finite element equations in the matrix notation.

Linearization of equation (5.26), as well as simplifications: 1) to 3) generate a certain error of equilibrium

$$E_{RROR} = {}^{t+\Delta t}R - \int_{{}^{t+\Delta t}V} {}^{t+\Delta t}_{t+\Delta t}\sigma_{ij}^{(k)}\, \delta^{t+\Delta t}_{t+\Delta t} e_{ij}^{(k)}\, {}^{t+\Delta t}dV \tag{5.32}$$

where: σ_{ij} - Cauchy stress tensor, e_{ij} - linear part of the Almansi strain tensor.

The above integral is defined at the updated (in each iteration) configuration of the body. The numbers in the parenthesis stand for the iterations which should be performed to eliminate error in the following iterative process

$$\int_{{}^{t}V} {}_{t}D_{ijrs}\Delta_{t}e_{rs}^{(k)}\delta_{t}e_{ij}\,{}^{t}dV+\int_{{}^{t}V}\delta_{ij}\,\alpha\,\Delta_{t}p\delta_{t}e_{ij}\,{}^{t}dV+\int_{{}^{t}V}({}_{t}^{t}S_{ij}^{//}+\delta_{ij}\alpha_{t}^{t}p)\delta\Delta_{t}\eta_{ij}^{(k)}\,{}^{t}dV=$$

$$={}^{t+\Delta t}R-\int_{{}^{t+\Delta t}_{(k-1)}V}({}_{t+\Delta t}^{t+\Delta t}\sigma_{ij}^{//\,(k-1)}+\delta_{ij}\alpha\,{}_{t+\Delta t}^{t+\Delta t}p^{(k-1)})\delta\,{}_{t+\Delta t}^{t+\Delta t}e_{ij}^{(k-1)}\,{}_{(k-1)}^{t+\Delta t}dV+ \tag{5.33}$$

$$+\int_{{}^{t}V} {}_{t}D_{ijrs}\Delta_{t}e_{rs}^{vp}\delta_{t}e_{ij}\,{}^{t}dV+$$

In each iteration step the displacements should be accumulated

$${}^{t+\Delta t}u_i^{(k)}={}^{t+\Delta t}u_i^{(k-1)}+\Delta u_i^{(k)} \tag{5.34}$$

and the configuration should be updated

$${}^{t+\Delta t}x_i^{(k)}={}^{t+\Delta t}x_i^{(k-1)}+\Delta u_i^{(k)} \tag{5.35}$$

Neither the building of a new stiffness array nor the decomposition of the global equation system is required in the above iterative process (5.33) to (5.35). After elimination of the kinematically admissible displacement field δu_i and extraction of the unknown increments of the displacement $\Delta u_i^{(k)}$ out of the integrals, the matrices resulting from the left hand side of (5.33) are constant during the whole iterative process. The iterative process will lead to the elimination of the errors resulting from the linearization of the strain tensor field. The iterative process should be continued so long as the convergence criterion is satisfied.

The attention should be paid to the fact that the conjugated strain and stress tensors are used in the current configuration of the body, which guarantees the energy balance. Starting from the virtual principle (5.4), it can easily be proved, Bathe (1982), Dłużewski (1988) that the Cauchy stress tensor is conjugated with the linear part of the Almansi strain tensor in the Eulerian description.

5.2 Pore Water

The continuity equation for the mass balance is the fundamental principle for the flow problem in a porous medium. The two-phase medium is considered. Pores in the skeleton are assumed to be fully saturated. The continuity equation takes the following form Lewis and Schrefler (1998)

$$\alpha\frac{\partial\varepsilon_v}{\partial t}+\frac{\alpha-n}{K_s}\frac{\partial p}{\partial t}+\frac{n}{K_w}\frac{\partial p}{\partial t}+\frac{\partial v_i}{\partial x_j}=0 \tag{5.36}$$

The three first terms of the left hand side of the equation defines the rate of fluid accumulation which consists of a few terms. The contributors which are taken into account are listed below:

a) the rate of change of the total volume strain

$$\frac{\partial \varepsilon_v}{\partial t}; \tag{5.37}$$

where $$\dot{\varepsilon}_v = \dot{e}_x + \dot{e}_y + \dot{e}_z \tag{5.38}$$

b) the rate of change of the grain volume due to pressure changes

$$\frac{\alpha - n}{K_s}\frac{\partial p}{\partial t}; \qquad n \text{ - porosity} \tag{5.39}$$

c) the rate of change of the water volume due to the change of the pore pressure

$$\frac{n}{K_w}\frac{\partial p}{\partial t} \tag{5.40}$$

The Darcy's flow rule for a certain porosity is assumed

$$v_i = -\frac{k_{ij}}{\mu}\left(-\frac{\partial p}{\partial x_j} + \rho g \frac{\partial z}{\partial x_j}\right) \tag{5.41}$$

where: μ - viscosity of the fluid, ρ - density, g - gravity, z - hydraulic head, k_{ij} - permeability tensor.

The relative (with respect to the solid phase) fluid flow velocity is governed by the Darcy's law (5.41). Both equations are defined in the material coordinate system. The terms of the left hand side of the continuity equation (5.36) depends either on the porosity, volumetric changes or the velocity field. The equation is defined in the current configuration of the body. Applying the updated Lagrange procedure, both increments and total strains are calculated, thus the volumetric changes can be summed by increments. The increment of the volumetric changes is calculated at the same level of the approximation as the constitutive equation (5.27). The exact definitions of the volumetric changes ε_v in both descriptions (Eulerian and Lagrangian) are given in the Appendix **B**. The calculated strain increments between time t and t+Δt are related to the current configuration. The summing up of the increments leads to the logarithmic strains, i.e. stretches (but not defined at the principal directions).

Herein the permeability is assumed to be dependent on the current porosity, so it can be written

$$k_{ij} = k_{ij}(n) = k_{ij}(e) \tag{5.42}$$

where $n = \dfrac{e}{1+e}$ is the porosity and e is the void ratio. The relation (5.42) has a different meaning for various types of materials. Considering the clay type of soil the relation as (5.42) is taken into account rarely. On the other hand side for the organic soils as peat or gyttja the proper description of relation (5.42) is very important. The permeability k can be of two orders of origin smaller due to closing of the peat pores during the building of the embankment, what very much affects time of consolidation. The results of the laboratory tests in the form of the relation between the void ratio and the permeability drawn in the semilogarithmic scale, after Chaciński and Dłużewski (1995), are depicted in Fig. 5.2.

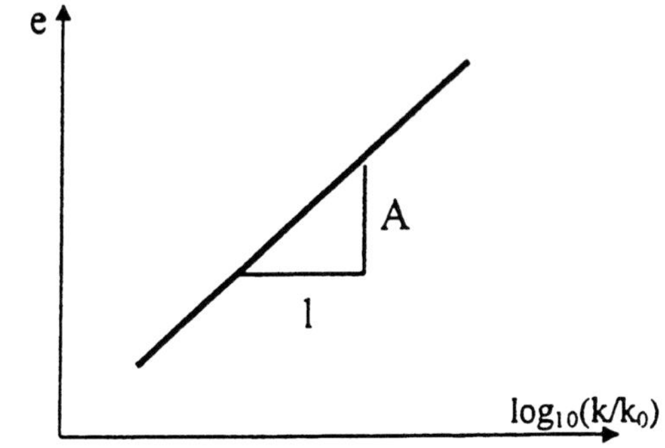

a) Graphical interpretation of equation (5.43)

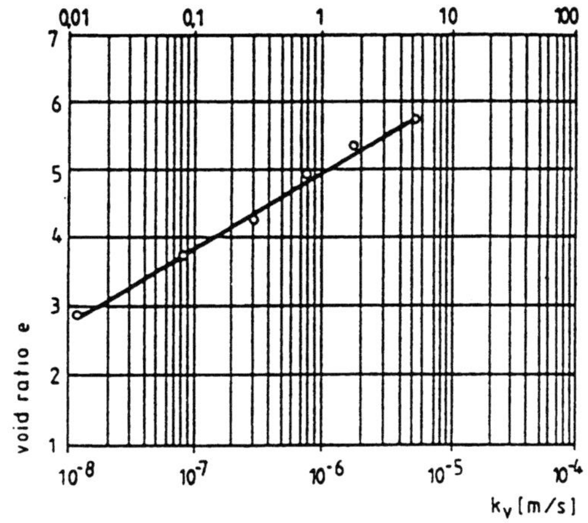

b) Permeability kv in vertical direction in relation to void ratio e.
Test results for peat after Chaciński and Dłużewski (1995)

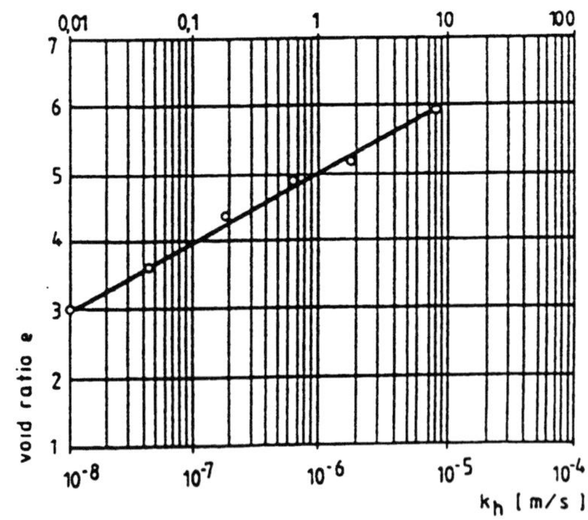

c) Permeability kh in horizontal direction in relation to void ratio e.
Test results for peat after Chaciński and Dłużewski (1995)

Figure 5.2. Permeability in relation to the void ratio

To describe the permeability changes, the following relation is used

$$\log_{10}\left(\frac{k_{ij}}{k_{ij}^{0}}\right) = \frac{e - e_0}{A} \tag{5.43}$$

where k_{ij}^{0} is the tensor of the initial permeability corresponding to the initial void ratio e_0. A is the material parameter with geometrical interpretation as tangent of the slope curve (5.43) in the semilogarithmic scale, Fig. 5.2.

The introduction of relation (5.43) is important. It is worth emphasizing that the changes of the orthotropic permeability for peat can be defined with the use of one material parameter A. The relation (5.43) defines changes of the permeability in orthotropic directions with the help of a single parameter only, what is in good agreement with experimental evidence. The relation has been successfully applied into finite element code to model consolidation process, Dłużewski (1997, 1998, 1999).

Monte and Kritzen (1976) proposed bilinear relation (in the semilogarithmic scale) for the permeability in relation to the void ratio e in the following form

$$\ln\left(\frac{k}{k_c}\right) = \frac{e - e_c}{A}; \tag{5.44}$$

$$A = A_1 \quad \text{for} \quad e < e_c \qquad A = A_2 \quad \text{for} \quad e > e_c \tag{5.45}$$

where k_c, e_c are the values of the permeability and the void ratio at which the slope changes. A_1 and A_2 are the material parameters.

It can be easily shown that the straight line drawn in one semilogarithmic scale is also straight in other semilogarithmic scales (for different bases of logarithms).

Kuklik et al. (1996, 1998) propose the relation between the permeability and the current void ratio e in the following form

$$k = k_0 \left(\frac{e}{e_0}\right)^{m_1+m_2} \left(\frac{1}{S\, r_0}\right)^{m_2} \tag{5.46}$$

where m_1, m_2, r_0 are the material parameters.

5.3 Finite Element Formulation for a Two-Phase Medium

Soil skeleton. The notation used in the paper is adopted after Zienkiewicz and Taylor (1991) (considering the basic equations) and after Lewis and Schrefler (1998) (considering the consolidation). The attention is paid to the fact that **S** means either the II Piola-Kichhoff stress tensor (written in the vector form) or the array responsible for the compressibility of the body. This collision stems from the consequences of applying the most common notations.

The considerations are confined to one element only, having in mind that assembling rules of the matrices are known. The basic equations can be written as follows

$$\mathbf{u}^e = \mathbf{N}\mathbf{u}, \qquad \mathbf{e} = \mathbf{B}\mathbf{u} \tag{5.47}$$

$$\mathbf{p}^e = \overline{\mathbf{N}}\mathbf{p} \qquad \mathbf{x}^e = \overline{\overline{\mathbf{N}}}\mathbf{x} \tag{5.48}$$

where: $\mathbf{N}$ - shape function array for displacement approximation, $\mathbf{B}$ - geometrical array containing the linear parts of the strain tensor, $\overline{\mathbf{N}}$ - shape function array for pore-water pressure approximation, $\overline{\overline{\mathbf{N}}}$ - shape function array for coordination approximation for 'inconsistent' formulation.

The forms of $\mathbf{N}, \overline{\mathbf{N}}, \overline{\overline{\mathbf{N}}}$ depend on the type of the finite element, type of the shape function applied to the approximation of displacements, pore-water pressure and coordinates. In consolidation problems the 'consistent' and 'inconsistent' formulations are considered very often. According to the decomposition of the stress (5.12), after Terzaghi, into effective one and pore-water pressure the 'consistent' approximation level of the stress and pore-water pressure is recommended. The effective stress or its rate, via constitutive equations, are calculated from strain, stemming from the differentials of shape functions in the form of $\partial N_i / \partial x_j$. Since the approximation of pore-water pressure is one order higher than strain (to obtain the same approximation level of stresses and pore-water pressures), the displacements should be approximated by the shape functions which are one order higher in relation to the pore-water pressure. Such formulation is called 'consistent' $N \neq \overline{N}$ and the polynomial N is one order higher in relation to $\overline{N}$. Most often the eight-noded quadrilaterals or six-noded triangles are used with displacements in all nodes and pore-water pressure at the corner nodes.

In 'inconsistent' formulation most often $N = \overline{N}$ and in all nodes the displacements and pore-water pressures are the independent degrees of freedom. What are the advantages and disadvantages of 'consistent' and 'inconsistent' formulation? The advantages of 'consistent' formulation are: the same order of the approximation of the pore-water pressure and stresses, smaller dimension of problem for the same mesh, shorter critical time step integration, what can be found in the thesis by Song (1990). The disadvantages of 'consistent' formulation are: limitation in types of elements, more difficult software development. The advantage of 'inconsistent' formulation is a much simpler numerical implementation. For 'inconsistent' formulation elements do exist (e.g. eight-noded quadrilateral) which give very accurate results in the pore-water pressure at the level of Gauss points even for very short time steps, when the oscillations are observed at nodes. The disadvantage is that the approximation levels of pore-water pressure and stresses are different but it does not yield considerable errors. The 'inconsistent' formulation is employed in this chapter.

To formulate the finite element equations for consolidation problem the stiffness matrix $\mathbf{K}_T$ derived in the Chapter 3 is used, equation (3.24). For a two phase medium we have

$$\mathbf{K}_T \frac{d\mathbf{u}}{dt} + \mathbf{L}\frac{d\mathbf{p}}{dt} = \frac{d\mathbf{P}}{dt} \tag{5.49}$$

In the above equation the following notation is accepted: **P** - vector of load, **u** - vector of displacement, **p** - vector of pore-water pressure, **L** – coupling array. The stiffness array is the sum of four terms $\mathbf{K}_0$, $\mathbf{K}_S$, $\mathbf{K}_{\sigma 1}$, $\mathbf{K}_{\sigma 2}$, defined by equations (3.25), (3.26), (3.29), (3.31).

$$\mathbf{L} = \int_V \mathbf{B}^T \alpha\, \mathbf{m}\, \overline{\mathbf{N}}\, dv \tag{5.50}$$

B - linear geometrical array, $\overline{\mathbf{N}}$ - shape function array for pore-water pressure approximation, **m** - one-zero vector (zero for shearing stresses, one for the remaining ones).

The matrix equation (5.50) contains the coupling between skeleton and water via the second term of the left hand side. Let us assume that the undrained situation is considered. In such situation the water outflow is very slow comparing to load increment, thus the pore-water pressure increment caused by deformation (at the level of the Gauss point) is equal to

$$\Delta p = 3K_w \mathbf{m}^T \Delta\mathbf{e} \tag{5.51}$$

Rewriting equation (2.44) in the matrix notation, we get

$$\Delta\boldsymbol{\sigma} = \Delta\boldsymbol{\sigma}^{//} + \mathbf{m}\alpha\Delta p \tag{5.52}$$

Introducing the linearized constitutive equation, written in the matrix notation, into equation (5.52), we have

$$\Delta\boldsymbol{\sigma} = (\mathbf{D}_T + 3K_w \mathbf{m}\mathbf{m}^T)\Delta\mathbf{e} \tag{5.53}$$

For the undrained analysis the set of equations (5.49) is reduced to the form as follows

$$(\mathbf{K}_{OU} + \mathbf{K}_S + \mathbf{K}_{\sigma 1} + \mathbf{K}_{\sigma 2})\Delta\mathbf{u} = \Delta\mathbf{R} \tag{5.54}$$

$$\mathbf{K}_{OU} = \int_{{}^tV} \mathbf{B}^T \mathbf{D}_{TU}\, \mathbf{B}\, {}^t dV; \tag{5.55}$$

The array $\mathbf{D}_{TU}$ depends on the water compressibility and is defined as follows

$$\mathbf{D}_{TU} = \mathbf{D}_T + 3K_w \alpha\, \mathbf{m}\mathbf{m}^T \tag{5.56}$$

The displacement increments determined from equation (5.54) and all other values obtained from $\Delta\mathbf{u}$ contain the errors discussed in detail in the previous section. The errors can be minimized in the iterative process during the incremental procedures

$$\mathbf{K}_T^{(k-1)} \Delta\mathbf{u}^{(k)} = \Delta\mathbf{R}^{(k-1)} \tag{5.57}$$

$$\Delta\mathbf{R}^{(k-1)} = \int\limits_{t+\Delta t_V} \mathbf{N}^T \mathbf{f}^{\,t+\Delta t} dV + \int\limits_{t+\Delta t_S} \mathbf{N}^T \mathbf{t}^{\,t+\Delta t} dS \; - \\ - \int\limits_{{}^{(k-1)}_{t+\Delta t}V} \mathbf{B}_{(k-1)}^T \left({}^{t+\Delta t}_{t+\Delta t}\sigma^{//\,(k-1)} + \mathbf{m}\,\alpha\; {}^{t+\Delta t}_{t+\Delta t}p^{(k-1)} \right) {}^{t+\Delta t}_{(k-1)}dV \tag{5.58}$$

If the pseudo-viscoplastic approach is used the following changes should be introduced: $\mathbf{D}_T$ should be replace by $\mathbf{D}_e$ in the stiffness array $\mathbf{K}_T$ in the equation (5.54) and the additional term will appear in the right hand side of the equation (5.58)

$$\int\limits_{{}^{(k-1)}_{t+\Delta t}V} \mathbf{B}_{(k-1)}^T \mathbf{D}_e \left({}^{t+\Delta t}_{t+\Delta t}\Delta\varepsilon_j^{vp\,(k-1)} \right) {}^{t+\Delta t}_{(k-1)}dV \tag{5.59}$$

where j denotes pseudo-viscoplastic iteration number. The configuration should be updated in each iterative step

$${}^{t+\Delta t}\mathbf{x}^{(k)} = {}^{t+\Delta t}\mathbf{x}^{(k-1)} + \Delta\mathbf{u}^{(k)} \tag{5.60}$$

until the convergence criterion is fulfilled. The above formulation (equation (5.57)) can be modified in such way that the building and decomposing the stiffness array $\mathbf{K}_T$ can be done after a few iteration steps.

If the assumption of the undrained analysis is not acceptable, the full consolidation process with the distribution of the excess pore-water pressure is required.

Pore water. The continuity equation (5.36) serves as the source for further formulation of the finite element equations for the liquid phase. The Galerkin method is applied to solve (5.36). Let us introduce the trial function in the form $\mathbf{p}_e = \overline{\mathbf{N}}\mathbf{p}$ where N_i are the elements of the array. The inconsistent formulation is applied (i.e. the same shape functions are used for the approximation of displacements and the pore-water pressure). The application of the consistent formulation (i.e. the shape functions one order higher for the approximation of the displacement field in relation to the pore-water pressure field) does not create any theoretical difficulties. Introducing the matrix notation and applying the Galerkin method to solve equation (5.36), we get

$$\mathbf{Hp} + \mathbf{S}\frac{d\mathbf{p}}{dt} + \mathbf{L}^T \frac{d\mathbf{u}}{dt} = \mathbf{Q} \tag{5.61}$$

where the forms of arrays are listed below

$$\mathbf{H} = \int\limits_V (\nabla\overline{\mathbf{N}})^T \frac{\mathbf{k}}{\mu} \nabla\overline{\mathbf{N}}\, dV \tag{5.62}$$

$$\mathbf{S} = \int\limits_V \overline{\mathbf{N}}^T s\, \overline{\mathbf{N}}\, dV \tag{5.63}$$

$$s = \frac{\alpha - n}{K_s} + \frac{n}{K_w} \tag{5.64}$$

$$\mathbf{L}^T = \int_V \overline{\mathbf{N}}^T \alpha \mathbf{m}^T \mathbf{B} dV \tag{5.65}$$

$$\mathbf{Q} = -\int_S \overline{\mathbf{N}}^T \mathbf{q} \; dS - \int_V (\nabla \overline{\mathbf{N}}^T)^T \frac{\mathbf{k}}{\mu} \nabla \rho g z dV \tag{5.66}$$

The following definitions are applied: **H** - array responsible for flow properties, **S** - array responsible for water compressibility, **L** - coupling array, **Q** - nodal discharges vector, ∇ - differential operators vector, $\overline{\mathbf{N}}$ - shape functions array, **k** - permeability array, **m** - one-zero vector, **q** - discharge at the boundary.
The more detail definitions of the arrays can be found in the book by Lewis and Schrefler (1998).

Consolidation. The nodal equilibrium equation (5.49) is coupled with the mass balance equation (5.61) via volumetric changes. Both equations can be linked in the following form

$$\begin{bmatrix} \mathbf{K}_T & \mathbf{L} \\ \mathbf{L}^T & -\mathbf{S} \end{bmatrix} \begin{bmatrix} \frac{d\mathbf{u}}{dt} \\ \frac{d\mathbf{p}}{dt} \end{bmatrix} = \begin{bmatrix} \mathbf{0} & \mathbf{0} \\ \mathbf{0} & -\mathbf{H} \end{bmatrix} \begin{bmatrix} \mathbf{u} \\ \mathbf{p} \end{bmatrix} + \begin{bmatrix} \frac{d\mathbf{P}}{dt} \\ \mathbf{Q} \end{bmatrix} \tag{5.67}$$

The above set of equations describes the behaviour of the two-phase medium in terms of the finite element approximation. To determine the displacements, and pore-water pressure the finite difference scheme is used and described in the Chapter 4.

In general case, the solution of the above matrix equation requires the iterative process in which the assembling of the left hand side matrix and determining of the unknown vector should be done in each iterative step. The array **H** depends on the current void ratio e, which changes during the deformation process and depends on the unknown $\Delta\mathbf{u}$, in other words $\mathbf{H}(\Delta\mathbf{u})$. Considering the material and geometrical nonlinearities, the iterative process is required.

The solution of the above equation system with taking into account current void ratio and its consequences, i.e. permeability k as a function of e and exact definition of volumetric changes according to the Appendix B, is very time consuming. The left hand side matrix of equation (5.67) has to be assembled and decomposed in each iteration step to determine $\Delta\mathbf{u}$ and $\Delta\mathbf{p}$ in the updated Lagrange procedure. In the iteration process both $\mathbf{K}_T$ and **H** depend on the unknowns. The iteration process takes the form as follows

$$\begin{bmatrix} \mathbf{K}_T^{(k-1)} & \lambda\mathbf{L} \\ \lambda\mathbf{L}^T & -\lambda^2(\mathbf{S}+\alpha\Delta t\mathbf{H}^{(k-1)}) \end{bmatrix}_t \begin{bmatrix} \Delta\mathbf{u}^{(k)} \\ \frac{1}{\lambda}\Delta\mathbf{p}^{(k)} \end{bmatrix} = \begin{bmatrix} \mathbf{0} & \mathbf{0} \\ \mathbf{0} & -\lambda\Delta t\mathbf{H}^{(k-1)} \end{bmatrix}_t \begin{bmatrix} \mathbf{u} \\ \mathbf{p} \end{bmatrix}_t + \begin{bmatrix} \Delta\mathbf{R}^{(k-1)} \\ \lambda\Delta t\,\mathbf{Q} \end{bmatrix}_t \tag{5.68}$$

where (5.69)

$$\Delta\mathbf{R}^{(k-1)} = \int_{^{t+\Delta t}V} \mathbf{N}^T \mathbf{f}^{\,t+\Delta t} dV + \int_{^{t+\Delta t}S} \mathbf{N}^T \mathbf{t}^{\,t+\Delta t} dS - \int_{^{(k-1)}_{t+\Delta t}V} \mathbf{B}^T_{(k-1)} ({}^{t+\Delta t}_{t+\Delta t}\sigma^{//\,(k-1)} + \mathbf{m}\alpha\, {}^{t+\Delta t}_{t+\Delta t}\mathbf{p})\, {}^{t+\Delta t}_{(k-1)}dV$$

If the pseudo-viscoplastic approach is used again the $\mathbf{D}_T$ should be replace by $\mathbf{D}_e$ in the stiffness array $\mathbf{K}_T$ and the additional term (5.59) will appear at the right hand side of the equation (5.69)
The coordinates are updated in each iteration step

$$ {}^{t+\Delta t}\mathbf{x}^{(k)} = {}^{t+\Delta t}\mathbf{x}^{(k-1)} + \Delta\mathbf{u}^{(k)} \tag{5.70} $$

The described iteration process is extremely time consuming, i.e. the following simplifications can be assumed. To make the time stepping process more adjusted to the updated Lagrange procedure, the assumption that during certain time step the values of kij(e) are constant is done. These values are changed with a new time step Δt. Such algorithm suits the Eulerian integration method. In other words, the updating of the configuration according to all changes combined with the deformation field, as well as the void ratio and the permeability, is performed at the end of the time step. According to such assumption in a certain time step *m* we have:
1) e - void ratio takes the value e_m
2) $k_{ij}(e)$ - permeability takes the value $k_{ij}(e_m)$.

According to the updated Lagrange procedure the iteration of the global matrix system in every time step (and load increment in time) is performed to minimise the global error of equilibrium. At the end of each time step the updating of the configuration is carried out

$$ \begin{bmatrix} \mathbf{K}_T & \lambda\mathbf{L} \\ \lambda\mathbf{L}^T & -\lambda^2(\mathbf{S}+\alpha\Delta t\mathbf{H}) \end{bmatrix}_t \begin{bmatrix} \Delta\mathbf{u}^{(k)} \\ \frac{1}{\lambda}\Delta\mathbf{p}^{(k)} \end{bmatrix} = \begin{bmatrix} \mathbf{0} & \mathbf{0} \\ \mathbf{0} & -\lambda\Delta t\mathbf{H} \end{bmatrix}_t \begin{bmatrix} \mathbf{u} \\ \mathbf{p} \end{bmatrix}_t + \begin{bmatrix} \Delta\mathbf{R}^{(k-1)} \\ \lambda\Delta t\,\mathbf{Q} \end{bmatrix}_t \tag{5.71} $$

where $\Delta\mathbf{R}^{(k-1)}$ is defined by equation (5.69). As long as the geometrical nonlinearites are solved the error of equilibrium can be improved in further numerical process, but error introduced by changes of the porosity is accumulated and influences the time of consolidation. Obviously, at the end of each time step the updating of all values combined with volumetric changes is done.

The above algorithm has been developed in HYDRO-GEO computer code, and applied to solve test problems shown in the next chapter. The code is developed at Warsaw University of Technology. To model the elasto-plastic behaviour of the skeleton, the pseudo-viscous iteration procedure was adopted, Zienkiewicz and Cormeau (1974), Cormeau (1975, 1976). The procedure takes its roots from the Perzyna (1966) model, and it was also applied to solve geometrically non-linear problems by Kanchi (1978).

5.4 Examples

One Dimensional Consolidation. One-dimensional consolidation problem is analysed. The layer of 10 m thickness founded on the impermeable basement is considered. The finite element mesh, boundary conditions and material data are shown in Fig.5.3. Eight-noded isoparametric elements are used. The whole loading is applied during the first time step $\Delta t = 2 \bullet 10^{+5}$ [s]. The time stepping process is continued until the excess of pore-water pressure is redistributed. 29 time steps are applied: $\Delta t = 5x(2 \bullet 10^{+5}), 9x(1 \bullet 10^{+6}), 5x(2 \bullet 10^{+6}), 6x(1 \bullet 10^{+7}), 5x(2 \bullet 10^{+7})$ [s].

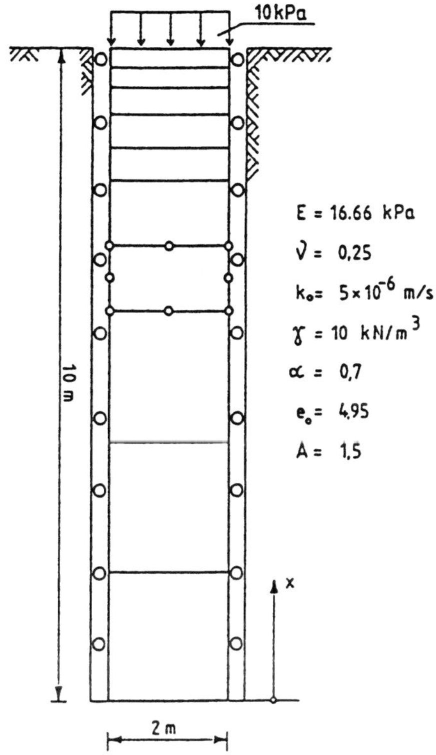

Figure 5.3. Mesh, boundary conditions, (bottom impermeable) loading, material data, time integration parameter α, initial permeability, initial void ratio and A parameter for equation (5.43).

The inconsistent formulation of consolidation process is used (here in every node two displacements and excess pore-water pressure are calculated). The analysis is carred out: first for small strains, next the geometrically nonlinear effects are taken into account. To determine the current permeability of the soil, the procedures described in the previous sections, based on equation (5.76) and relation (5.43) are engaged.

Numerical results are shown in Figs. 5.4-5.7. The value of the load intensity is chosen in such a way that the displacements could be equal half of the height of the soil layer. The obtained displacements versus time both for small and large strains are shown in Fig. 5.4. The settlements for the large strain case are smaller comparing the same steps for the small strains. The final displacements reached 5 m for small strains, and 3.83 m for large strains.

The excess pore-water pressure versus time is shown in Fig. 5.5. The redistribution of the excess pore-water pressure is much slower for the case of large strain comparing with the small strain results. Taking into account the closing of pores, the decay of pressure is much slower. After 29 time steps (time $1.7 \bullet 10^{+7}$ s) the excess of pore-water pressure reached -0.052 kPa. For small strains with constant permeability the redistribution of excess pore-water pressure is obtained at time $2 \bullet 10^{+6}$s, it is ten times faster. The comparison of the excess of pore-water pressure between various solutions is shown in Fig. 5.6. The analytical solution for small strain is compared with numerical results for small and large strains. The distribution of the excess pore-water pressure in the vertical cross-section is shown for various time periods. The divergence in the excess pore-water pressure between large and small strains is increasing at the beginning of the time stepping process and then is decreasing to reach 0 in the end.

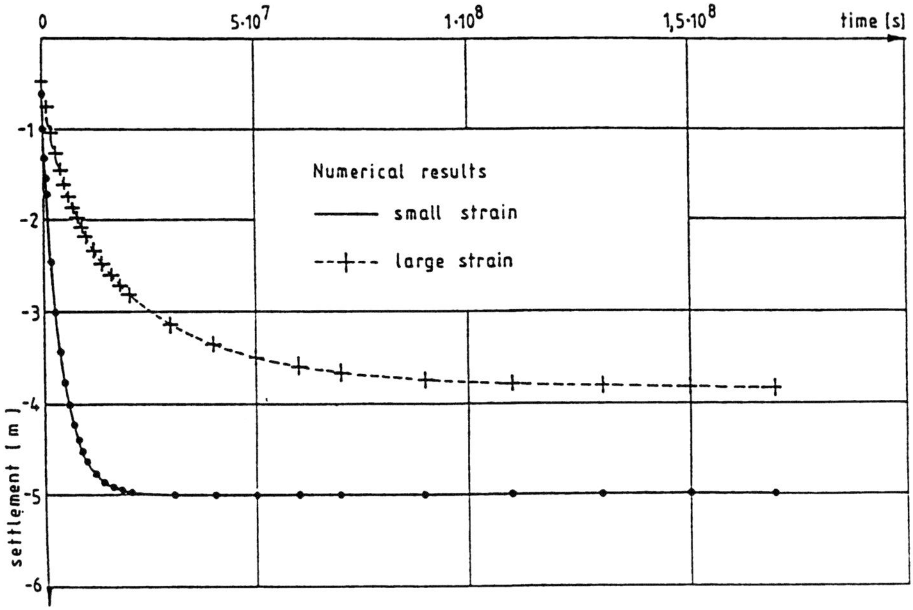

Figure 5.4. Settlement of the ground level during consolidation process (max. for small strain equals 5 m, for large strains 3.83 m).

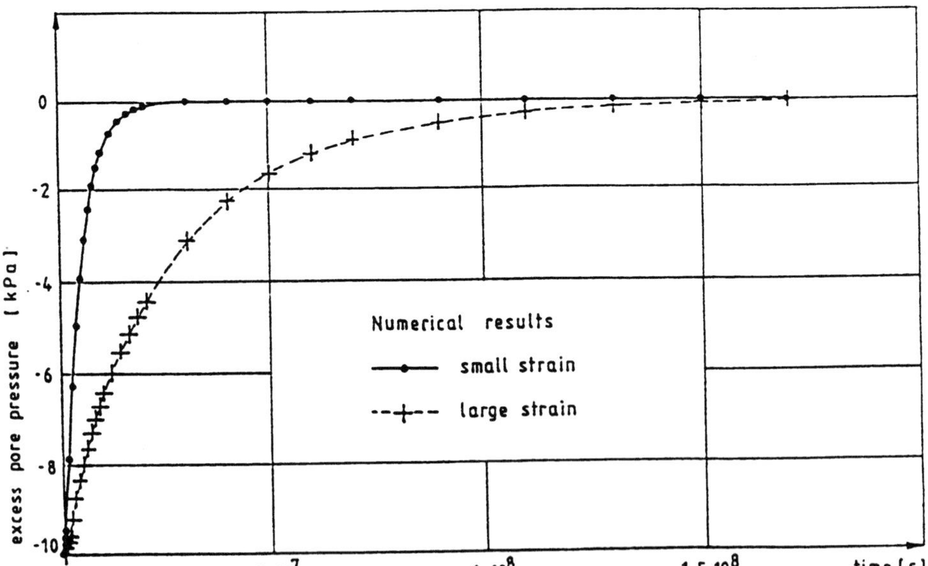

Figure 5.5. Comparison of the excess pore-water pressure versus time during consolidation process. Excess pore-water pressure is compared for point at the level x=0 m. (see Fig. 5.3).

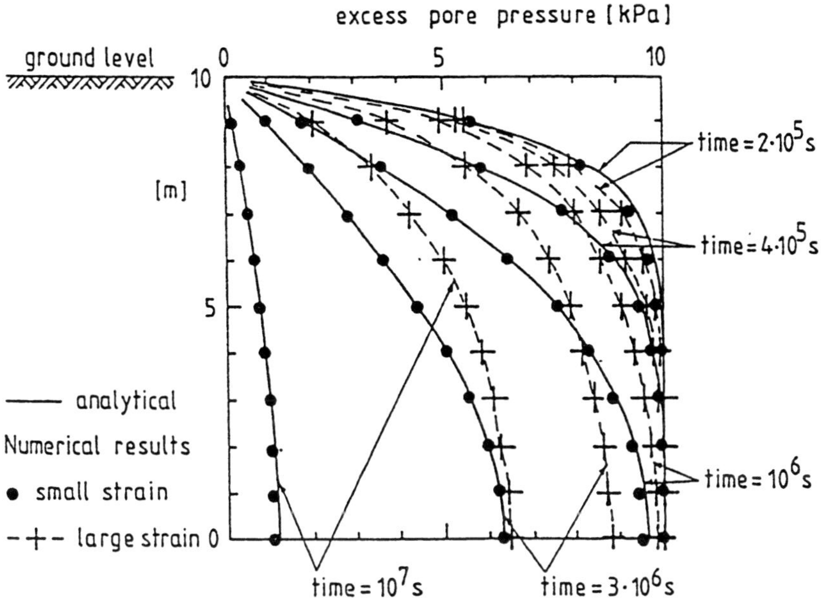

Figure 5.6. Comparison of excess pore-water pressure in the layer for various time periods.

Embankment on o Peaty Subsoil. The engineering problem is approached by taking full advantage of the methods and algorithms described in the previous sections. The tailing reservoir of the sugar factory has been risen on the multilayer subsoil. The subsoil is composed of the peat layer (3.1 m thick), mud layer (1.2 m thick) and sand foundation below, Fig. 5.7.

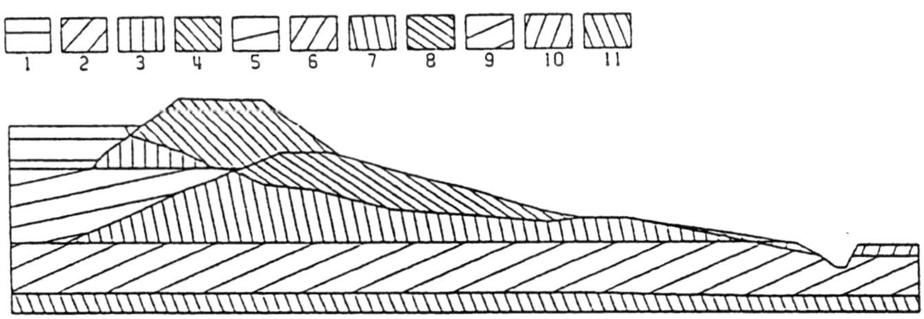

Figure 5.7. Materials of the embankment and subsoil. Phreatic line divides the embankment into saturated and not fully saturated zones what influences the weight of the material.

The sand foundation is not taken into account. The dam has been built in stages starting from June 1979. The reservoir has been filled up with the wet wastes, then raised and filled up again. The history of the building of the tailing dam and filling of the reservoir is as follows: 180 days of the dam building, 4320 days of filling, 90 days of the dam heightening, 1080 days of filling, 360 days of no activity. In numerical modelling the process is divided into stages which are drawn in Fig. 5.8. The stages with time periods and loading are given in Table 5.1. The elasto-plastic model is used based on Coulomb-Mohr yield condition with non-associated flow rule. The strain hardening is introduced for peat and mud. The hardening rule is defined for cohesion in the following way: $c = c_o + H\,\varepsilon^i_p$, where c_o is the initial cohesion, H is the hardening modulus, ε^i_p is the intensity of the plastic strain (based on the second invariant of the plastic strain tensor). The cohesion can take the value from the interval $c_o \le c \le c_r$, where c_r is the maximum cohesion in the hardening process. The material parameters are given in Table 5.2. In large strain the permeability for peat and mud changes in relation to the current void ratio, according to equation (5.43). Large compressibility of the peat has been found during the laboratory tests by Chaciński and Dłużewski (1995), Fig. 5.2.

Table 5.1. Stages of the embankment raising

State	Stage number	Time increment [days]	Load multiplier
Introduction of initial stresses into subsoil	1	-	1
Raising of first embankment	2	22.5	0.5
		22.5	0.5
	3	22.5	0.5
		22.5	0.5
	4	22.5	0.5
		22.5	0.5
	5	22.5	0.5
		22.5	0.5
Filling of reservoir	6	30	0.007
		330	0.076
		360	0.083
		720	0.168
		1440	0.333
		1350	0.312
		90	0.021
Raising of second embankment	7	90	1
Filling of reservoir	8	90	0.083
		90	0.083
		180	0.168
		360	0.333
		360	0.333
No activity	9	360	0.

The calculations are performed for two cases; small strains and large strains. For the large strain changes of the permeability are described according to (5.43). First the initial stresses are introduced into subsoil. Next the heightening of the dam and filling of the reservoir is modelled according to Table 5.1. The raising of the dam and filling of the reservoir is modelled by adding the elements. The boundary conditions are updated in each stage. Since no saturation problem is studied, the changes of the phreatic line is modelled by means of the weight of the soil and wastes. Such modelling is efficient and realistic provided the material of the dam and the fill is much more permeable compering with the subsoil. Otherwise the excess pore-water pressure can be generated in the non-saturated zones.

Table 5.2. Material parameters

Soil	No	E [kPa]	ν	γ [kN/m³]	c_o [kPa]	φ [°]	H [kPa]	c_r [kPa]	k_H [m/d]	k_V [m/d]	n	A
wastes 1	1, 2	1500.	0.35	15.	15.	5.	-	-	1000.	1000.	-	-
wastes 2	3, 4	3500.	0.35	19.	27.	15.	-	-	1000.	1000.	-	-
wastes 3	5, 6	2000.	0.35	18.	27.	9.	-	-	1000.	1000.	-	-
sand + wastes 4	7, 8	2500.	0.25	17.5	1.	35.	-	-	1000.	1000.	-	-
peat	9, 10	90.	0.30	11.	8.	12.	50.	20.	0.7	0.36	0.75	1.
mud	11	120.	0.35	10.5	10.	13.	50.	20.	0.086	0.086	0.75	1.

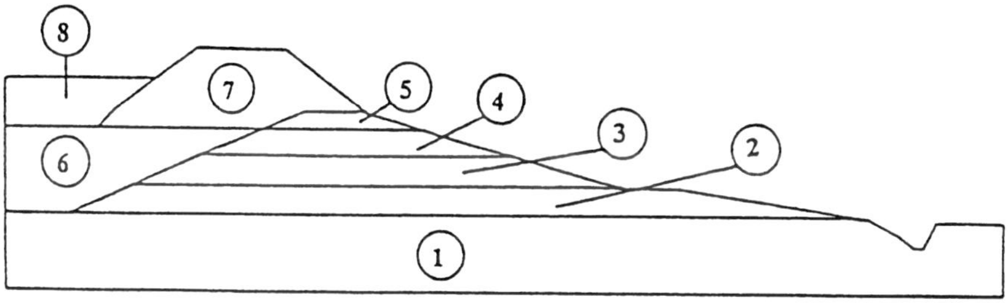

Figure 5.8. Stages of the embankment heightening and filling up of the reservoir.

During the process of the dam building and filling of the reservoir three types of loading are applied: weight, hydrostatic lift and flow pressure according to the current phreatic line which is assumed to be known at each stage, thus the finite element mesh, Fig. 5.9, should meet this requirement. The calculations are performed for small and large strains. The finial deformations for both cases are shown in Fig. 5.10. The peat and mud layers cause large settlement of the subsoil. The maximum displacement obtained for small strains reached 4.27 m and for the large strains 2.61 m respectively. The isolines of the vertical displacements in the subsoil are shown in Fig. 5.11. The isolines are drawn every 0.5 m. The difference in displacements for small and large strains approaches 100%. The range of the final plastic zones is shown in Fig. 5.12. Large plastic zone appeared in the subsoil and dam, two local small zones appeared close to the slope surface. The settlement of the point A (located at the ground level, Fig. 5.9) are depicted in Fig.5.13. The process has been traced from 1979. The effect of the decreasing permeability due to pore-water closing is shown in Fig. 5.14. The excess pore-water pressure is redistributed relatively fast. Decreasing of the permeability leads to the generation of a generate greater excess pore-water pressure during the heightening of the dam. The redistribution of the excess pore-water pressure is smaller with respect to the constant permeability case.

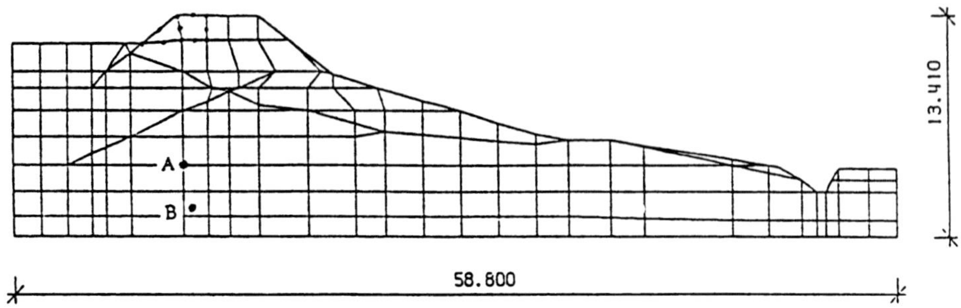

Figure 5.9. Finite element mesh.

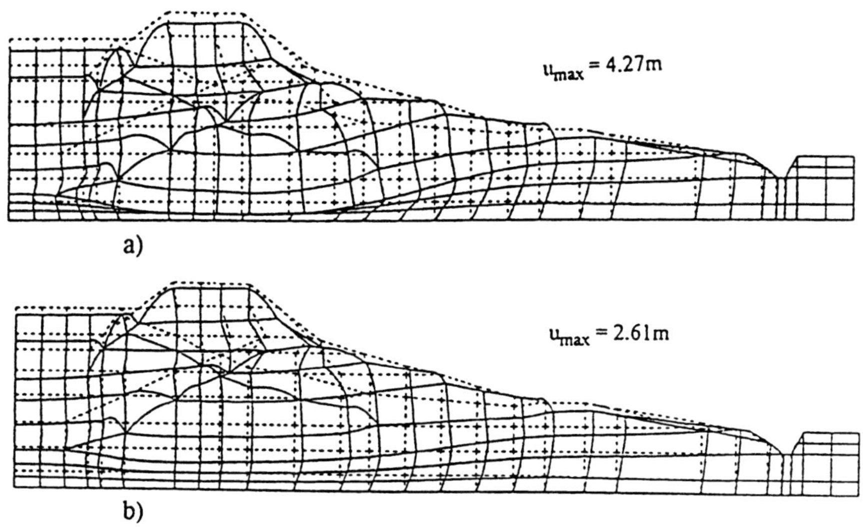

Figure 5.10. Deformed mesh at the end of the process; a) small strains, b) large strains.

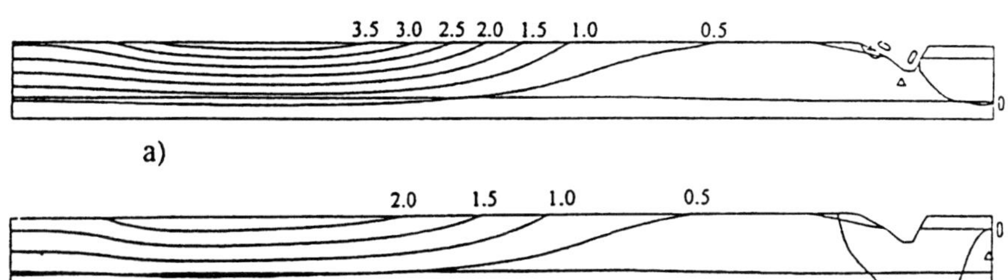

Figure 5.11. Isolines of the vertical displacements for the subsoil drawn every 0.5 m: a) small strains, b) large strains.

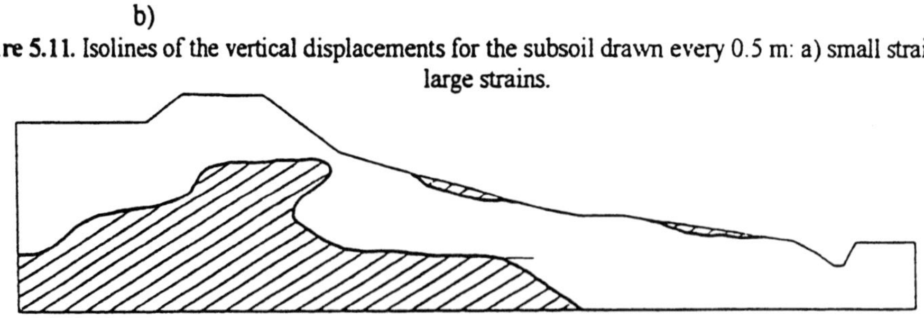

Figure 5.12. Plastic zones created during the process.

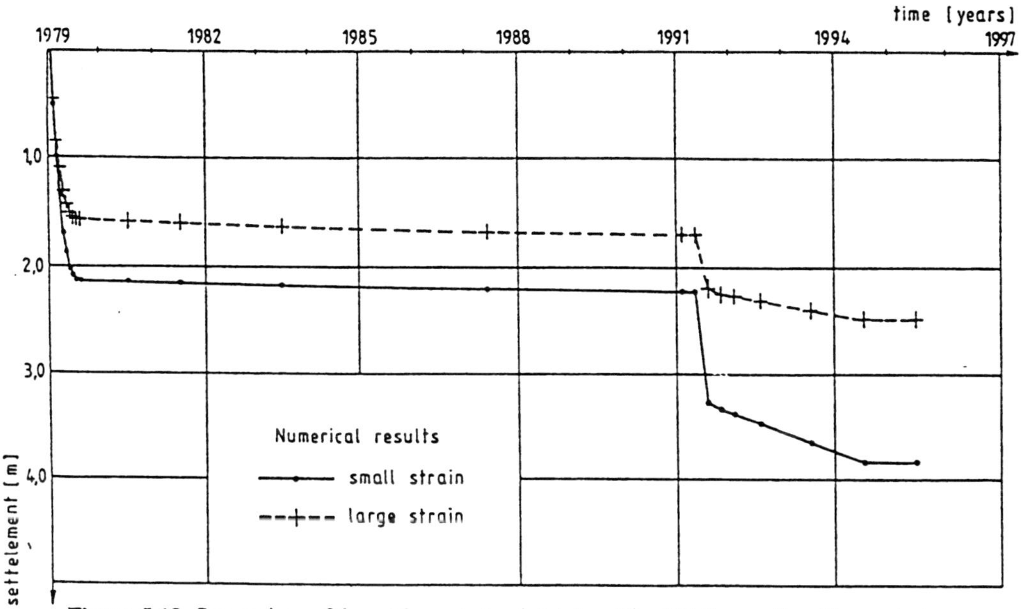

Figure 5.13. Comparison of the settlement at point A (see Fig. 5.9) for small and large strains.

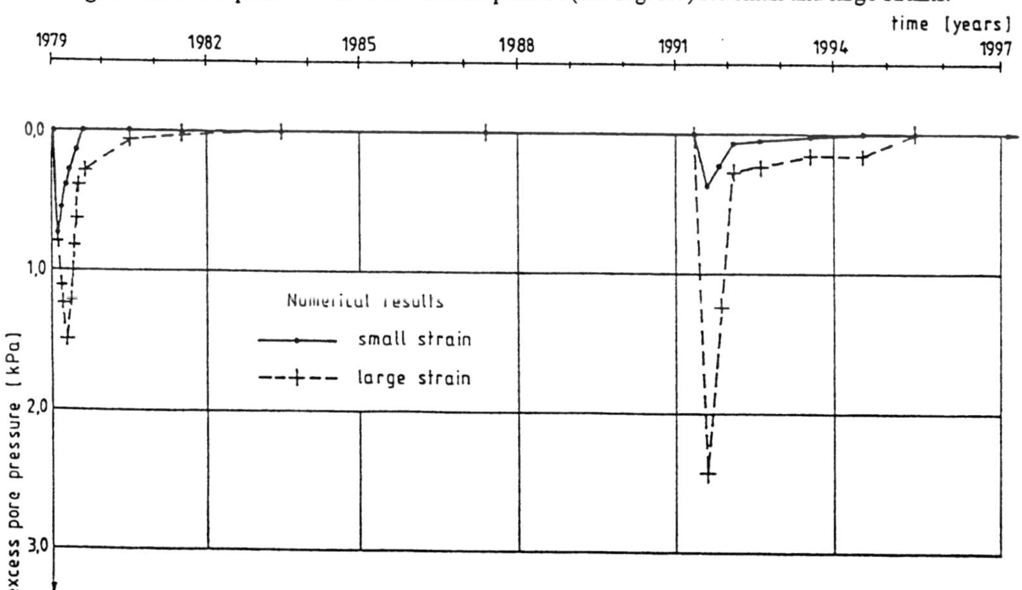

Figure 5.14. Comparison of the redistribution of the excess pore-water pressure at point B (see Fig. 5.9) for small and large strains.

The updated Lagrange procedure extended for the two-phase medium can successfully be applied to model consolidation process. The numerical results show the differences between small and large strain analyses. The algorithm for solving large strain consolidation process is found efficient. The pseudo-viscoplastic iteration method can efficiently be applied to solve physically and geometrically nonlinear problems for consolidation process. The iterative process is convergent for the shown examples.

Acknowledgments

The author acknowledges gratefully the support provided by Mr. Krzysztof Ciuhak in preparing the manuscript.

Appendix A

The arrays for six-noded interface element for plane problem are defined in the Appendix A.

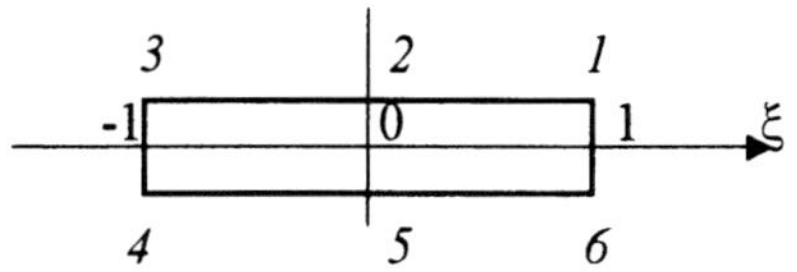

Figure A.1 Node numbers for six noded interface element

1) The pore pressure inside the element is calculated as follows

$$p = \mathbf{N}_p \mathbf{R} \, \mathbf{p}_e \, , \tag{A.1}$$

where

$$\mathbf{N}_p = [N_1, N_2, N_3] \tag{A.2}$$

Nodes 1 and 6, 2 and 5, 3 and 4 are the coupled pairs which occupy the same coordinate x. The same shape functions are used for the coupled nodes, i.e.

$$\begin{aligned} N_1 &= N_6 = 0.5\,\xi\,(1+\xi), \\ N_2 &= N_5 = 1-\xi^2 \\ N_3 &= N_4 = 0.5\,\xi\,(1-\xi), \end{aligned} \tag{A.3}$$

$$\mathbf{R} = \begin{bmatrix} \frac{1}{2} & 0 & 0 & 0 & 0 & \frac{1}{2} \\ 0 & \frac{1}{2} & 0 & 0 & \frac{1}{2} & 0 \\ 0 & 0 & \frac{1}{2} & \frac{1}{2} & 0 & 0 \end{bmatrix} \tag{A.4}$$

$$\mathbf{p}_e^T = [p_1, p_2, p_3, p_4, p_5, p_6]. \tag{A.5}$$

2) The jump in the pore pressure is defined as follows

$$\Delta p = \mathbf{N}_p \mathbf{S} \, \mathbf{p}_e \, , \tag{A.6}$$

where $\mathbf{N}_p$ is defined by (A.2), $\mathbf{p}_e$ by (A.5), **S** takes the following form

$$\mathbf{S}=\begin{bmatrix}1 & 0 & 0 & 0 & 0 & -1\\ 0 & 1 & 0 & 0 & -1 & 0\\ 0 & 0 & 1 & -1 & 0 & 0\end{bmatrix} \tag{A.7}$$

3) The jumps in displacements are calculated as follows

$$\Delta\mathbf{u}_{s-n}=\mathbf{c}\mathbf{N}_u\mathbf{T}\mathbf{u}_e, \tag{A.8}$$

where

$$\Delta\mathbf{u}_{s-n}=\begin{bmatrix}\Delta u_s\\ \Delta u_n\end{bmatrix} \tag{A.9}$$

where: Δu_s and Δu_n are the jumps in the displacements in the interface in the s- and n-directions. The rotation array **c** and the shape function array $\mathbf{N}_u$ are defined as follows

$$\mathbf{c}=\begin{bmatrix}\cos\alpha - & \sin\alpha\\ \sin\alpha & \cos\alpha\end{bmatrix}, \tag{A.10}$$

$$\mathbf{N}_u=\begin{bmatrix}N_1 & 0 & N_2 & 0 & N_3 & 0\\ 0 & N_1 & 0 & N_2 & 0 & N_3\end{bmatrix}. \tag{A.11}$$

T takes the following form

$$\mathbf{T}=\begin{bmatrix}1&0&0&0&0&0&0&0&0&0&-1&0\\ 0&1&0&0&0&0&0&0&0&0&0&-1\\ 0&0&1&0&0&0&0&0&-1&0&0&0\\ 0&0&0&1&0&0&0&0&0&-1&0&0\\ 0&0&0&0&1&0&-1&0&0&0&0&0\\ 0&0&0&0&0&1&0&-1&0&0&0&0\end{bmatrix} \tag{A.12}$$

The vector of the nodal displacements for the six-noded interface has the ordinary form

$$\mathbf{u}_e^T=\begin{bmatrix}u_x^1, & u_y^1, & u_x^2, & u_y^2, & u_x^3, & u_y^3, & u_x^4, & u_y^4, & u_x^5, & u_y^5, & u_x^6, & u_y^6\end{bmatrix} \tag{A.13}$$

4) Volumetric changes defined by the equations (4.31) are the products of the below defined arrays

$$\varepsilon_v^s=\frac{1}{D}\mathbf{l}^T\,\mathbf{c}\,\mathbf{B}_\partial\mathbf{u}_e \tag{A.14}$$

where

$$\mathbf{l}=\begin{bmatrix}1\\ 0\end{bmatrix} \tag{A.15}$$

$$\mathbf{B}_{\partial}=\frac{1}{2}\begin{bmatrix} \frac{\partial N_1}{\partial \xi} & 0 & \frac{\partial N_2}{\partial \xi} & 0 & \frac{\partial N_3}{\partial \xi} & 0 & \frac{\partial N_3}{\partial \xi} & 0 & \frac{\partial N_2}{\partial \xi} & 0 & \frac{\partial N_1}{\partial \xi} & 0 \\ 0 & \frac{\partial N_1}{\partial \xi} & 0 & \frac{\partial N_2}{\partial \xi} & 0 & \frac{\partial N_3}{\partial \xi} & 0 & \frac{\partial N_3}{\partial \xi} & 0 & \frac{\partial N_2}{\partial \xi} & 0 & \frac{\partial N_1}{\partial \xi} \end{bmatrix} \tag{A.16}$$

The nodal displacement vector $\mathbf{u}_e$ is defined by (A.13).

The ε_v^n is defined as follows

$$\varepsilon_v^n=\frac{1}{d}\mathbf{n}^T\mathbf{c}\mathbf{B}_u\mathbf{u}_e \tag{A.17}$$

where $\mathbf{B}_u$ is expressed by

$$\mathbf{B}_u=\mathbf{N}_u\mathbf{T} \tag{A.18}$$

$\mathbf{n}$ is defined by (4.3), while d is the thickness of the element.

5) The flow array defined in local coordinate system equals

$$\mathbf{H}=\int_{\Omega}\overset{*}{\mathbf{B}}_f^T\overset{*}{\mathbf{k}}\overset{*}{\mathbf{B}}_f D d\xi \tag{A.19}$$

is composed of the following arrays

$$\overset{*}{\mathbf{B}}_f=\begin{bmatrix} \frac{1}{2D}\frac{\partial N_1}{\partial \xi}; & \frac{1}{2D}\frac{\partial N_2}{\partial \xi}; & \frac{1}{2D}\frac{\partial N_3}{\partial \xi}; & \frac{1}{2D}\frac{\partial N_3}{\partial \xi}; & \frac{1}{2D}\frac{\partial N_2}{\partial \xi}; & \frac{1}{2D}\frac{\partial N_1}{\partial \xi}; \\ N_1 & N_2 & N_3 & -N_3 & -N_2 & -N_1 \end{bmatrix} \tag{A.20}$$

$$\overset{*}{\mathbf{k}}=\begin{bmatrix} \frac{k_s d}{\gamma_w} & 0 \\ 0 & \frac{k_n}{\gamma_w d} \end{bmatrix} \tag{A.21}$$

where d is the thickness of the element; k_s, k_n are the permeabilities; D is the Jacobian of the transformation from global to the local coordinates

$$D=\sqrt{\left(\frac{\partial x}{\partial \xi}\right)^2+\left(\frac{\partial y}{\partial \xi}\right)^2} \tag{A.22}$$

For a straight element the Jacobian reduces itself to the half of the element length

$$D=\frac{1}{2}l \tag{A.23}$$

Appendix B

The basic relations between the porosity, void ratio and volume changes in relation to the initial and current configurations are defined in the appendix. According to Fig. B.1 the following relations are valid:

$$V_0 = V_{P0} + V_{S0}, \quad V = V_P + V_S \tag{B.1}$$

$$V_P = V_{P0} + \Delta V_P, \quad V_S = V_{S0} + \Delta V_S, \quad \Delta V = \Delta V_P + \Delta V_S \tag{B.2}$$

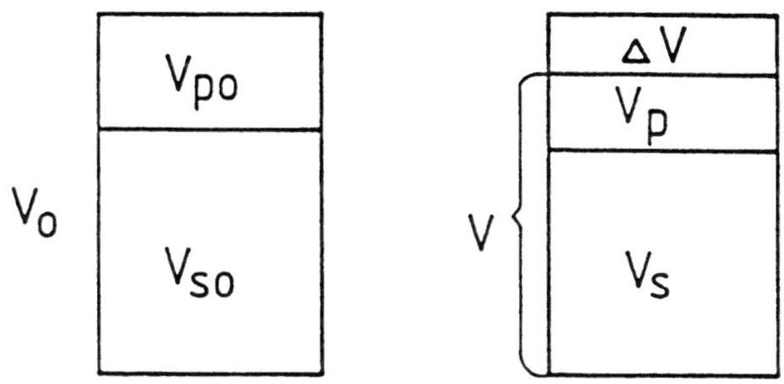

Figure B.1. Changes in volumes of pores and pore-water pressure.

Assuming the grain incompressibility

$$\Delta V_S = 0; \ \Delta V_P = \Delta V; \ V_S = V_{S0} \tag{B.3}$$

Void ratios are defined as follows

$$e_0 = \frac{V_{P0}}{V_{S0}}; \quad e = \frac{V_P}{V_S} \tag{B.4}$$

Introducing equations (B.1), (B.2) into (B.4) with the assumptions (B.3), we can define the volume changes by the initial and current void ratios

$$\varepsilon_{VE} = \frac{e - e_0}{1 + e}; \quad \varepsilon_{VL} = \frac{e - e_0}{1 + e_0} \tag{B.5}$$

The first above equation describes volumetric changes in the Eulerian description, the second one in the Lagrangian description. In the whole paper the compression is negative according to continuum mechanics notation.

Since the mass balance equation is being simultaneously solved with the equilibrium equation, two various definitions of volume changes are useful depending on the configuration and deformation field

$$\varepsilon_{VE} = \frac{\Delta V}{V}; \text{ or } \varepsilon_{VL} = \frac{\Delta V}{V_o} \tag{B.6}$$

Void ratio depends on the total volume changes ε_{VE} or ε_{VL} defined in the Eulerian or Lagrangian description respectively. The current void ratio can be defined in relation to volume changes (B.5).

$$e = \frac{e_0 + \varepsilon_{VE}}{1 - \varepsilon_{VE}} \text{ or } e = e_0 + \varepsilon_{VL}(1 + e_0) \tag{B.7}$$

The proper definition of the volumetric changes is very important for the consolidation process. The volumetric changes (B.6) depend on three basic strain tensor invariants, Eringen (1974).

Depending on the deformation description, the invariants are defined at the basis of the different strain tensors. In the Lagrangeian description the volume changes are defined in terms of three basic invariants of the Green strain tensor

$$\varepsilon_{VL} = \frac{\Delta V}{V_o} = -1 + \sqrt{1 + 2I_L + 4II_L + 8III_L} \tag{B.8}$$

where I_L, II_L, III_L are the basic invariants of the Green strain tensor. In the Eulerian description the volumetric changes by means of the Almansi strain tensor

$$\varepsilon_{VE} = \frac{\Delta V}{V} = 1 - \sqrt{1 - 2I_E + 4II_E - 8III_E} \tag{B.9}$$

where I_E, II_E, III_E are the three basic invariants of the Almansi strain tensor. The numerical tests showing the differences between the volumetric changes defined using (B.9) in relation to ε_v for small strain definition can be found in Dłużewski (1998).

Appendix C - List of Symbols

Vectors and matrices:

Chapter 4

σ - stresses

σ' - effective stresses

$\mathbf{t}$ - traction at the discontinuity lines

$\mathbf{t}'$ - effective traction at the discontinuity lines

ε - strains

$\mathbf{f}$ - body forces

$\mathbf{s}$ - surface traction

$\Delta\mathbf{u}$ - relative displacements

$\Delta\mathbf{u}^p$ - plastic relative displacements

$\mathbf{D}^e$ - elastic material array

$\mathbf{D}_T$ - tangent material array

$\mathbf{m},\mathbf{n},\mathbf{l}$	- zero-one vectors
$\mathbf{c}$	- rotation matrix
$\mathbf{N}_p$	- shape function vector for pore pressures at the interface
$\mathbf{R}$	- zero-1/2 array to pick up pore pressures
$\mathbf{p}_e$	- pore pressures vector at nodes of element
$\mathbf{N}_u$	- shape function array for displacements at interface
$\mathbf{T}$	- zero-one-minus-one array to pick up proper displacements
$\mathbf{S}$	- zero-one-minus-one array to pick up proper pore pressures
$\mathbf{K}_{in}$	- stiffness array for the interface element
$\mathbf{L}_{inI}$	- the first coupling array
$\mathbf{L}_{inII}$	- the second coupling array
$\mathbf{F}$	- nodal load vector
$\mathbf{u}$	- nodal displacements vector
$\mathbf{u}^e$	- nodal displacements vector for element
$\mathbf{p}$	- vector of excess pore pressure at nodes
$\nabla_{s\text{-}n}$	- differential operator
$\mathbf{v}_{s\text{-}n}$	- flow velocity
$\mathbf{k}_{s\text{-}n}$	- permeabilities array
$\mathbf{l}$	- zero-one vector
$\mathbf{B}_\partial$	- array of differentials of shape functions
$\mathbf{q}$	- flux of liquid
$\mathbf{B}_f$	- array for flow definition for interface element
$\mathbf{H}_{in}$	- flow array for interface
$\mathbf{S}_{in}$	- compressibility array of the liquid
$\mathbf{Q}$	- vector of nodal discharges

Chapter 5

$\mathbf{N}$	- shape function array for displacement approximation
$\mathbf{B}$	- geometrical array containing the linear parts of the strain tensor
$\overline{\mathbf{N}}$	- shape function array for pore-water pressure approximation
$\overline{\overline{\mathbf{N}}}$	- shape function array for coordination approximation
$\Delta\mathbf{R}$	- vector of load increment
$\Delta\mathbf{u}$	- vector of displacement increment
$\Delta\mathbf{p}$	- vector of pore-water pressure increment
$\mathbf{B}$	- linear geometrical array
$\mathbf{D}$	- constitutive array
$\mathbf{B}_N$	- array stemming from non-linear part of strain tensor
$\mathbf{S}$	- II Piola-Kirchhoff stresses tensor
$\mathbf{E}$	- Green strain tensor
σ	- Cauchy stress tensor
σ''	- Cauchy effective stresses
$\mathbf{N}$	- shape function array for pore-water pressure approximation
$\mathbf{m}$	- one-zero vector (zero for shearing stresses, 1 for the remaining ones)

H	- array responsible for flow properties
S	- array responsible for water compressibility
L	- coupling array
Q	- nodal discharges vector
∇	- differential operators vector
$\overline{\mathbf{N}}$	- shape functions array
k	- permeability array
m	- one-zero vector
q	- discharge at the boundary

Scalars:

Chapter 4

σ_n	- normal stress at the discontinuity
τ	- shearing stress at the discontinuity
p_t	- total pore pressure
p	- excess pore pressure
d_s	- shear modulus at the discontinuity
d_n	- bulk modulus at the discontinuity
ν	- Poisson's ratio
G	- shear modulus
K	- bulk modulus
l	- length of the interface element
α	- rotation angle
μ	- multiplier
c	- cohesion
ϕ	- friction angle
ψ	- dilatation angle
f(t')	- yield function
g(t')	- plastic potential
φ	- porosity
K_w	- bulk modulus of water
β	- water compressibility
ε_v	- volumetric changes
γ_w	- water density
Δt	- time step
θ	- time integration coefficient
k_x	- permeability in the x-direction
k_y	- permeability in the y-direction
Δh	- distance between two nodes
c_v	- consolidation coefficient
P	- pore-water pressure

References

Atkinson, J. H., and Salfors, G. (1991). Experimental determination of stress-strain-time characteristics in laboratory and in situ tests, General report, *Proceedings 10th ECSMFE*, Firence, Vol. 3, 915-956.

Asaoka, A., Noda, T., and Fernando, G.S.K. (1997). Effects of changes in geometry on the linear elastic consolidation deformation, *Soils and Foundations, Japanese Society of Soil Mechanics and Foundation Engineering*, Vol. 37, No. 1. 29-39.

Asaoka, A., Nakano, M., and Noda, T. (1997). Soil-water coupled behaviour of heavily overconsolidated clay near/at critical state, *Soils and Foundations, Japanese Society of Soil Mechanics and Foundation Engineering*, Vol. 37, No. 1., 13-28.

Asaoka, A., Noda, T., Fernando, G.S.K., and Yamada, E. (1998). Creep-like delayed failure of clayey ground after the end of embankment construction, *Proceedings of the 4th European Conference on Numerical Methods in Geotechnical Engineering*, Udine, Italy, 243-252.

Bakker, K. J., and Vermeer P. A. (1986). Finite element analysis of sheetpile walls, *Proceedings of the 2nd International Symposium on Numerical Models in Geomechanics*, Ghent, Belgium, 409-416.

Bakker, K. J., and Brinkgreve R. B. J. (1990). The use of hybrid beam elements to model sheetpile behaviour in two dimensional deformation analysis. *Proceedings 2nd European Specialty Conference on Numerical Methods in Geotechnical Engineering*, Santander, Spain, 559-571.

Bathe, K. J. (1982). *Finite element procedures in engineering analysis*. Prentice-Hall, Inc. New Jersey.

Biot, M. A. (1941). General theory of three-dimensional consolidation. *Journal of Applied Physics*, Vol. 12, 155-164.

Biot, M. A. (1956). General solution of the equations of elasticity and consolidation for porous materials. *Journal of Applied Mechanics*, Vol. 23, 91-96.

Booker, J. R., Small, J. S. (1975). An investigation of the stability of numerical solutions of Biot's equations of consolidation. *International Journal Solids and Structures*, Vol. 11, 907-917.

Burland, J. B. (1967). *Deformation of soft clay*. Ph.D. Thesis, University Cambridge.

Chaciński, Z., Dłużewski, J. M. (1995). Badanie charakterystyk przepływu w procesie konsolidacji gruntów organicznych. *II Seminarium Instytutu Zaopatrzenia w Wodę i Budownictwa Wodnego*, Warszawa. (Study of the permeability relation for soft organic soils, *Proceedings of 2nd Seminar Institute of Water Supply and Hydrotechnical Structures*, Warsaw, Poland), 69-74.

Cormeau, C. (1975). Numerical stability in quasi static elasto-visco-plasticity. *International Journal of the Numerical Methods Engineering*, Vol. 9, 109-127.

Cormeau, C. (1976). *Viscoplasticity and plasticity in the finite element methods*. Ph.D. Thesis, Swansea.

Crisfield, M. A. (1996). *Non-linear Finite Element Analysis of Solids and Structures*. Vol. 1, John Wiley & Sons, New York.

Crisfield, M. A. (1997). *Non-linear Finite Element Analysis of Solids and Structures*. Vol. 2, Advanced Topics, John Wiley & Sons, New York.

Day, R. A., and Poots, D. M. (1994). Zero thickness interface elements - numerical stability and application. *International Journal for Numerical and Analytical Methods in Geomechanics*, Vol 18, 689-708.

Dafalias, Y. F., and Herrmann, L. R. (1980). A bounding surface soil plasticity model. *Proceedings of International Symposium Soils under Cyclic Trans. Load.*, Swansea, 335-345.

Derski, W., Izbicki, R., Kisiel, I., and Mroz Z. (1988). *Rock and Soil Mechanics*. PWN-Elsevier, Warszawa-Amsterdam.

Desai, C. S., and Zhang, D. (1987). Viscoplastic model for geologic materials with generalized flow rule. *International Journal for Numerical and Analytical Methods In Geomechanics*, Vol.11, 38-45.

Desai, C. S., Zaman, M. M., Lightner, J. G., Siriwardane, H. J. (1984). Thin-layer element for interfaces and joints. *International Journal for Numerical and Analytical Methods in Geomechanics*, Vol. 8, 19-43.

Dienes, J. K. (1979). On the analysis of rotation and stress rate in deforming bodies. *Acta Mechanica*, Vol. 32, 217-232.

Dłużewski, J. M. (1988). Total formulation for large strains in soils. *International Journal Computers and Geotechnics*, Vol. 5, No. 3, 197-211.

Dłużewski, J. M., Termaat, R. J. (1990). Consolidation by finite element method in engineering problems. *Proceedings 2nd European Speciality Conference on Numerical Methods in Geotechnical Engineering*, Santander, Spain, 213-222.

Dłużewski, J. M. (1991). Soil structure interactions in consolidation problems. *Proceedings 7th Conference of International Associations for Computer Methods and Advances in Geomechanics*, Carinas, Australia, May, 1141-1146.

Dłużewski, J. M. (1993). *Numerical modelling of soil structure interactions in consolidation problems*. Warsaw University of Technology Publications, Civil Engineering, Vol. 123

Dłużewski, P., and Antunez, H. (1994). Finite element simulation of dislocation field movement. *Computer. Assist. Mechanics Engineering. Science*. 2, 141-145.

Dłużewski, J. M. (1997). *HYDRO-GEO Program elementów skończonych dla geotechniki, hydrotechniki i inżynierii środowiska*. Oficyna Wydawnicza Politechniki Warszawskiej, Warszawa. (*HYDRO-GEO – finite element program for hydrotechnics, geotechnics and environmental engineering*. Warsaw University of Technology Publications)

Dłużewski, J. M. (1997). Nonlinear consolidation in finite element modelling. *Proceedings of the IX International Conference on Computer Methods and Advances in Geomechanics*, Wuhan, China, 1089-1094.

Dłużewski, J. M. (1998). Large strain consolidation for elasto-plastic materials. *Proceedings of the 4th European Conference on Numerical Methods in Geotechnical Engineering*, Udine, Italy, 473-482.

Dłużewski, J.M., Popielski, P., Sternik, K., and Gryczmański, M. (1999). Consolidation of soft soils in terms of finite element modelling. *Proceedings of the Seventh International Conference on Numerical Models in Geomechanics (NUMOG VII)* Graz, Austria, September, 569-572.

Eringen, A. C. (1974). *Basic Principles in "Continuum Physics"*. Vol. 2, Academic Press, NY.

Fung, Y. C. (1965). *Foundations of solid mechanics*. Prentice Hall.

Ghaboussi, J., Wilson E. L., and Isenberg, J. (1973). Finite element for rock joints and interfaces. *Journal of the Soil Mechanics and Foundations Division*, ASCE, 99, SM10, October, 833-848.

Gibson, R. E., Schiffman, R. L., and Pu S. L. (1970). Plane strain and axially symmetric consolidation of a clay layer on a smooth impervious base. *Quart. Journal Mechanics and Applied Mathematics.*, Vol. XXIII, Pt. 4, 505 – 520.

Goodman, R. E., Taylor R. L., and Brakke T. L. (1968). A model for the mechanics of jointed rock. *Journal of the Mechanics Division*, ASCE, 94, SM3, 637-659.

Goodman, R. E., and Dubois J. (1972). Duplication of dilatancy in analysis of jointed rocks. *Journal of the Soil Mechanics and Foundations Division*, ASCE, 98, SM4, April, 399-422.

Gryczmański, M. (1994). Analytical and numerical subsoil models for soil-foundation interaction problems. Studia Geotechnika et Mechanica, Vol. XVI, No. 3-4, 29-72.

Gryczmański, M. (1995). Próba klasyfikacji modeli konstytutywnych gruntów. *Zeszyty Naukowe Politechniki Śląskiej, Budownictwo*, 81. (Trial of the constitutive soil models classification. *Scientific Note, Silesian University of Technology, Civil Engineering, 81*, 433-446

Gryczmański, M. (1995). *Wprowadzenie do opisu sprężysto-plastycznych modeli gruntów*. Studia z zakresu inżynierii, 40, KILiW, IPPT PAN, Warszawa. (*Introduction to elasto-plastic soil models*. Civil Engineering Study No. 40, Polish Academy of Science, Institute of Fundamental Technical Research, Warsaw).

Hashiguchi, K. (1986). A mathematical description of elastic-plastic deformation in normal-yield and sub-yield states. *Proceedings of the International Symposium on Numerical Models in Geomechanics „NUMOG 2"*, Ghent, 17-24.

Hibbit, H. D., Marcal, P. V., and Rice, J. R. (1970). A finite element formulation for problems of large strain and large displacement. *International Journal of Solids and Structures*, Vol. 6, 1069-186.

Hill, R. (1968). On constitutive inequalities for simple materials. I, II, *Journal Mechanics Physics*. Solids, Vol. 16, 229-242.

Hill, R. (1978). Aspects of invariance in solid mechanics. *Advances in Applied Mechanics*, Vol. 18, Academic Press, 1-75.

Hohberg, J. M. (1990). A note on spurious oscillations in FEM joint elements. *Earthq. Engineering Struct. Dynamic*, Vol.19

Huyakorn, P. S., and Pinder, G. F. (1983). *Computational Methods in Subsurface Flow*. Academic Press, NY.

Kanchi, M. B., Zienkiewicz, O. C., and Owen, P.R.J. (1978). The visco-plastic approach to problems of plasticity and creep involving geometric nonlinear effects. *International Journal for the Numerical Methods in Engineering* Vol.12, 169-181.

Kleiber, M. (1975). Lagrangian and Eulerian finite element formulation for large strain elasto-plasticity. *Bull. Acad. Polon. Sci. Ser. Techn.* 23, 117-126.

Koiter, W. T. (1960). General theorem for elastic-plastic solids. *Progress in Solid Mechanics*, Vol.1 (eds. I.N. Sneddon and R.Hill), North-Holland Publishing Co., Amsterdam. 165-221

Kuklik, P., Sejnoha, M., and Mares, J. (1998). Dimensional reduction applied to specific problems of consolidation. *Proceedings of the IV European Conference on Numerical Methods in Geotechnical Engineering – NUMGE98*, Udine, Italy, 337-346.

Kuklik, P., and Zalesky, J. (1995). Analysis of some experiments carried on the Cam-clay model. WORKSHOP 95, Praque, 357-358.

Kuklik, P. (1998). Veryfication of Cam-clay model in description of consolidation. SS CC98, Durban, South Africa, 281-285.

Lee, E. H. (1969). Elasto-plastic deformation at finite strain. *Journal of Applied Mechanics 36*, 1-6.

Lee, E. H., and McMeeking, R. M. (1980). Concerning elastic and plastic components of deformations. *International Journal Solids and Structures*, Vol. 18, 715-721.

Lewis, R. W., and Schrefler, B. A. (1998). *The finite element method in the static and dynamic deformation and consolidation of porous media.* John Wiley, Second edition, New York.

Malvern, L. E., *Introduction to the Mechanics of a Continuos Medium*, Prentice Hall, Englewood Cliffs NY

Mandel, J. (1972). *Plasticite classique et viscoplasticite.* Lecture Notes International Centre for Mechanical Sciences, Udine, Springer-Verlag, Berlin

McMeeking, R. M., and Rice, J. R. (1975). Finite-element formulations for problems of large elasto-plastic deformation. *International Journal Solids and Structures*, Vol. 11, 601-616.

Meijer, K. L. (1985). *Computation of stresses and strains in saturated soil.* Delft University of Technology, Ph.D. thesis, The Netherlands.

Meroi, E. A., Schrefler, B. A., and Zienkiewicz, O. C. (1995). Large strain static and dynamic semisaturated soil behaviour. *International Journal for Numerical and Analytical Methods in Geomechanics*, Vol. 19, 81-106.

Mikasa, M. (1965). The consolidation of soft clay - A new consolidation theory and its application. *Civil Engineering in Japan*, 21-26.

Monte, J. L., and Kritzen, R. J. (1976). One-dimensional mathematical model for large-strain consolidation. *Geotechnique* 26 (3), 495-510.

Mróz, Z. (1985) *Non-linear continuum mechanics of solids.* Lecture notes, Polish Academy of Science, Institute of Fundamental Technical Research, Warsaw.

Norris, V. A., and Lewis, R. W. (1979). A finite element procedure for consolidation with non-linear constitutive law. *International Journal for Numerical and Analytical Methods in Geomechanics* Vol. 1.

Pande, G. N., and Sharma, K. G. (1979). On joint/interface elements and associated problems of numerical ill-conditioning, *International Journal for Numerical and Analytical_Methods in Geomechanics*, 3, 293-300.

Perzyna, P. (1966). Fundamental problems in visco-plasticity. *Advances in Applied Mechanics*, Vol. 9, Academic Press, New York, 243-377.

Reed, M. B. (1993). Incorporation of strain-hardening in the implicit elasto-viscoplasticity algorithm. *Communication in Numerical Methods in Engineering*, Vol. 9, 331-336.

Rice, J. M. (1971). Inelastic constitutive relations for solids: an interval variable theory and its application to metal plasticity. *Journal of Mechanics and Physics of Solids*, 19, 433-455.

Russell, D. (1992). An element to model thin, highly permeable materials in two dimensional finite element consolidation analysis. *Proceedings of the 2nd European Speciality Conference on Numerical Methods in Geotechnical Engineering*, Santander, Spain, 303-310.

Sandhu, R. S. (1968). Fluid flow in saturated porous elastic media. Ph.D. thesis, University of California, Berkeley.

Sandhu, R. S., and Wilson, E. L. (1969). Finite element analysis of seepage in elastic media. *ASCE Journal* Vol. 95, EM5, 641-652.

Schofield, A. N., and Wroth,. C. P. (1968). *Critical State Soil Mechanics.* Mc Graw-Hill, London.

Song Er-Xiang. (1990). *Elasto-plastic consolidation under steady and cyclic loads*. PhD. thesis, Delft University of Technology, The Netherlands.

Stolle, D. F. E. (1991). An interpretation of initial stress and strain methods, and numerical stability. *International Journal for Numerical and Analytical, Methods in Geomechanics* Vol. 15, 399-416.

Swoboda, G., and Lai X.Y. (1994). Simulation of arch dam-foundation interaction with a new friction interface element. *International Journal for Numerical and Analytical Methods in Geomechanics*, Vol. 17, 601-617.

Szefer, G. (1977). Non-linear problems of consolidation theory. Symposium Franco-Polonais Problems non-lineares de mecanique, Kraków, Poland.

Szymański, A. (1991). *The factors determining the deformations analysis of organic subsoil under embankment. (in Polish), habilitation thesis*, Warsaw Agricultural University Press, Treaties and Monographs.

Tan, A. and Scott, R. F. (1988). Finite strain consolidation, A study of convection. *Soil and Foundation*, Vol. 28, No.3, Sept, 64-75.

Teodosiu, C. (1970). A dynamic theory of dislocations and its applications to the theory of the elastic plastic continuum, In *Fundamental Aspects of Dislocation Theory* ed. A. Simmonols et al., NBSSP, 317, II, 837-876.

Terzaghi, K. (1923). Die Berechnung der Durchlassigkeitsziffer des Tones aus dem Verlauf der hydrodynamischen Spannungserscheinungen. Akademie der Wissenschaften in Wien, Sitzungsberichte, *Mathematisch-naturwissenschaftiche Klasse, IIa*, 132. (3/4), 125-138

Terzaghi, K. (1943). *Theoretical Soil Mechanics*. New York, J. Wiley & Sons.

Tezduya, T. E., and Ganjoo, D. K. Petrov-Galerkin formulations with weighting functions dependent upon spatial and temporal discretizations: Applications to transient convection-diffusion problems", *Computers Methods. Applied Mechanics Engineering* 59, 49-71.

Truesdell, C., and Toupin, R. A. (1960). *The classical field theory*. Handbuch der Physik, III/1 Springer.

Truesdell, C. And Noll, W. (1965). *The non-linear field theories of mechanics*. Handbuch der Physik, III/3 Springer.

Van Langen, H., (1991). *Numerical analysis of soil-structure interaction*. Delft University of Technology, Ph.D. thesis.

Van Langen, H., Vermeer, P. A. (1991). Interface elements for singular plasticity points. *International Journal for Numerical and Analytical Methods in Geomechanics*, Vol. 15, 1-15.

Vermeer, P., and Verruijt, A. (1981). An accuracy condition for consolidation by finite elements. *International Journal for Numerical and Analytical Methods in Geomechanics*, Vol 5, 1-14.

Vermeer, P., Bonnier, P. G., Brand, P. J. W., Brinkgreve, R. B. J., Dłużewski, J. M., and Van Langen, H. (1990). *PLAXIS - Finite Element Code for Soil and Rock Plasticity*. Balkema.

Verruijt, A. (1977). Generation and dissipation of pore - water pressures. *Finite Elements in Geomechanics*, Chapter 9, ed. G. Gudehus, Wiley, London, 293-317.

Zienkiewicz, O.C., and Cormeau, I. C. (1974). Visco-plasticity–plasticity and creep in elastic solids - a unified numerical solution approach. *International Journal for Numerical. Methods in Engineering*, Vol. 8, 821-845.

Zienkiewicz, O.C. (1982). Basic formulation of static and dynamic behaviour of soil and other porous media. In *Numerical Methods in Geomechanics* ed. J. B. Martin, 39-57.

Zienkiewicz, O.C., and Shiomi, T. (1984). Static and dynamic behaviour of saturated porous media: the generalized Biot formulation and its numerical solution. *International Journal for the Numerical and Analytical Methods in Geomechanics*, Vol. 8, 71-96.
Zienkiewicz, O.C., Chan, A. H. C., Pastor, M., Paul, D. K., and Shiomi, T. (1990). Static and dynamic behaviour of soils: a rational approach to quantitative solutions. I – Fully saturated problems, *Proceedings of Royal Society* London, A, 429, 285-309.
Zienkiewicz, O.C., Xie, Y. M., Schrefler, B. A., Ledesma, A., and Bicanic, N. (1990). Static and dynamic behaviour of soils: a rational approach to quantitative solutions. II – Semi-saturated problems, *Proceedings of Royal Society* London, A, 429, 311-321.
Zienkiewicz, O. C., and Taylor R. (1991). *The finite element method.* Vols. 1 and 2. McGraw - Hill, New York.

Stability analysis of highly variable soils by elasto-plastic finite elements

D.V. Griffiths

Colorado School of Mines, Golden, Colorado, U.S.A.

Abstract. Geotechnical analyses involving bearing capacity, excavation and slope stability of highly variable soils have been performed using elasto-plastic finite elements. The finite element analyses are validated against well-known solutions where possible, before being applied to more complicated problems in which the classical approaches are at best inconvenient, and at worst may lead to misleading results. The types of heterogeneity considered in this work include simple layering due to stratification, water table effects, voids due to excavation and variability based on a statistical description of the material properties. In the statistical approach, the results lead to a probabilistic interpretation of design parameters that have traditionally been dealt with using "average" soil properies in conjunction with a "Factor of Safety".

1 Slope Stability Analysis by Finite Elements

The majority of slope stability analyses performed in practice still use traditional limit equilibrium approaches involving methods of slices that have remained essentially unchanged for decades. This was not the outcome envisaged when Whitman and Bailey (1967) set criteria for the then emerging methods to become readily accessible to all engineers.

The finite element method represents a powerful alternative approach for slope stability analysis which is accurate, versatile and requires fewer *a priori* assumptions, especially regarding the failure mechanism. Slope failure in the finite element model occurs "naturally" through the zones in which the shear strength of the soil is insufficient to resist the shear stresses. This section, which is based on a paper by Griffiths and Lane (1999), describes several examples of finite element slope stability analysis with comparison against other solution methods, including the influence of a layering and a free surface on slope and dam stability. Graphical output is included to illustrate deformations and mechanisms of failure.

It is argued that the finite element method of slope stability analysis is a more powerful alternative to traditional limit equilibrium methods and its widespread use should now be standard in geotechnical practice.

1.1 Introduction

Elasto-plastic analysis of geotechnical problems using the finite element (FE) method has been widely accepted in the research arena for many years, however its routine use in geotechnical practice for slope stability analysis still remains limited. The reason for this lack of acceptance is not entirely clear, however advocates of finite element techniques in academe must take some responsibility. Practicing engineers are often sceptical of the need for such complexity especially in view of the poor quality of soil property data often available from routine site investigations. Although this scepticism is often warranted, there are certain types of geotechnical problem for which the finite element approach offers real benefits. The challenge for an experienced engineer is to know which kind of problem would benefit from a finite element treatment and which would not.

In general, linear problems such as the prediction of settlements and deformations, the calculation of flow quantities due to steady seepage or the study of transient effects due to consolidation are all highly amenable to solution by finite elements. Traditional approaches involving charts, tables or graphical methods will often be adequate for routine problems but the finite element approach may be valuable if awkward geometries or material variations are encountered which are not covered by traditional chart solutions.

The use of nonlinear analysis in routine geotechnical practice is harder to justify because there is usually a significant increase in complexity which is more likely to require the help of a modelling specialist. Nonlinear analyses are inherently iterative in nature, because the material properties and/or the geometry of the problem are themselves a function of the "solution". Objections to nonlinear analysis on the grounds that they require excessive computational power however have been largely overtaken by developments in, and falling costs of, computer hardware. A desktop computer with a standard processor is now capable of performing nonlinear analyses such as those described in this report in a reasonable time span - minutes rather than hours or days.

Slope stability represents an area of geotechnical analysis in which a nonlinear finite element approach offers real benefits over existing methods. As this report will show, slope stability analysis by elasto-plastic finite elements is accurate, robust and simple enough for routine use by practicing engineers. Perception of the finite element method as complex and potentially misleading is unwarranted and ignores the real possibility that misleading results can be obtained with conventional "slip circle" approaches. The graphical capabilities of FE programs also allow better understanding of the mechanisms of failure, simplifying the output from reams of paper to managable graphs and plots of displacements.

1.2 Traditional Methods of Slope Stability Analysis

Most textbooks on soil mechanics or geotechnical engineering will include reference to several alternative methods of slope stability analysis. In a survey of equilibrium methods of slope stability analysis reported in Duncan's 1996 'State of the Art' review, the characteristics of a large number of methods were summarised including the Ordinary Method of Slices (Fellenius 1936), Bishop's Modified Method (1955), Force Equilibrium Methods (e.g. Lowe and Karafiath 1960), Janbu's Generalised Procedure of Slices (1968), Morgenstern and Price's Method (1965) and Spencer's Method (1967).

Although there seems to be some consensus that Spencer's Method is one of the most reliable, textbooks continue to describe the others in some detail and the wide selection of available methods is at best confusing to the potential user. For example, the controversy was recently revisited by Lambe and Silva (1995) who maintained that the Ordinary Method of Slices had an undeservedly bad reputation.

A difficulty with *all* the equilibrium methods is that they are based on the assumption that the failing soil mass can be divided into slices. This in turn necessitates further assumptions relating to side force directions between slices, with consequent implications for equilibrium. The side force assumption is one of the main characteristics distinguishing one limit equilibrium method from another, yet the concept of side forces is entirely artificial.

1.3 Finite Element Methods for Slope Stability Analysis

Duncan's review of finite element analysis of slopes concentrated mainly on deformation rather than stability analysis of slopes, however, attention was drawn to some important early papers in which elasto-plastic soil models were used to assess stability. Smith and Hobbs (1974) reported results of $\phi_u = 0$ slopes and obtained reasonable agreement with Taylor's (1937) charts. Zienkiewicz *et al* (1975) considered a c', ϕ' slope and obtained good agreement with slip circle solutions. Griffiths (1980) extended this work to show reliable slope stability results over a wide range of soil properties and geometries as compared with charts of Bishop and Morgenstern (1960). Subsequent use of the finite element method in slope stability analysis has added further confidence in the method, (e.g. Griffiths 1989, Potts *et al* 1990, Matsui and San 1992). Duncan mentions the potential for improved graphical results and reporting utilising FE but cautions against artificial accuracy being assumed when the input parameters themselves are so variable.

Wong (1984) gives a useful summary of potential sources of error in the finite element modelling of slope stability, although recent results, including those presented in this paper, indicate that better accuracy is now possible.

Advantages of the Finite Element Method. The advantages of a finite element approach to slope stability analysis over traditional limit equilibrium methods can be summarised as follows:

1. No assumption needs to be made in advance about the shape or location of the failure surface. Failure occurs "naturally" through the zones within the soil mass in which the soil shear strength is unable to resist the applied shear stresses.
2. Since there is no concept of slices in the finite element approach there is no need for assumptions about slice side forces. The finite element method preserves global equilibrium until "failure" is reached.
3. If realistic soil compressibility data is available, the finite element solutions will give information about deformations at working stress levels.
4. The finite element method is able to monitor progressive failure up to and including overall shear failure.

Brief Description of the Finite Element Model. The programs used in this paper are based closely on Program 6.2 in the text by Smith and Griffiths (1998)–the main difference being the ability to model more general geometries and soil property variations including variable water levels and pore pressures. Further graphical output capabilities have also been added. The programs are for 2-d plane strain analysis of elastic-perfectly plastic soils with a Mohr-Coulomb failure criterion utilising 8-node quadrilateral elements with reduced integration (4 Gauss-points per element) in the gravity loads generation, the stiffness matrix generation and the stress redistribution phases of the algorithm. The soil is initially assumed to be elastic and the model generates normal and shear stresses at all Gauss-points within the mesh. These stresses are then compared with the Mohr-Coulomb failure criterion. If the stresses at a particular Gauss-point lie within the Mohr-Coulomb failure envelope then that location is assumed to remain elastic. If the stresses lie on or outside the failure envelope, then that location is assumed to be yielding. Yielding stresses are redistributed throughout the mesh utilising the visco-plastic algorithm (Perzyna 1966, Zienkiewicz and Cormeau 1974). Overall shear failure occurs when a sufficient number of Gauss-points have yielded to allow a mechanism to develop.

The analyses presented in this paper do not attempt to model tension cracks. Although "no-tension" criteria can be incorporated into elasto-plastic finite element analyses (see e.g. Naylor and Pande 1981), this additional constraint on stress levels complicates the algorithm, and in addition, there is still some debate as to how "tension" should properly be defined. Further research in this area is warranted,

Soil Model. The soil model used in this study consists of six parameters as shown in Table 1.1

Table 1.1. Six parameter soil model

ϕ'	Friction angle
c'	Cohesion
ψ	Dilation angle
E'	Young's modulus
ν'	Poisson's ratio
γ	Unit weight

The dilation angle ψ affects the volume change of the soil during yielding. It is well known that the actual volume change exhibited by a soil during yielding is quite variable. For example a medium dense material during shearing might initially exhibit some volume decrease ($\psi < 0$) followed by a dilative phase ($\psi > 0$), leading eventually to yield under constant volume conditions ($\psi = 0$). Clearly this type of detailed volumetric modelling is beyond the scope of the elastic-perfectly plastic models used in this study where a constant dilation angle is implied.

The question then arises as to what value of ψ to use. If $\psi = \phi'$ then the plasticity flow rule is "associated" and direct comparisons with theorems from classical plasticity can be made. It is also the case that when the flow rule is associated, the stress and velocity characteristics coincide, thus closer agreement can be expected between failure mechanisms predicted by finite elements and critical failure surfaces generated by limit equilibrium methods.

In spite of these potential advantages of using an associated flow rule, it is also well known that associated flow rules with frictional soil models predict far greater dilation than is ever observed in reality. This in turn leads to increased failure load prediciton, especially in "confined" problems such as bearing capacity (Griffiths 1982). This shortcoming has led some of the more successful constitutive soil models to incorporate non-associated plasticity (e.g. Molenkamp 1981, Griffiths *et al* 1982, Hicks and Boughrarou 1998).

Slope stability analysis is relatively unconfined, thus the choice of dilation angle is less important. As the main objective of the current study is the accurate prediction of slope factors of safety, a compromise value of $\psi = 0$, corresponding to a non-associated flow rule with zero volume change during yield, has been used throughout this report. It will be shown that this value of ψ enables the model to give reliable factors of safety and a reasonable indication of the location and shape of the potential failure surfaces.

The parameters c' and ϕ' refer to the effective cohesion and friction angle of the soil. Although a number of failure criteria have been suggested for modelling the strength of soil (see e.g. Griffiths 1990) , Mohr-Coulomb's criterion remains the one most widely used in geotechnical practice and has been used throughout this paper. In terms of principal stresses and assuming a compression-negative sign convention, the criterion can be written as follows:

$$F = \frac{\sigma_1' + \sigma_3'}{2} \sin\phi' - \frac{\sigma_1' - \sigma_3'}{2} - c' \cos\phi' \tag{1}$$

where σ_1' and σ_3' are the major and minor principal effective stresses.

The failure function F can be interpreted as follows:

$F < 0$ stresses inside failure envelope (elastic)
$F = 0$ stresses on failure envelope (yielding)
$F > 0$ stresses outside failure envelope (yielding and must be redistributed)

The elastic parameters E' and ν' refer to Young's modulus and Poisson's ratio of the soil. If a value of Poisson's ratio is assumed (typical drained values lie in the range $0.2 < \nu' < 0.3$), the value of Young's modulus can be related to the compressibility of the soil as measured in a 1-d oedometer (see e.g. Lambe and Whitman 1969) ,

$$E' = \frac{(1+\nu')(1-2\nu')}{m_v(1-\nu')} \tag{2}$$

where m_v is the coefficient of volume compressibility.

Although the actual values given to the elastic parameters have a profound influence on the computed deformations prior to failure, they have little influence on the predicted factor of safety in slope stability analysis. Thus, in the absence of meaningful data for E' and ν', they can be given nominal values (e.g. $E' = 10^5$ kN/m^2 and $\nu' = 0.3$).

The total unit weight γ assigned to the soil is proportional to the nodal self-weight loads generated by gravity and will be discussed further in the next section.

In summary, the most important parameters in a finite element slope stability analysis are the same as they would be in a traditional limit equilibrium approach, namely the total unit weight γ, the shear strength parameters c' and ϕ' and the geometry of the problem.

Gravity Loading. The forces generated by the self weight of the soil are computed using a standard gravity "turn-on" procedure involving integrals over each element of the form:

$$\mathbf{p}^{(e)} = \gamma \int_{V^e} \mathbf{N}^T \, dV^e \tag{3}$$

where $\mathbf{N}$ are the shape functions of the element and the superscript e refers to the element number. This integral evaluates the volume of each element, multiplies by the total unit weight of the soil and distributes the net vertical force consistently to all the nodes. These element forces are assembled into a global gravity force vector that is applied to the finite element mesh in order to generate the initial stress state of the problem.

The present work applies gravity in a single increment to an initially stress-free slope. Others have shown that under elastic conditions, sequential loading in the form of incremental gravity application or embanking, affects deformations but not stresses (see e.g. Clough and Woodward 1967). In nonlinear analyses, it is recognised that the stress paths followed due to sequential excavation may be quite different to those followed under a gravity "turn-on" procedure, however the factor of safety appears unaffected when using simple elasto-plastic models (see.e.g Borja *et al* 1989, Smith and Griffiths 1998).

In comparing results with limit equilibrium solutions which generally take no account of loading sequence, experience has shown that the predicted factor of safety is insensitive to the form of gravity application when using elastic-perfectly plastic Mohr-Coulomb models. An example of this insensitivity is demonstrated later in the report.

The factor of safety may be sensitive to loading sequence when implementing more complex constitutive laws, such as those that attempt to reproduce volumetric changes accurately in an undrained or partially drained environment. For example, Hicks and Wong (1988) showed that the effective stress path could have a big influence on the factor of safety of an undrained slope.

Determination of the Factor of Safety The Factor of Safety (FOS) of a soil slope is defined here as the factor by which the original shear strength parameters must be divided in order to bring the slope to the point of failure.[1] The factored shear strength parameters c'_f and ϕ'_f, are therefore given by:

$$c'_f = c'/SRF \tag{4}$$

[1] This definition of the factor of safety is exactly the same as that used in traditional limit equilibrium methods, namely the ratio of restoring to driving moments.

$$\phi'_f = \arctan\left(\frac{\tan\phi'}{SRF}\right) \tag{5}$$

where SRF is a "Strength Reduction Factor". This method is referred to as the "shear strength reduction technique" (e.g. Matsui and San 1992) and allows for the interesting option of applying different strength reduction factors to the c' and $\tan\phi'$ terms. In this paper however, the same factor is always applied to both terms. To find the "true" factor of safety FOS, it is necessary to initiate a systematic search for the value of SRF that will just cause the slope to fail. When this value has been found, $FOS = SRF$.

Definition of Failure. There are several possible definitions of failure e.g. some test of bulging of the slope profile (Snitbhan and Chen 1976); limiting of the shear stresses on the potential failure surface (Duncan and Dunlop 1969) or nonconvergence of the solution (Zienkiewicz and Taylor 1989). These are discussed in Abramson *et al* (1995) from the original paper by Wong (1984) but without resolution. In the examples studied here the non-convergence option is taken as being a suitable indicator of failure.

When the algorithm cannot converge within a user-specified maximum number of iterations, the implication is that no stress distribution can be found that is simultaneously able to satisfy both the Mohr-Coulomb failure criterion and global equilibrium. If the algorithm is unable to satisfy these criteria "failure" is said to have occurred. Slope failure and numerical non-convergence occur simultaneously, and are accompanied by a dramatic increase in the nodal displacements within the mesh. Most of the results shown in this section used an iteration ceiling of 1000 and present results in the form of a graph of SRF vs. $E'\delta_{max}/\gamma H^2$ (a dimensionless displacement), where δ_{max} is the maximum nodal displacement at convergence and H is the height of the slope. This graph may be used alongside the displaced mesh and vector plots to indicate both the factor of safety and the nature of the failure mechanism.

1.4 Slope Stability Examples and Validation

Several examples of finite element slope stability analysis are now presented with validation against traditional stability analyses where possible. Initial consideration will be given to slopes containing no pore pressures in which total and effective stresses are equal. This is followed by examples of layered slopes. Finally, submerged and partially submerged slopes are considered in which pore pressures are taken into account.

Example 1: Homogeneous slope with no foundation layer (D–1). The homogeneous slope shown in Figure 1 has the following soil properties:

$$\phi' = 20^o$$
$$c'/\gamma H = 0.05$$

The slope is inclined at an angle of 26.57^o (2:1) to the horizontal and the boundary conditions are given as vertical rollers on the left boundary and full fixity at the base.

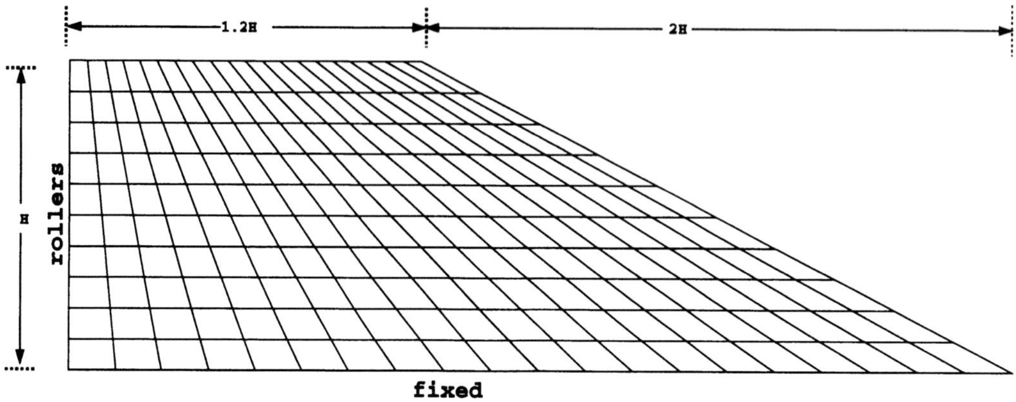

Figure 1. Example 1: Mesh for a homogeneous slope with a slope angle of 26.57° (2:1), $\phi' = 20°$, $c'/\gamma H = 0.05$

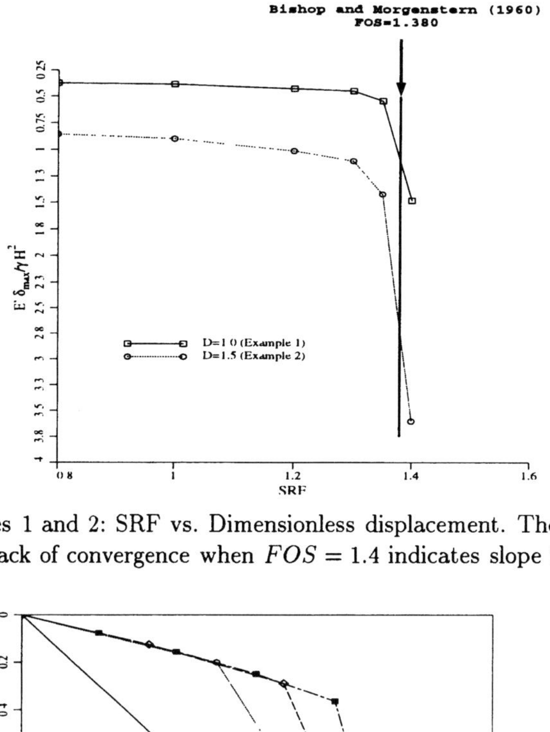

Figure 2. Examples 1 and 2: SRF vs. Dimensionless displacement. The rapid increase in displacement and the lack of convergence when $FOS = 1.4$ indicates slope failure.

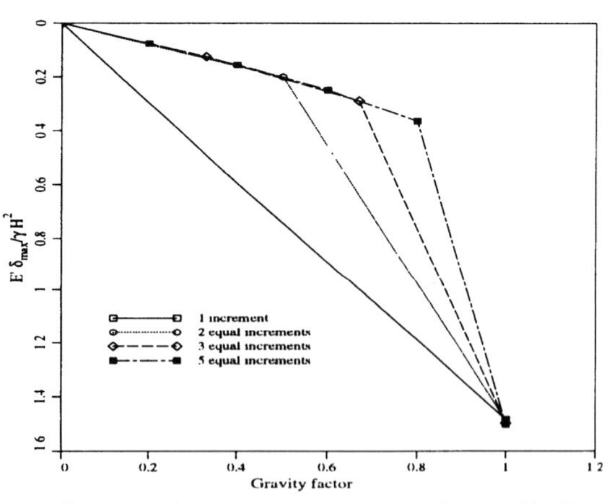

Figure 3. Influence of gravity increment size on maximum displacement at failure (FOS=1.4) from Example 1.

Gravity loads were applied to the mesh and the strength reduction factor (SRF) gradually increased until convergence could not be achieved within the iteration limit as shown in Table 1.2.

Table 1.2. Results from Example 1

SRF	$E'\delta_{max}/\gamma H^2$	Iterations
0.80	0.379	2
1.00	0.381	10
1.20	0.422	20
1.30	0.453	41
1.35	0.544	792
1.40	1.476	1000

The table indicates that six trial strength reduction factors were attempted ranging from 0.8 to 1.4. Each value represented a completely independent analysis in which the soil strength parameters were scaled by SRF as indicated in equations 4 and 5. Some efficiencies are possible in that the gravity loads and global stiffness matrix are the same in each analysis and are therefore generated once only.

The "Iterations" column indicates the number of iterations for convergence corresponding to each SRF value. The algorithm has to work harder to achieve convergence as the "true" FOS is approached. When $SRF = FOS = 1.4$, there is a sudden increase in the dimensionless displacement $E'\delta_{max}/\gamma H^2$, and the algorithm is unable to converge within the iteration limit. Figure 2 shows a plot of the data from Table 1.2 and indicates close agreement between the finite element result and the factor of safety given for the same problem by the charts of Bishop and Morgenstern (1960).

Figure 3 shows the influence of gravity loading increment size on displacements in Example 1. With a "failure" strength reduction factor of $SRF = 1.4$ applied to the soil properties, the four graphs correspond to the maximum displacement obtained when gravity was applied in a single increment as compared with that obtained with 2, 3 or 5 equal increments. The figure demonstrates that the displacement obtained with full gravity loading is barely affected by the increment size.

Figures 4a and 4b give the nodal displacement vectors and the deformed mesh corresponding to the unconverged situation with $SRF = FOS = 1.4$. The deformed mesh corresponding to this unconverged solution gives a rather diffuse indication of the failure mechanism. This is due to the relatively crude finite element mesh which must remains continuous even at "failure". Continuum finite element analysis is unable to model gross discontinuities along potential failure surfaces although techniques have been described for enhancing the visualisation of the failure surfaces (see e.g. Griffiths and Kidger 1995). More advanced finite element methods for modelling shear bands in conjunction with adaptive mesh refinement techniques have been described by Loret and Prevost (1991) and Zienkiewicz *et al* (1995).

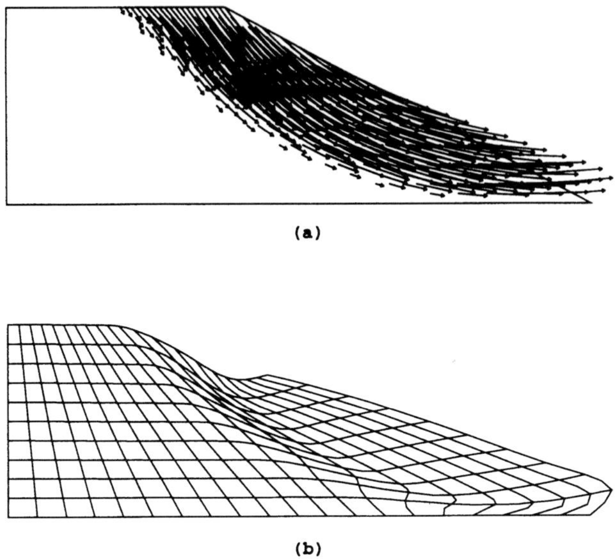

Figure 4. Example 1: Deformed mesh plots corresponding to the unconverged solution with FOS=1.4. (a) Nodal displacement vectors, (b) Deformed mesh.

Example 2: Homogeneous slope with a foundation layer (D=1.5). In this example, a foundation layer of thickness $H/2$ has been added to the base of the slope of Example 1 with all other properties and geometry remaining the same.

The initial mesh and the deformed mesh at failure are shown in Figures 5a and 5b respectively. It is clear from Figure 5b, that a mechanism of the "toe failure" type has been obtained. Figure 2 indicates that the critical factor of safety is essentially unchanged from Example 1 at $FOS = 1.4$ although the displacements are increased due to the greater volume of compressible soil.

This finite element result confirms that the addition of the foundation layer has *not* led to any perceptible change in the Factor of Safety of the slope. Bishop and Morgenstern (1960) give $FOS = 1.752$ as one possible solution for this example ($D = 1.5, c'/\gamma H = 0.05, \phi' = 20^o$, 2:1 slope) although it is important to check the alternative solution corresponding to D=1.0 to verify which gives the lower FOS. The charts of Cousins (1978) essentially agree with the finite element result and indicate that with a foundation layer, the critical circular mechanism at its lowest point passes fractionally below the base of the slope and gives a slightly lower factor of safety than when there is no foundation layer present.

Solving this example using a proprietary slip circle program also found the possible "result" of $FOS = 1.7$ when a failure circle tangent to the base of the foundation was assumed. It was necessary to force the slip circle to pass through the toe to obtain the "correct" $FOS = 1.37$.

This example demonstrates one of the main advantages of finite element slope stability analysis over conventional methods. The FE approach requires no a *priori* assumption of the location or shape of the critical surface. Failure occurs "naturally" within the zones of the soil mass where the shear strength of the soil is insufficient to resist the shear stresses.

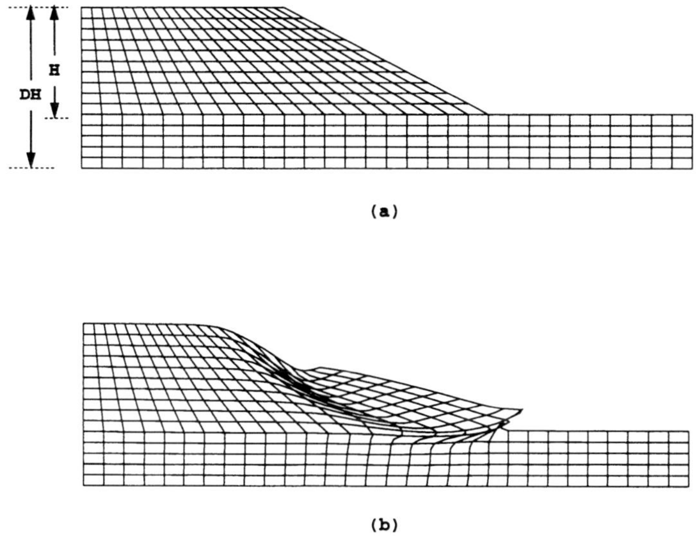

Figure 5. Example 2: Homogeneous slope with a foundation layer. Slope angle 26.57° (2:1), $\phi' = 20°$, $c'/\gamma H = 0.05$, $D = 1.5$. (a) Undeformed mesh, (b) Mesh corresponding to unconverged solution with $FOS = 1.4$.

The use of a limit equilibrium methods at the very least, requires some experience and care on the part of the user in order to initiate appropriate search procedures which avoid the possibility of homing in on the wrong "critical" circle.

Example 3: An undrained clay slope failure with a thin weak layer. Figure 6 shows a slope on a foundation layer ($D = 2$) of undrained clay. The slope includes a thin layer of weaker material which initially runs parallel to the slope, then horizontally in the foundation and finally outcrops at an angle of 45° beyond the toe.

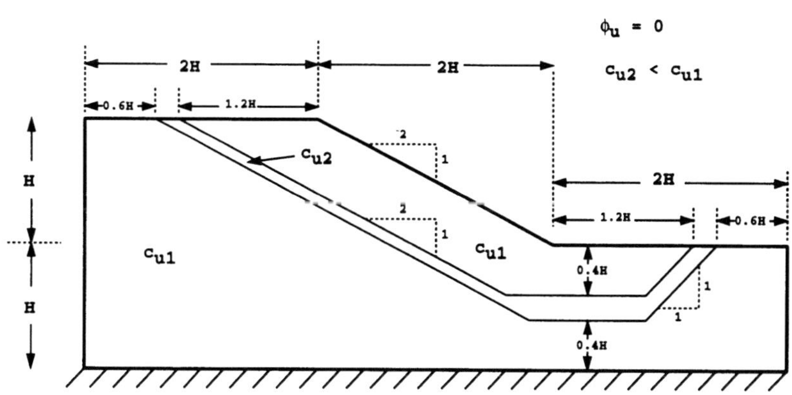

Figure 6. Example 3: Undrained clay slope with a foundation layer including a thin weak layer ($D = 2$, $c_{u1}/\gamma H = 0.25$)

Although this example may seem contrived, it is not unlike the situation of a thin, "slippery" liner within a landfill system. The factor of safety of the slope was estimated by finite element analysis for a range of values of the undrained shear strength of the thin layer (c_{u2}) while maintaining the strength of the surrounding soil at $c_{u1}/\gamma H = 0.25$.

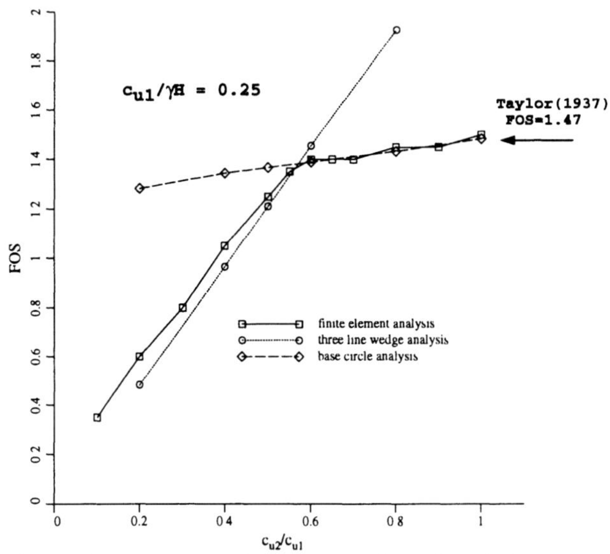

Figure 7. Example 3: Computed factor of safety (FOS) for different values of c_{u2}/c_{u1}.

The finite element results shown in Figure 7 give the computed factor of safety expressed to the nearest 0.05. For a homogeneous slope ($c_{u2}/c_{u1} = 1$), the computed factor of safety was close to the Taylor solution (Taylor 1937) of $FOS = 1.47$ and gave the expected circular base failure mechanism. As the strength of the thin layer was gradually reduced, a distinct change in the nature of the results was observed when $c_{u2}/c_{u1} \approx 0.6$.

Also shown on this figure are limit equilibrium solutions obtained using Janbu's method assuming both circular (base failure) and three line wedge mechanisms following the path of the weak layer. The discontinuity when $c_{u2}/c_{u1} \approx 0.6$ clearly represents the transition between the circular mechanism and the non-circular mechanism governed by the weak layer. For $c_{u2}/c_{u1} > 0.6$ the (circular) base failure mechanism governs the behaviour and the factor of safety is essentially unaffected by the strength of the weaker thin layer. For $c_{u2}/c_{u1} < 0.6$ the (non-circular) thin layer mechanism takes over and the factor of safety falls linearly.

This behaviour is explained more clearly in Figure 8 which shows the deformed mesh at failure for three different values of the ratio c_{u2}/c_{u1}. Figure 8a, corresponding to the homogeneous case ($c_{u2}/c_{u1} = 1$), indicates an essentially circular failure mechanism tangent to the firm base as predicted by Taylor. Figure 8c, in which the strength of the thin layer is only 20% of the surrounding soil ($c_{u2}/c_{u1} = 0.2$), indicates a highly concentrated non-circular mechanism closely following the path of the thin weak layer. Figure 8b, in which the strength of the thin layer is 60% of the surrounding soil ($c_{u2}/c_{u1} =$

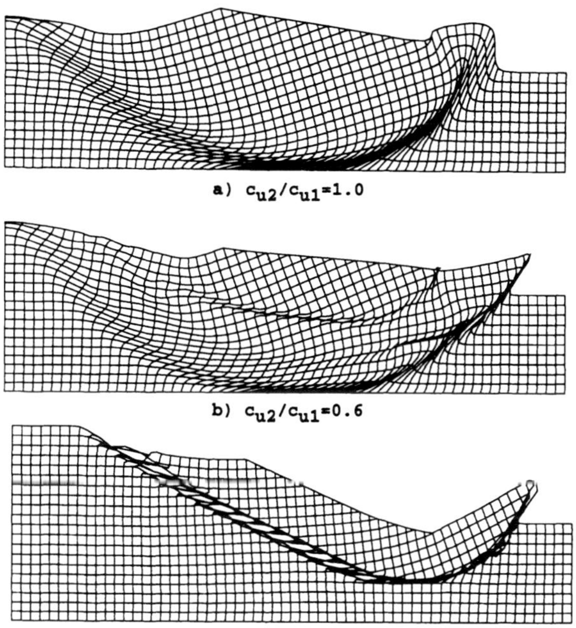

Figure 8. Example 3: Deformed meshes at failure corresponding to the unconverged solution for three different values of c_{u2}/c_{u1}.

0.6) indicates considerable complexity and ambiguity. At least two conflicting mechanisms are apparent. Firstly there is a base failure mechanism merging with the weak layer beyond the toe of the slope and secondly there is a mechanism running along the weak layer parallel to the slope and outcropping at the toe.

Without prior knowledge of the two alternative mechanisms, a traditional limit equilibrium search could overestimate the factor of safety. This is illustrated in Figure 7 where, for example, a circular mechanism with $c_{u2}/c_{u1} = 0.2$ would indicate $FOS = 1.3$ when the correct factor of safety is closer to 0.6.

Example 4: An undrained clay slope with a weak foundation layer. In this case, the same slope geometry and finite element mesh as in the previous example has been used but with a different type of inhomogeneity as shown in Figure 9. The shear strength of the slope material has been maintained at a constant value of $c_{u1}/\gamma H = 0.25$ while the shear strength of the foundation layer has been varied. The relative size of the two shear strengths has again been expressed as the ratio c_{u2}/c_{u1}. Figure 10 shows the computed factor of safety for a range of c_{u2}/c_{u1} values together with classical solutions of Taylor for the two cases when $c_{u2} = c_{u1}$ and $c_{u2} >> c_{u1}$.

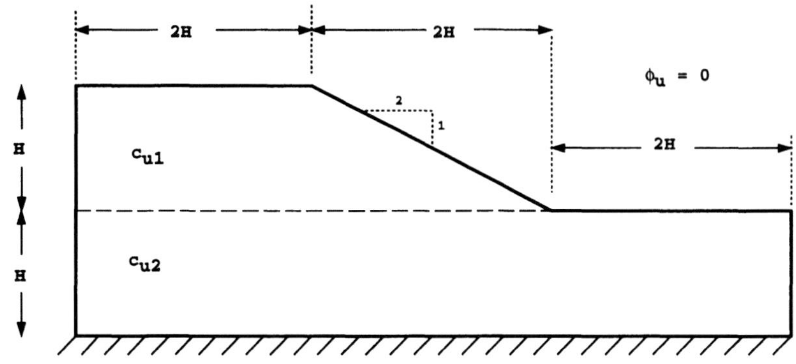

Figure 9. Example 4: Undrained clay slope with a weak foundation layer ($D = 2$, $c_{u1}/\gamma H = 0.25$)

There is clearly a change of behavior occurring at $c_{u2}/c_{u1} \approx 1.5$ as indicated by the flattening out of the curve. Also shown on this figure are limit equilibrium solutions for both toe and base circle mechanisms. The discontinuity corresponding to $c_{u2}/c_{u1} \approx 1.5$ obviously represents the transition between these two fundamental mechanisms.

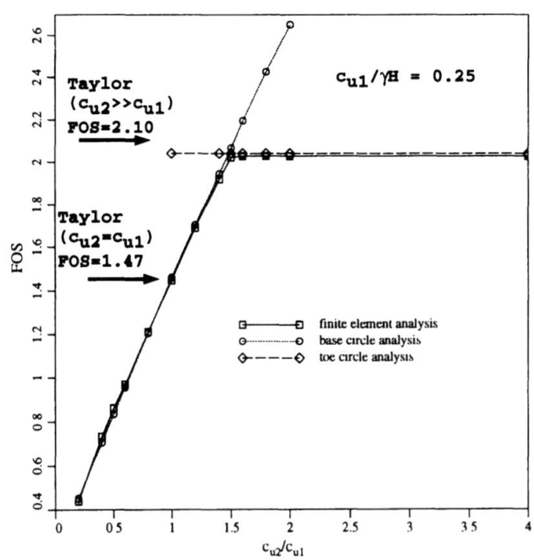

Figure 10. Example 4: Computed factor of safety (FOS) for different values of c_{u2}/c_{u1}.

This transition is clearly demonstrated by the finite element failure mechanisms shown in Figure 11. When $c_{u2} \ll c_{u1}$ (Figure 11a) a deep-seated base mechanism is observed, whereas a shallow "toe" mechanism is seen when $c_{u2} \gg c_{u1}$ (Figure 11c). The result corresponding to the approximate transition point at $c_{u2} \approx 1.5c_{u1}$ (Figure 11b) shows an ambiguous situation in which both mechanisms are trying to form at the same time. It is interesting to note that the lower soil must be at least 50% stronger than the upper soil before the toe mechanism becomes the most critical. The slip circle program STABR (Duncan and Wong, 1985) when applied to this example with $c_{u2}/c_{u1} = 1.46$ gave identical Factors of Safety of 2.02 for circles tangent to the base of the slope and the base of the foundation.

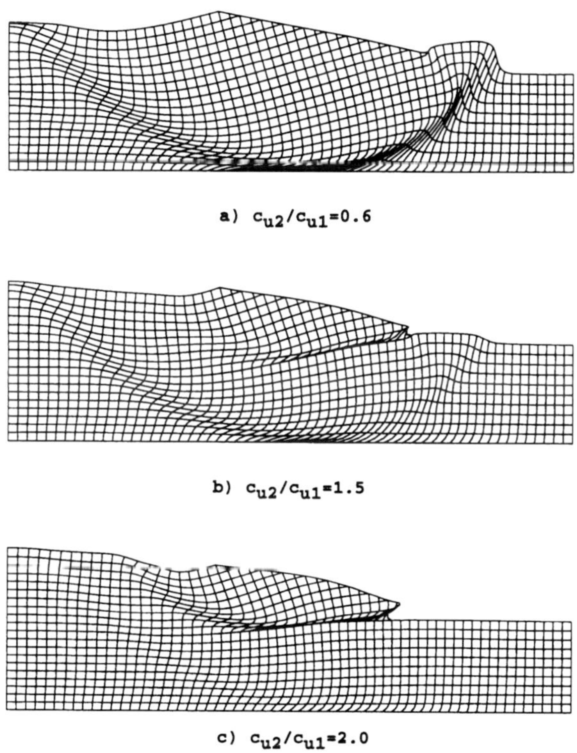

Figure 11. Example 4: Deformed meshes at failure corresponding to the unconverged solution for three different values of c_{u2}/c_{u1}

The previous two examples have shown that in even quite simple cases of soil heterogeneity in the form of layering, complex interactions can occur between conflicting mechanisms which are detected by the finite element approach. For more complicated stability problems involving several soil property groups, such as a zoned earth embankment, numerical approaches are arguably the *only* robust way of generating the minimum factor of safety and indicating the location and shape of the critical mechanism.

1.5 Influence of Free Surface and Reservoir Loading on Slope Stability

We now consider the influence of a free-surface within an earth slope and reservoir loading on the outside of a slope as shown in Figure 12.

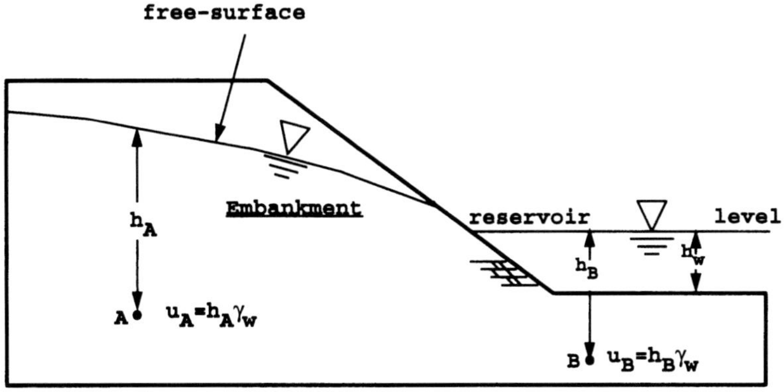

Figure 12. Slope with free-surface and reservoir loading.

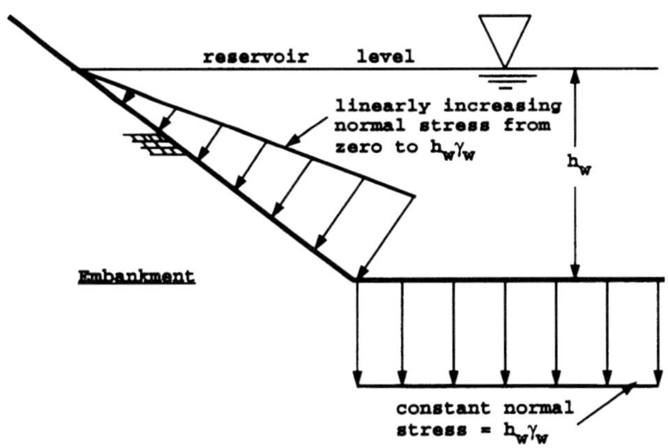

Figure 13. Detail of submerged area of slope beneath free-standing reservoir water showing stresses to be applied to the surface of the mesh as equivalent nodal loads.

Regarding the role of the free-surface, a rigorous approach would firstly involve obtaining a good quality flow net for free-surface flow through the slope, enabling pore pressures to be accurately estimated at any point within the flow region. For the purposes of slope stability analysis however, it is usually considered sufficiently accurate and conservative to estimate pore pressure at a point as the product of the unit weight of water (γ_w) and the *vertical* distance of the point beneath the free-surface. In Figure 12 the pore pressures at two locations, A and B, have been calculated using this assumption.

In the context of finite element analysis, the pore pressures are computed at all submerged (Gauss) points as described above, and subtracted from the *total* normal stresses computed at the same locations following the application of surface and gravity loads. The resulting *effective* stresses are then used in the remaining parts of the algorithm relating to the assessment of Mohr-Coulomb yield and elasto-plastic stress redistribution. Note that the gravity loads are computed using *total* unit weights of the soil.

The external loading due to the reservoir is modelled by applying a normal stress to the face of the slope equal to the water pressure. Thus, as shown in Figure 13, the applied stress increases linearly with water depth and remains constant along the horizontal foundation level. These stresses are converted into equivalent nodal loads on the finite element mesh (see e.g. Appendix I in Smith and Griffiths 1998) and added to the initial gravity loading.

Example 5: Homogeneous slope with horizontal free-surface. Figure 14 shows a similar slope to that analysed in Example 1, but with a horizontal free surface at a depth L below the crest. Using the method described above, the factor of safety of the slope has been computed for several different values of the drawdown ratio (L/H) which has been varied from -0.2 (slope completely submerged with the water level 0.2H above the crest), to 1.0 (water level at the base of the slope). The problem could be interpreted as a "slow" drawdown problem in which a reservoir, initially above the crest of the slope (!), is gradually lowered to the base, with the water level within the slope maintaining the same level. A constant total unit weight of 20kN/m^3 has been assigned to the entire slope, both above and below the water level.

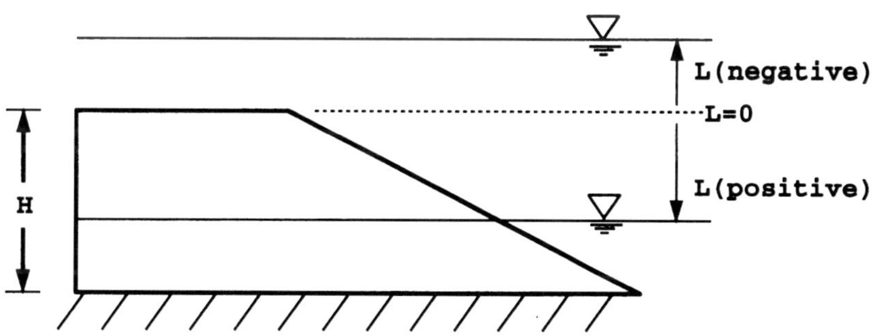

Figure 14. Example 5: "Slow" drawdown problem. Homogeneous slope with a horizontal free surface. Slope angle 26.57° (2:1), $\phi' = 20°$, $c'/\gamma H = 0.05$ (above and below free surface).

The interesting result shown in Figure 15 indicates that the factor of safety reaches a *minimum* of $FOS \approx 1.3$ when $L/H \approx 0.7$. A limit equilibrium solution shown on the same figure indicates a similar trend (see e.g. Cousins 1978). The special cases corresponding to $L/H = 0$ and $L/H = 1$ agree well with chart solutions given, respectively, by Morgenstern (1963) (F=1.85), and Bishop and Morgenstern (1960) ($FOS = 1.4$). The fully submerged

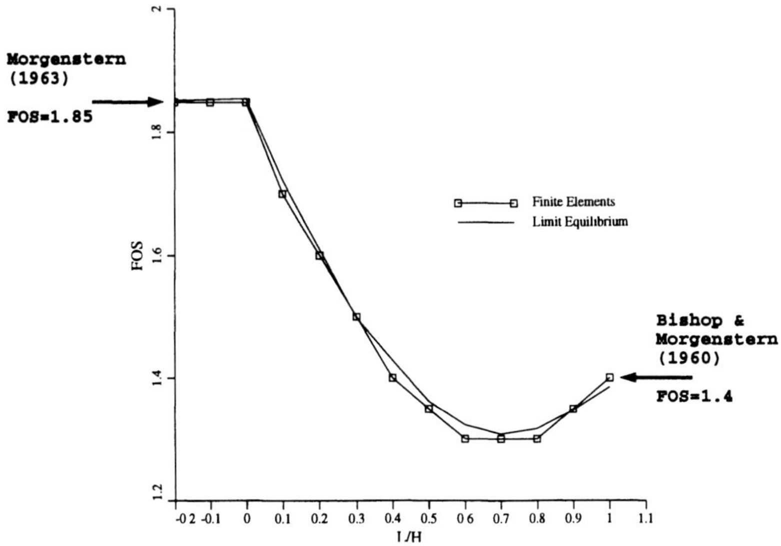

Figure 15. Factor of safety in a "slow" drawdown problem for different values of the drawdown ratio L/H.

slope ($L/H \leq 0$) is more stable than the "dry" slope ($L/H \geq 1$) as indicated by a higher factor of safety.

An explanation of the observed minimum is due to the cohesive strength of the slope (which is unaffected by buoyancy) and the trade-off between soil weight and soil shear strength as the drawdown level is varied. In the initial stages of drawdown ($L/H < 0.7$), the increased weight of the slope has a proportionately greater destabilising effect than the increased frictional strength and the factor of safety falls. At higher drawdown levels ($L/H > 0.7$) however, the increased frictional strength starts to have a greater influènce than the increased weight and the factor of safety rises. Other results of this type have been reported by Lane and Griffiths (2000) for a slope which was stable ($FOS > 1$) when "dry" or fully submerged, but became unstable ($FOS < 1$) at a critical value of the drawdown ratio L/H. It should also be pointed out from the horizontal part of the graph in Figure 15, corresponding to $L/H \leq 0$, that the factor of safety for a fully submerged slope is unaffected by the depth of water above the crest.

Excellent agreement with Morgernstern (1963) for "rapid drawdown" problems has also been demonstrated for a range of slopes using this approach by Lane and Griffiths 2000.

Example 6: Two-sided earth embankment. The example given in Figure 16 is of an actual earth dam cross section including a free surface which slopes from the reservoir level to foundation level on the downstream side (Torres and Coffman 1997). For the purposes of this example, the material properties have been made homogeneous. Figure 17 shows the finite element model used for the slope stability analysis (Paice 1997). The boundary conditions consist of vertical rollers on the faces at the left and right ends of the foundation layer with full fixity at the base. It should be noted that the downstream slope of the embankment is slightly steeper than the upstream slope.

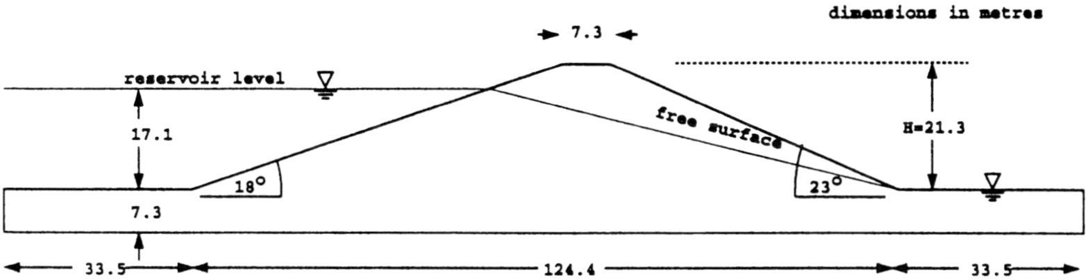

Figure 16. Example 6: Two-sided earth embankment with a sloping free surface. $\phi' = 37^o$, $c = 13.8\ \mathrm{kN/m^2}$, $\gamma = 18.2\ \mathrm{kN/m^3}$ (above and below WT).

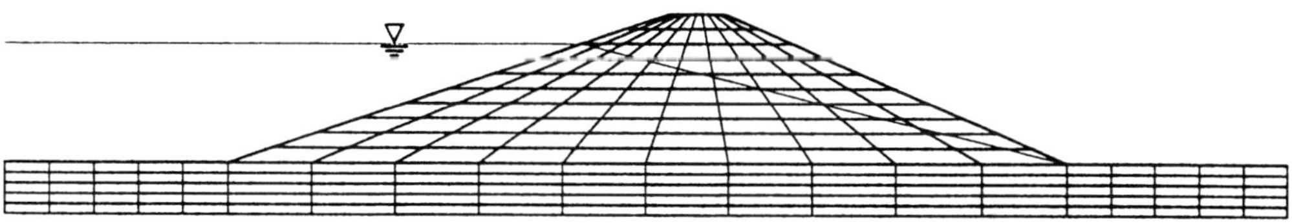

Figure 17. Example 6: Finite element mesh.

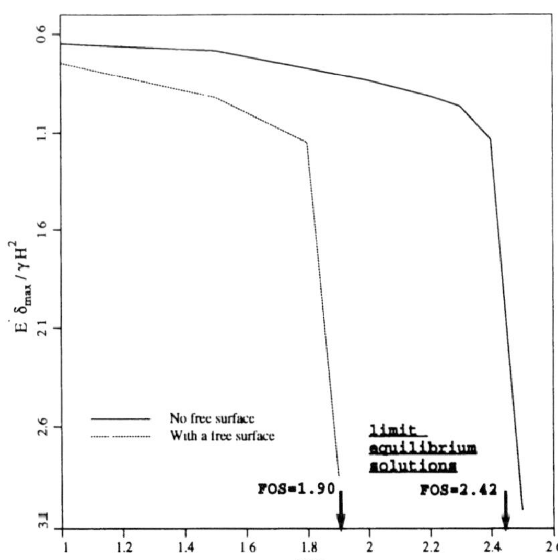

Figure 18. Example 6: SRF vs. Dimensionless displacement.

A second analysis was also performed with no free-surface corresponding to the embankment before the reservoir was filled. Finite element slope stability analysis led to the results shown in Figure 18. Both cases were also solved using a conventional limit equilibrium approach which gave $FOS = 1.90$ with a free surface and $FOS = 2.42$ without a free surface. The limit equilibrium and finite element factors of safety values were in close agreement in both cases.

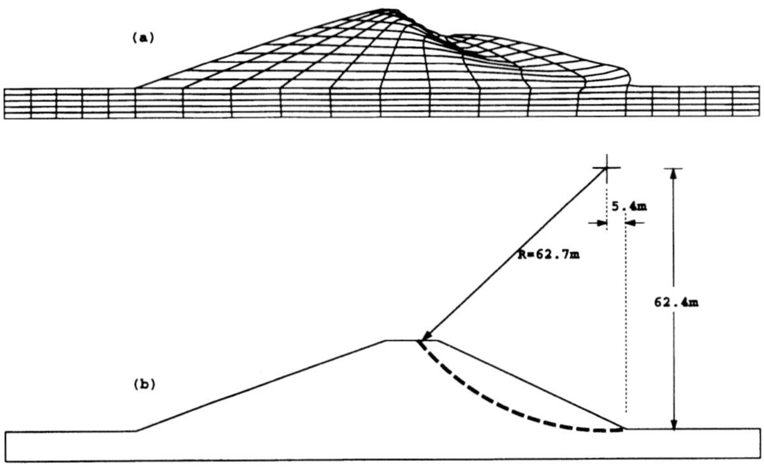

Figure 19. Example 6 with no free surface: (a) Deformed mesh corresponding to the unconverged solution by finite elements, (b) the critical slip circle by limit equilibrium. Both methods give FOS=2.4.

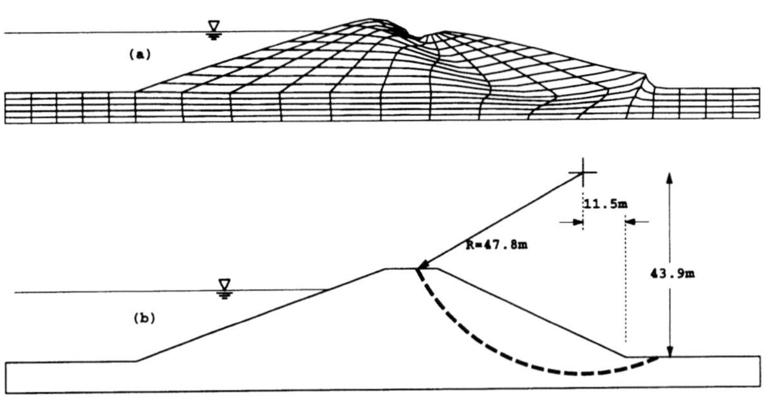

Figure 20. Example 6 with a free surface: (a) Deformed mesh corresponding to the unconverged solution by finite elements, (b) the critical slip circle by limit equilibrium. Both methods give FOS=1.9.

Regarding the critical mechanisms of failure, Figures 19 and 20 show the deformed mesh corresponding to the unconverged finite element solution as compared with the slip circle that gave the lowest factors of safety from the limit equilibrium approach. As expected, the lowest factor of safety occurs on the steeper, downstream side of the embankment in both cases. It should also be noted that both the finite element and limit equilibrium results indicate a toe failure for the case with no free-surface (Figure 19), and a deeper mechanism extending into the foundation layer for the case with a free surface (Figure 20). Figure 21 shows the corresponding displacement vectors from the finite element solutions. Reasonably good agreement between the locations of the failure mechanisms indicated by both types of analysis is indicated.

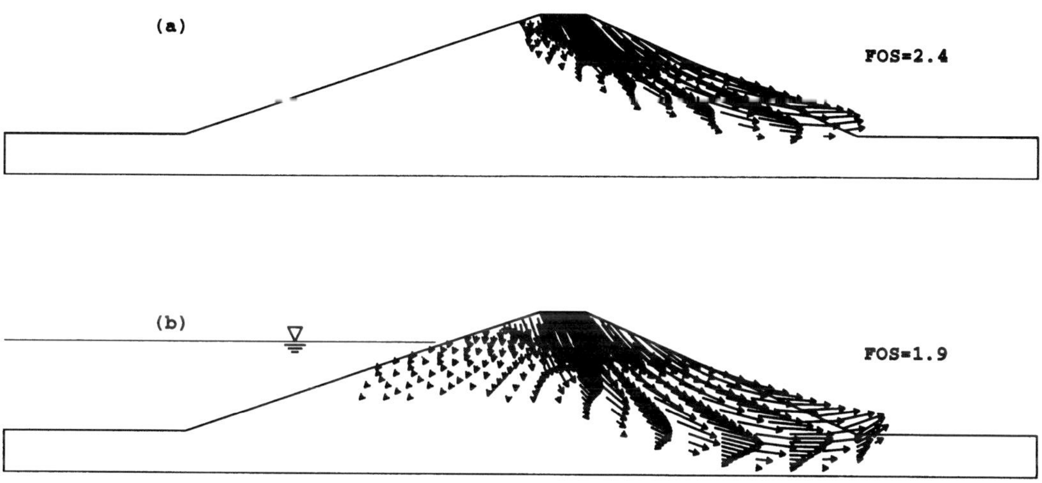

Figure 21. Example 6: Displacement vectors corresponding to the unconverged solution by finite elements, (a) no free surface, (b) with a free surface.

1.6 The Criteria for Computer Aided Analysis

In their 1967 paper Whitman and Bailey looked forward to the future of computer aided analysis for engineers and set criteria by which it could be judged. Their comments were originally addressed to the automation of limit equlibrium methods but they also commented on the then emerging numerical analysis techniques.

They judged that the system must be sufficiently accurate for confidence in its use and appropriate for the parameters being input. Finite element analysis meets these criteria with a degree of accuracy decided by the engineer in designing the model.

It should be possible, in a realistic timescale, to do sufficient trials to examine all the key modes of behaviour; to consider different times in the life of the structure and to vary parameters during design to test options for cost and efficiency. All this is now possible with FE methods.

Finally the method of human-machine communication must be be user-friendly and readily accessible. This is partly a matter of program design but easily achieved. Graphical output greatly enhances the process of design and analysis over and above that from the numerical results.

Similarly, Chowdhury (1981) in his discussion of Sarma (1979) commented on the perceived reluctance to develop alternatives to limit equilibrium methods for practice when the tools to do so were already available. Since then numerous applications and experience have verified the possibilities offered by finite elements.

The key issues of cost and turnaround time have been overtaken by the falling cost of powerful hardware and processor speeds which now make the FE method available to engineers at less than the cost of their CAD systems. What remains is the concern of powerful tools used wrongly. That is no more true of finite elements after years of application than of limit equilibrium methods which can themselves produce seriously misleading results. Engineering judgement is still essential whichever method is being used.

1.7 Concluding Remarks

The finite element method in conjunction with an elastic-perfectly plastic (Mohr-Coulomb) stress strain model has been shown to be a reliable and robust method for assessing the factor of safety of slopes. One of the main advantages of the finite element approach, is that the factor of safety emerges naturally from the analysis without the user having to comit to any particular form of the mechanism *a priori*.

The finite element approach for determining the factor of safety of slopes has satisfied the criteria for effective computer aided analysis. The widespread use of this method should now be seriously considered by geotechnical practitioners as a more powerful alternative to traditional limit equilibrium methods.

2 A Probabilistic Approach to Slope Stability using Finite Elements

An investigation has been performed into the stability of an undrained clay slope having spatially randomly varying shear strength. The results of the study lead to a direct comparison between the probability of slope failure and the traditional Factor of Safety for a range of statistically defined input shear strength properties. The results highlight the influence of the spatial correlation length, a variable which is routinely omitted from conventional probabilistic studies in geotechnics.

2.1 Introduction

The section presents results obtained using a program developed by Gordon Fenton of Dalhousie University and the author (Griffiths and Fenton 2000), which merges non-linear elasto-plastic finite element analysis with random field theory (e.g. Fenton 1990,

Vanmarcke 1984). Some initial work using this approach has been reported by Paice and Griffiths (1997), however the problem to be considered in this paper is an undrained clay slope ($\phi_u = 0$) of height H with a gradient of 2:1 resting on a foundation layer, also of depth H. A typical finite element mesh is shown in Figure 22.

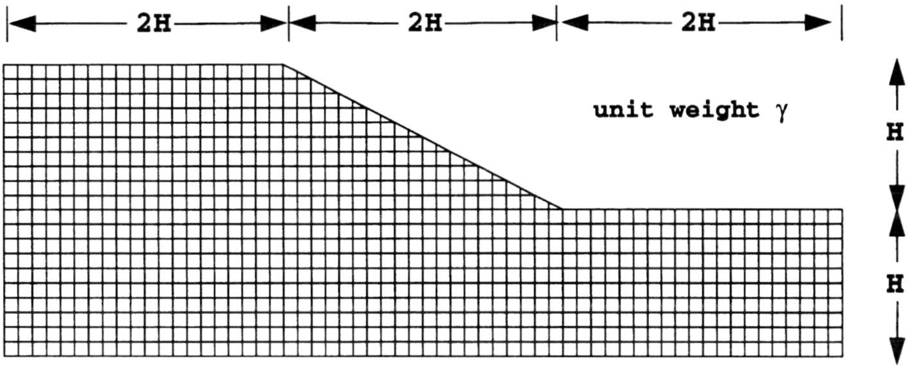

Figure 22. Mesh used for slope stability analyses.

In this study, the variability of the undrained shear strength (c_u) is assumed to be characterized by a lognormal distribution [2] with the parameters shown in Table 2.1.

Table 2.1. Statistical Parameters

		Units
Mean	μ_{c_u}	Stress
Standard Deviation	σ_{c_u}	Stress
Spatial Correlation Length	$\theta_{\ln c_u}$	Length

The mean and standard deviation can conveniently be expressed in terms of the dimensionless coefficient of variation defined as

$$C.O.V._{c_u} = \frac{\sigma_{c_u}}{\mu_{c_u}} \tag{6}$$

Rearranged the following par the actual undrained shear field is assumed lognormally distributed, taking its logarithm yields an "underlying" normally distributed (or Gaussian) field. The spatial correlation length is measured with respect to this underlying field, that is, with respect to $\ln c_u$. In particular, the spatial correlation length ($\theta_{\ln c_u}$) describes the distance over which the spatially random values will tend to be significantly correlated in the underlying Gaussian field. Thus, a large value of $\theta_{\ln c_u}$ will imply a smoothly varying field, while a small value will imply a ragged field. The spatial correlation length can be estimated from a set of shear strength data taken over some spatial

[2] A more detailed discussion of the lognormal distribution is given in Section 5.2 of this report.

region simply by performing the statistical analyses on the log-data. In practice, however, $\theta_{\ln c_u}$ is not much different in magnitude from the correlation length in real space and, for most purposes, θ_{c_u} and $\theta_{\ln c_u}$ are interchangeable given their inherent uncertainty in the first place. In the current study, the spatial correlation length has been non-dimensionalized by dividing it by the height of the embankment H, introducing to the variable $\Theta_{c_u} = \theta_{\ln c_u}/H$.

It should be emphasised that the spatial correlation length is rarely taken into account in routine probabilistic studies relating to geotechnical engineering. In the majority of these cases, a Single Random Variable approach (e.g. Harr 1987, Duncan 2000) is used, which is equivalent to setting $\Theta_{c_u} = \infty$.

It has been suggested (see e.g. Lee *et al* 1983, Kulhawy *et al* 1991 and Duncan 2000) that typical $C.O.V._{c_u}$ values for the undrained shear strength lie in the range 0.1-0.5, however the spatial correlation length is less well documented, especially in the horizontal direction, and may well exhibit anisotropy. While the analysis tools used in this study have the capability of modeling an anisotropic spatial correlation field, all the results presented in this report assume that Θ_{c_u} is isotropic. This is not a severe restriction, since the geometry can often be scaled to achieve the desired spatial correlation structure

2.2 Brief description of the finite element method used

The slope stability analyses use an elastic-perfectly plastic stress-strain law with a Tresca failure criterion similar to that described in the first section of this report. Plastic stress redistribution is accomplished using a viscoplastic algorithm which uses 8-node quadrilateral elements and reduced integration in both the stiffness and stress redistribution parts of the algorithm.

In brief, the analyses involve the application of gravity loading, and the monitoring of stresses at all the Gauss points. If the Tresca criterion is violated, the program attempts to redistribute those stresses to neighboring elements that still have reserves of strength. This is an iterative process which continues until the Tresca criterion and global equilibrium are satisfied at all points within the mesh under quite strict tolerances.

In this study, "failure" is said to have occurred if, for any given realization, the algorithm is unable to converge within 500 iterations. Following a set of 1000 realizations of the Monte-Carlo process the probability of failure is simply defined as the proportion of these realizations that required 500 or more iterations to converge.

While the choice of 500 as the iteration ceiling is subjective, Figure 23 confirms, for the case of $\Theta_{c_u} = 1$, that the probability of failure computed using this criterion is quite stable after about 200 iterations.

2.3 Parametric studies

In the parametric studies described in this section, the mean strength expressed in the form of a Stability Number

$$N_s = \mu_{c_u}/\gamma H \tag{7}$$

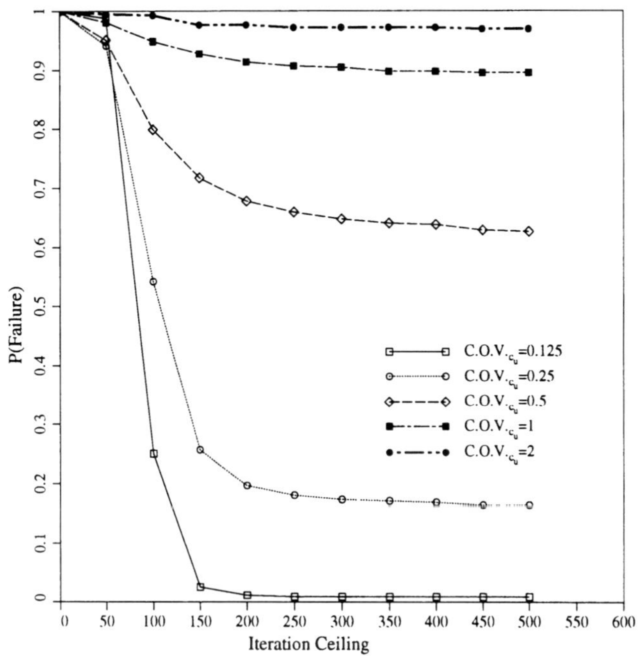

Figure 23. Probability of failure vs. Iteration Ceiling.

where γ is the unit weight, was given the values 0.15, 0.20, 0.25 and 0.30, and in each case, a range of $C.O.V._{c_u}$ and Θ_{c_u} values were investigated as follows:

$$\begin{aligned} \Theta_{c_u} &= 0.5,\ 1,\ 2,\ 4,\ 8,\ \infty \\ C.O.V._{c_u} &= 0.125,\ 0.25,\ 0.5,\ 1,\ 2,\ 4 \end{aligned} \tag{8}$$

To put the probabilistic results in context, Table 2.2 shows the Factor of Safety F from conventional limit equilibrium analysis, for the slope in Figure 22 assuming a homogeneous shear strength defined by the Stability Number N_s.

Table 2.2. Factors of Safety Assuming Homogeneous Soil

N_s	F
0.15	0.88
0.17	1.00
0.20	1.18
0.25	1.47
0.30	1.77

For each set of assumed statistical properties given by $C.O.V._{c_u}$ and Θ_{c_u}, Monte-Carlo simulations were performed, typically involving 1000 repetitions or "realizations" of the shear strength random field and the subsequent finite element analysis. Each realization

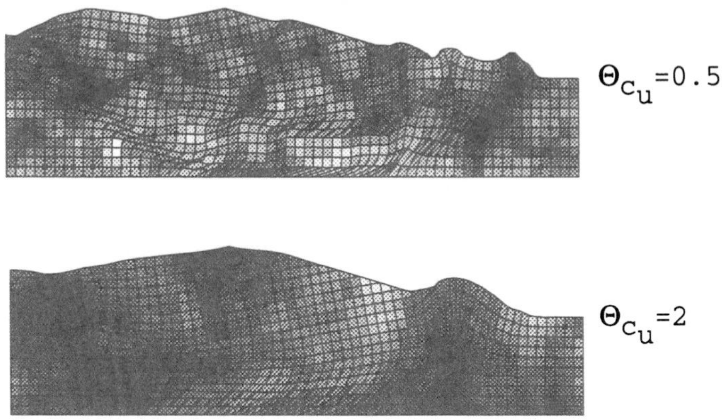

Figure 24. Typical random field realizations. Darker zones indicate weaker soil.

of the random field, while having the same underlying statistics, led to a quite different spatial pattern of shear strength values within the slope.

Figure 24 shows two typical random field realizations and associated failure mechanisms for slopes with $\Theta_{c_u} = 0.5$ and $\Theta_{c_u} = 2$. Notice how the higher Θ_{c_u} gives a more gradually varying shear strength over space and a smoother failure surface.

2.4 Single random variable approach

It is instructive to consider the special case of $\Theta_{c_u} = \infty$, which implies that each realization of the Monte-Carlo process gives a uniform strength, the same everywhere, but with the strength varying from one realization to the next according to the governing lognormal distribution. The probability of failure in such a case is simply equal to the probability that the Stability Number N_s will be below 0.17, the value that would give a Factor of Safety of unity.

For example, if $\mu_{c_u} = 0.25\gamma H$ and $\sigma_{c_u} = 0.125\gamma H$, corresponding to $C.O.V._{c_u} = 0.5$, the statistics of the Stability Number are therefore given by $\mu_{N_s} = 0.25$, $\sigma_{N_s} = 0.125$ and $C.O.V._{N_s} = 0.5$.

From standard relationships, the mean and standard deviation of the underlying *normal* distribution of the Stability Number are given by:

$$\sigma_{\ln N_s} = \sqrt{\ln\left\{1 + \left(\frac{\sigma_{N_s}}{\mu_{N_s}}\right)^2\right\}} \tag{9}$$

$$\mu_{\ln N_s} = \ln \mu_{N_s} - \frac{1}{2}\sigma^2_{\ln N_s} \tag{10}$$

hence $\mu_{\ln N_s} = -1.498$ and $\sigma_{\ln N_s} = 0.472$.

The probability of failure is therefore given by:

$$p(N_s < 0.17) = \Phi\left(\frac{\ln 0.17 - \mu_{\ln N_s}}{\sigma_{\ln N_s}}\right) \tag{11}$$

$$= 0.281 \tag{12}$$

where Φ is the cumulative normal distribution function. The relationship between the Factor of Safety (assuming a constant shear strength equal to μ_{c_u}) and the probability of failure assuming a Single Random Variable ($\Theta_{c_u} = \infty$) is summarized in Figure 25 for a range of $C.O.V._{c_u}$ values.

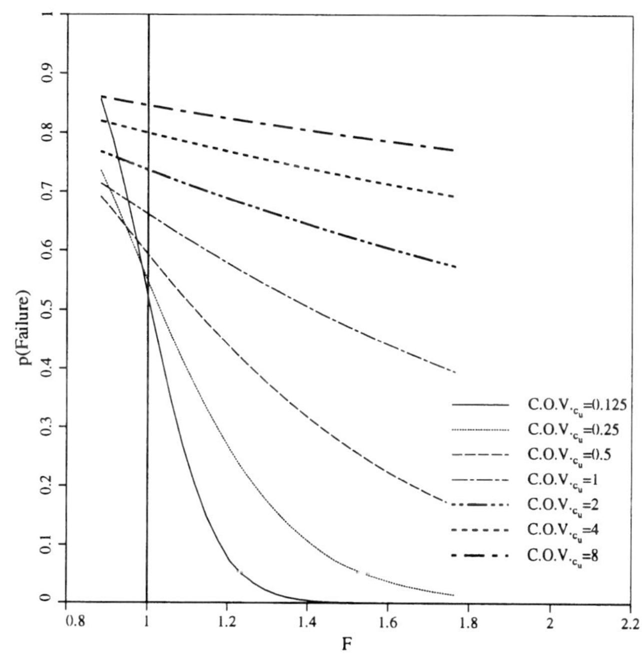

Figure 25. Factor of Safety vs. Probability of Failure. Single random variable approach, $\Theta_{c_u} = \infty$

Apart from the rather obvious conclusion that the probability of failure goes up as the Factor of Safety goes down, it is also clear that for the majority of cases, the probability of failure also goes up as the $C.O.V._{c_u}$ of the shear strength increases. This result is not necessarily intuitive, since soil with a higher $C.O.V._{c_u}$ contains elements that are much weaker *and* much stronger than the mean. The result indicates however, that the weaker elements dominate the stability calculation.

The only exception to this trend occurs when the mean strength indicates a Factor of Safety of *less* than unity. As shown in Figure 25, the probability of failure in such cases is understandably high, however the role of $C.O.V._{c_u}$ has the opposite effect to that described above, with lowest values of $C.O.V._{c_u}$ tending to give the highest values of the probability of failure.

It is interesting to note that using this approach, a slope with a Factor of Safety of 1.50, based on the mean strength, would have a probability of failure as high as 27% if $C.O.V._{c_u} = 0.5$, the upper limit of the recommended range of Lee *et al* (1983) and others.

2.5 Random field approach.

The code developed by the authors enables a random field of shear strength values to be generated and subsequently mapped onto the finite element mesh. In a random field, the value assigned to each cell (or finite element in this case) is itself a random variable, thus the mesh of Figure 22 which has 910 finite elements consists of 910 random variables. The random variables can be correlated to one another by controlling the spatial correlation length Θ_{c_u} as described previously, hence the single random variable approach discussed in the previous section can now be viewed as just a special case of a much more powerful analytical tool.

Figures 26 and 27 show the effect of the spatial correlation length Θ_{c_u} on a soil with a Factor of Safety of 1.47 (based on the mean strength) for a range of $C.O.V._{c_u}$ values. Figure 26 clearly indicates two branches relating to the value of $C.O.V._{c_u}$. For low values of $0 < C.O.V._{c_u} \leq 0.5$, the probability of failure increases as Θ_{c_u} increases, indicating that the Single Random Variable approach in which $\Theta_{c_u} = \infty$ is conservative. For high values of $1 \leq C.O.V._{c_u}$ quite the reverse trend is apparent, with the higher values of Θ_{c_u} tending to *underestimate* the probability of failure.

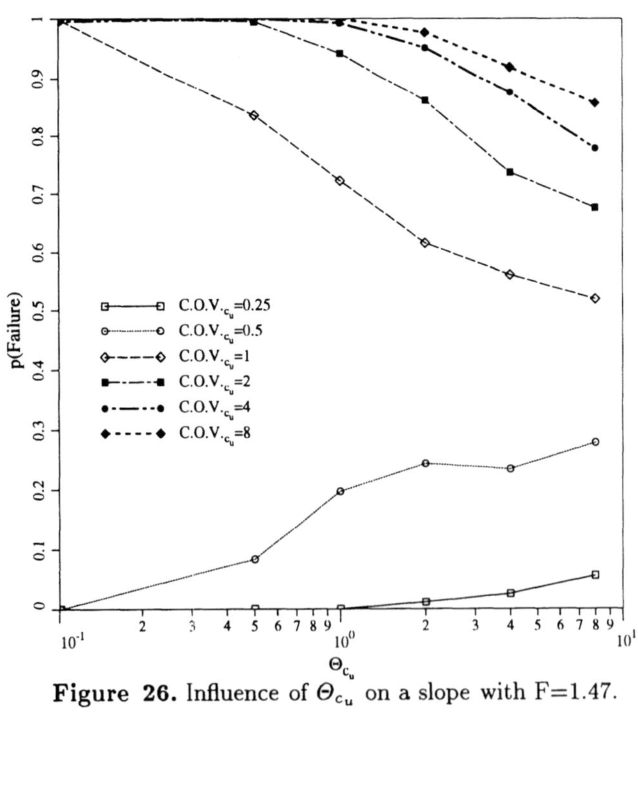

Figure 26. Influence of Θ_{c_u} on a slope with F=1.47.

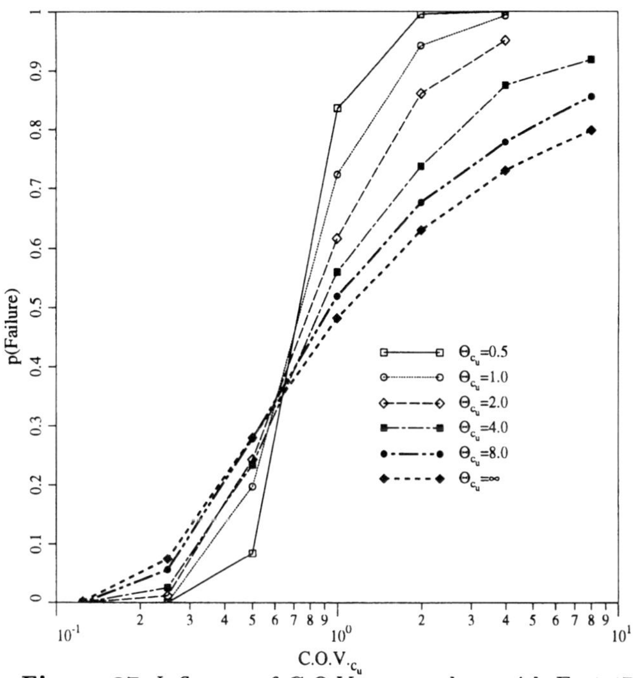

Figure 27. Influence of $C.O.V._{\cdot c_u}$ on a slope with F=1.47.

Figure 27 shows an alternative representation of the same data with $C.O.V._{\cdot c_u}$ plotted along the abscissa. This figure shows more clearly how $\Theta_{c_u} = \infty$ will tend to overestimate the probability of failure for low $C.O.V._{\cdot c_u}$ values and underestimate it for high values. It is also of interest to note the sensitivity of the probability of failure to the value of $C.O.V._{\cdot c_u}$ for low levels of correlation. For example, the line corresponding to $\Theta_{c_u} = 0.5$ rises steeply from zero to 100% probability of failure within the relatively narrow band of $0.25 < C.O.V._{\cdot c_u} < 2$. For even smaller values of Θ_{c_u} the rise was observed to be even more dramatic, although these results are not presented here. A further point of interest from Figure 27 is that all the lines appear to coincide at approximately the same value of $C.O.V._{\cdot c_u} \approx 0.65$, implying that at this level of shear strength variance, the probability of failure is independent of Θ_{c_u}.

The observations made with respect to Figures 26 and 27 were for the particular case of a mean shear strength that would have given a Factor of Safety of 1.47. The results from further analyses of a range of mean shear strength values corresponding to the Stability Numbers in Table 2.2 are shown in Figure 28. In order to reduce the number of variables, only the results assuming $C.O.V._{\cdot c_u} = 0.5$ are shown.

Figure 28 indicates another type of "cross over" with respect to the Factor of Safety. For the given value of $C.O.V._{\cdot c_u} = 0.5$, the Single Random Variable approach corresponding to $\Theta_{c_u} = \infty$ appears to overestimate the probability of failure for slopes with relatively high deterministic Factors of Safety ($F > 1.4$) and underestimate it for lower Factors of Safety ($F < 1.4$).

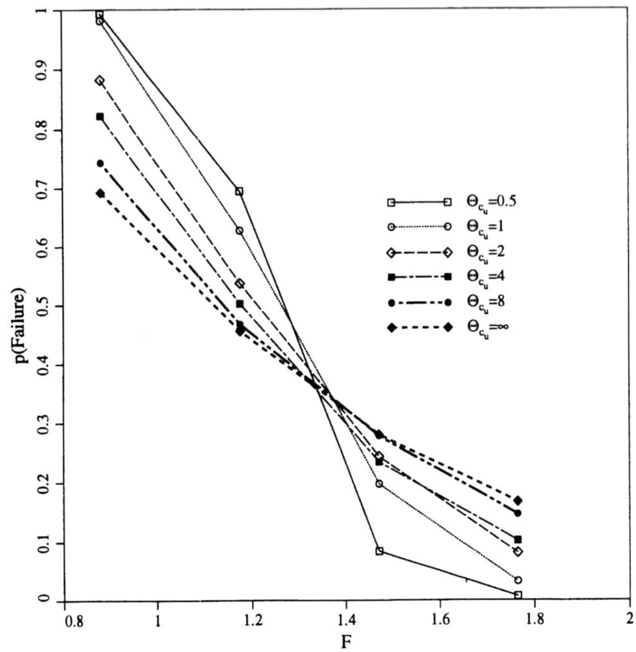

Figure 28. Influence of Θ_{c_u} on the Probability of Failure for a range of deterministic Factors of Safety ($C.O.V._{c_u} = 0.5$)

2.6 Concluding remarks

The section has shown that soil strength heterogeneity in the form of a spatially varying lognormal distribution can significantly affect the stability of a slope of undrained clay when viewed in a probabilistic context. In this paper, particular attention was paid to the validity of treating the heterogeneity as a Single Random Variable which was shown to be a special case of the authors' formulation corresponding to an infinite correlation length of $\Theta_{c_u} = \infty$.

The following more specific observations can be made from the results presented in this paper:

1. For the slope considered in this study with a Factor of Safety of 1.47 based on the mean strength, the Single Random Variable approach gave conservative estimates of the probability of failure for Coefficient of Variation values in the "typical" range of $0 < C.O.V._{c_u} < 0.5$. For higher values of $C.O.V._{c_u}$ however, the Single Random Variable approach gave unconservative estimates.
2. For the slope considered in this study with $C.O.V._{c_u} = 0.5$, the Single Random Variable approach gave conservative estimates of the probability of failure for higher Factors of Safety in the approximate range $F > 1.4$ and unconservative estimates for lower Factors of Safety when $F < 1.4$.

More work remains to be done in this area, but the implications of this study are that the Single Random Variable approach is an acceptable guide to probabilistic slope stability providing the mean strength indicates a relatively high Factor of Safety. For more critical cases, in which the mean strength indicates a Factor of Safety closer to unity, the Single Random Variable approach can give an unconservative estimate of the probability of failure, i.e. *lower* than the "true" value.

A final comment relates to the influence of the Coefficient of Variation of the soil shear strength. While increasing the value of $C.O.V._{c_u}$ introduces both stronger and weaker zones of soil into the slope, the weaker soil always dominates the overall performance leading to a *less* stable slope.

3 Excavation Software

Heterogeneity in geotechnical problems can also occur in the form of voids caused by tunneling and excavation.

All the slope stability analyses described in the first two sections of this report assumed the slope was initially weightless and then suddenly subjected to gravity loading in a single increment. While this is unrealistic, evidence was produced in Figure 3 showing that the factor of safety is insensitive to the rate of gravity loading (e.g. Lechman 2000).

The basic slope stability approach however takes no account of construction sequence in the form of "building up" or "excavating down". The text by Smith and Griffiths (1998) contains Program 6.9 (construction of an embankment) and Program 6.10 (excavation of an embankment). In this section, the excacation program 6.10 is briefly revisited, and a slightly modified version of the code is included. Some examples of tunneling type excavation are then presented. The interested reader is referred to Brown and Booker (1985) and Smith and Ho (1992) for more background on the excavation algorithm.

3.1 Excavation loads

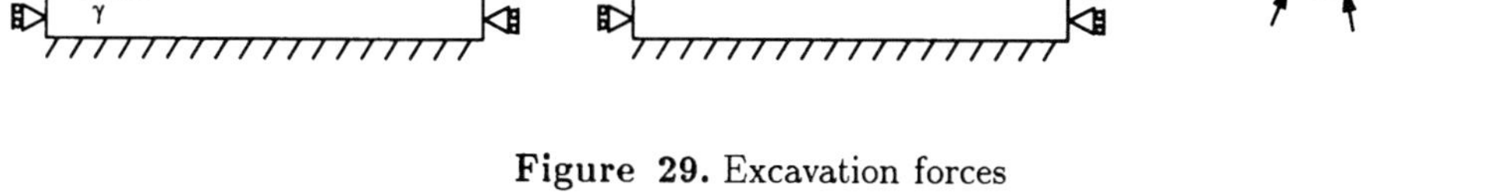

Figure 29. Excavation forces

The aim of an analysis is that when a portion of material is excavated, either in open excavations or an enclosed tunnel, forces must be applied along the excavated surface such that:

- The remaining material experiences the correct unloading effect.
- The new "free surface" is stress free.

Figure 29 shows a 2-d block of soil from which a zone of material is to be excavated. The soil to be removed is zone A and the surrounding soil that will remain after excavation is zone B. The initial stresses in the zone to be removed and the surrounding soil are labelled σ_{A0} and σ_{B0} respectively. Any self-weight and external loads are taken into account in forming these stresses prior to the removal of A. Since both bodies are in equilibrium, forces F_{AB} must be applied to body B by body A to maintin σ_{B0} and, similarly, F_{BA} must act on body A as shown in Figure 29. Forces F_{AB} and F_{BA} are equal and opposite. In general therefore, the forces to be applied to the excavated boundaries depend on the initial stress state in the excavated material and its self-weight. It can be shown that (e.g. Brown and Booker 1985):

$$F_{AB} = \int_{V_A} \mathbf{B}^T \sigma_{A0} dV_A + \gamma \int_{V_A} \mathbf{N}^T dV_A \tag{13}$$

where $\mathbf{B}$ is the strain-displacement matrix, $\mathbf{N}$ the element shape functions, and V_A and γ are respectively the volume and unit weight of the excavated material.

3.2 Listing and sample data for excavation program

The following listing represents a modified version of Program 6.10 from the text by Smith and Griffiths (1998). The modified program is called `p610_1.f90` and this, together with all other programs, libraries and data from the text, can be obtained from the author's web site at:

`http://www.mines.edu/fs_home/vgriffit/`

then under the heading "SOFTWARE FROM TEXTBOOKS", click on "Programming the finite element method" and click on "chap6".

The general approach for each excavation step is to compute the loads to be applied to the excavation boundary using equation (13). The properties of the elements in the mesh are then modified by setting Young's Modulus of excavated elements to zero. Although the total number of nodes in the mesh remains constant throughout, the introduction of "air" elements involves a re-numbering of the active freedoms within the mesh, and hence a full re-assembly of the global stiffness matrix at each excavation step. It should also be noted that to avoid zeros on the diagonal of the modified global stiffness matrix, any freedoms attached to nodes that become completely surrounded by "air", must be removed from the analysis and not assembled into the global stiffness matrix.

The reader is referred to the original program 6.10 for the meaning of most variable names, however the main changes in the current version may be summarised as follows:

- Only the excavated element numbers need to be given as data. The nodes affected by the excavation are computed automatically. New variables include:
  noexe–the number of elements to be removed at each excavation step,

ntote–a running total of the total number of elements excavated,
exele(noexe)–the element numbers to be removed at each excavation step,
totex(ntote)–all the element numbers removed.

- Subroutines **mesh_exc.f90**, **dis_exc.f90**, **vec_exc.f90** have been developed to generate PostScript output files corresponding to, repectively, the undeformed mesh, the deformed mesh and the nodal displacment vectors. All figures correspond to the situation following the final excavation step, whether converged or not. These subroutines have not been included in the listing that follows, but are appended to the end of the program available on the web as described above.
- The program is quite general, and reads the coordinates and connectivity of the initial mesh via data. The program also permits different material property groups in the style of Program 6.8.
- Some minor cosmetic changes have been made to allow data control of which specific nodal displacements are to be printed following each excavation step.

The data file that follows represents the same analysis presented in the text with Program 6.10, in which a vertical cut in undrained clay is excavated in two stages as shown in Figure 30. The rather crude mesh is used for demonstration purposes only. The nodes are indicated by numbers with circles around them and element numbers are indicated in the middle of each element. The side boundary conditions are rollers (vertical movement only) and the base is fixed. The meanings of the input variable names are defined in Table 3.1.

Table 3.1. Input variable names

nels	number of elements
nn	number of nodes
limit	iteration ceiling
tol	iteration tolerance
epk0	"at rest" earth pressure coefficient
nprops	number of properties (6)
np_types	number of property groups
prop(nprop,np_types)	properties (ϕ, c, ψ, γ, E, ν)
nouts	number of output nodes
no(nouts)	output node numbers
g_coord(2,nn)	nodal coordinates (x, y)
g_num(8,nels)	element node numbers
nr	number of restrained nodes
k,nf(:,k)	node number and fixity (0 or 1), (nr-times)
layers	number of excavations
incs	increments per excavation
noexe	number of elements to be removed
exele(noexe)	element numbers to be removed

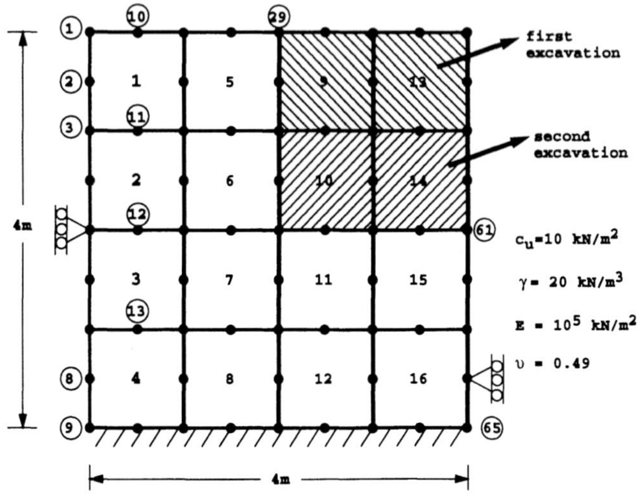

Figure 30. Test excavation problem

Data File p610_1.dat

```
nels  nn  limit  tol    epk0
 16   65   250  .0001   1.

nprops  np_types
  6        1

prop
0.0  10.0  0.0   20.0  1.e5  0.49

nouts
2

no
29 61

g_coord
0.0000E+00  0.0000E+00
0.0000E+00 -0.5000E+00
0.0000E+00 -0.1000E+01
0.0000E+00 -0.1500E+01
0.0000E+00 -0.2000E+01
0.0000E+00 -0.2500E+01
0.0000E+00 -0.3000E+01
0.0000E+00 -0.3500E+01
0.0000E+00 -0.4000E+01
0.5000E+00  0.0000E+00
0.5000E+00 -0.1000E+01
```

```
0.5000E+00 -0.2000E+01
0.5000E+00 -0.3000E+01
0.5000E+00 -0.4000E+01
0.1000E+01  0.0000E+00
0.1000E+01 -0.5000E+00
0.1000E+01 -0.1000E+01
0.1000E+01 -0.1500E+01
0.1000E+01 -0.2000E+01
0.1000E+01 -0.2500E+01
0.1000E+01 -0.3000E+01
0.1000E+01 -0.3500E+01
0.1000E+01 -0.4000E+01
0.1500E+01  0.0000E+00
0.1500E+01 -0.1000E+01
0.1500E+01 -0.2000E+01
0.1500E+01 -0.3000E+01
0.1500E+01 -0.4000E+01
0.2000E+01  0.0000E+00
0.2000E+01 -0.5000E+00
0.2000E+01 -0.1000E+01
0.2000E+01 -0.1500E+01
0.2000E+01 -0.2000E+01
0.2000E+01 -0.2500E+01
0.2000E+01 -0.3000E+01
0.2000E+01 -0.3500E+01
0.2000E+01 -0.4000E+01
0.2500E+01  0.0000E+00
0.2500E+01 -0.1000E+01
0.2500E+01 -0.2000E+01
0.2500E+01 -0.3000E+01
0.2500E+01 -0.4000E+01
0.3000E+01  0.0000E+00
0.3000E+01 -0.5000E+00
0.3000E+01 -0.1000E+01
0.3000E+01 -0.1500E+01
0.3000E+01 -0.2000E+01
0.3000E+01 -0.2500E+01
0.3000E+01 -0.3000E+01
0.3000E+01 -0.3500E+01
0.3000E+01 -0.4000E+01
0.3500E+01  0.0000E+00
0.3500E+01 -0.1000E+01
0.3500E+01 -0.2000E+01
0.3500E+01 -0.3000E+01
0.3500E+01 -0.4000E+01
0.4000E+01  0.0000E+00
0.4000E+01 -0.5000E+00
0.4000E+01 -0.1000E+01
0.4000E+01 -0.1500E+01
```

```
0.4000E+01 -0.2000E+01
0.4000E+01 -0.2500E+01
0.4000E+01 -0.3000E+01
0.4000E+01 -0.3500E+01
0.4000E+01 -0.4000E+01

g_num
 3   2   1  10  15  16  17  11
 5   4   3  11  17  18  19  12
 7   6   5  12  19  20  21  13
 9   8   7  13  21  22  23  14
17  16  15  24  29  30  31  25
19  18  17  25  31  32  33  26
21  20  19  26  33  34  35  27
23  22  21  27  35  36  37  28
31  30  29  38  43  44  45  39
33  32  31  39  45  46  47  40
35  34  33  40  47  48  49  41
37  36  35  41  49  50  51  42
45  44  43  52  57  58  59  53
47  46  45  53  59  60  61  54
49  48  47  54  61  62  63  55
51  50  49  55  63  64  65  56

nr
25

k,nf(:,k),i=1,nr
 1 0 1  2 0 1  3 0 1  4 0 1  5 0 1  6 0 1  7 0 1  8 0 1  9 0 0
14 0 0 23 0 0 28 0 0 37 0 0 42 0 0 51 0 0 56 0 0
57 0 1 58 0 1 59 0 1 60 0 1 61 0 1 62 0 1 63 0 1 64 0 1 65 0 0

layers incs
2       5

noexe
2
exele
9  13

noexe
2
exele
10  14
```

Program p610_1.f90

```
program p610_1
!-----------------------------------------------------------------------
!      A more general version of Program 6.10.
!      Plane strain of an elastic-plastic (Mohr-Coulomb) solid
!      using 8-node quadrilateral elements, viscoplastic strain method.
!      Construction of an excavation in stages introducing ''air'' elements
!
!      General program permitting any shape of initial mesh with
!      multiple property groups if required.
!-----------------------------------------------------------------------
use main
use geom
implicit none
integer::nels,neq,nband,nn,nr,nip=4,nodof=2,nod=8,nst=4,ndof,              &
         i,k,iel,iters,limit,incs,iy,ndim=2,layers,ii,iq,noexe,nprops,     &
         np_types,jj,nodex,modex,ncheck,ntote,nouts
logical::converged
character(len=15)::element='quadrilateral'
real::det,phi,psi,c,gama,dt,ddt,f,dsbar,dq1,dq2,dq3,lode_theta,sigm,pi,    &
      snph,e,v,epk0,tol
!----------------------------- dynamic arrays--------------------------------
real,allocatable::kb(:,:),exc_loads(:),points(:,:),bdylds(:),tot_d(:,:),   &
                  evpt(:,:,:),oldis(:),loads(:),dee(:,:),coord(:,:),       &
                  fun(:),jac(:,:),weights(:),der(:,:),deriv(:,:),bee(:,:),&
                  km(:,:),eld(:),eps(:),bload(:),eload(:),erate(:),        &
                  g_coord(:,:),prop(:,:),evp(:),devp(:),m1(:,:),m2(:,:),   &
                  m3(:,:),flow(:,:),tensor(:,:,:),gc(:),s(:)
integer,allocatable::nf(:,:),g(:),num(:),g_num(:,:),etype(:),solid(:),     &
                  lnf(:,:),exele(:),totex(:),no(:)
!-----------------------input and initialisation-----------------------------
open(10, file = 'fe90.dat')
open(11, file = 'fe90.res')
!
read(10,*)nels,nn,limit,tol,epk0
read(10,*)nprops,np_types
allocate(prop(nprops,np_types),etype(nels))
read(10,*)prop
etype=1
if(np_types>1)read(10,*)etype
read(10,*)nouts
ndof=nod*nodof
!
write(11,'(a,i5)')"The total number of elements is ",nels
allocate(nf(nodof,nn),points(nip,ndim),weights(nip),g_coord(ndim,nn),     &
         num(nod),dee(nst,nst),evpt(nst,nip,nels),coord(nod,ndim),        &
         fun(nod),solid(nels),jac(ndim,ndim),der(ndim,nod),               &
         deriv(ndim,nod),g_num(nod,nels),bee(nst,ndof),km(ndof,ndof),     &
```

```
         eld(ndof),eps(nst),totex(nels),bload(ndof),eload(ndof),          &
         erate(nst),evp(nst),devp(nst),g(ndof),m1(nst,nst),m2(nst,nst),   &
         m3(nst,nst),flow(nst,nst),s(nst),tot_d(nodof,nn),gc(ndim),       &
         tensor(nst,nip,nels),lnf(nodof,nn),no(nouts))
!------ nf is an index array of 1s and 0s : lnf is the local nf  -------------
 read(10,*)no
 solid=1
!--------------------- read geometry and connectivity -----------------------
 read(10,*)g_coord
 read(10,*)g_num
 nf=1
 read(10,*)nr
 if(nr>0)read(10,*)(k,nf(:,k),i=1,nr)
 lnf=nf
 call formnf(lnf)
 neq=maxval(lnf)
 write(11,'(a,i5)')"The total possible number of equations is:",neq
 pi=acos(-1.)
!-------- set up the global node numbers and global nodal coordinates ---------
!--------------- loop the elements to set starting stresses ------------------
 call sample(element,points,weights)
 elements_0: do iel=1,nels
   num=g_num(:,iel)
   coord=transpose(g_coord(:,num))
   gama=prop(4,etype(iel))
   gauss_pts_0: do i=1,nip
     call shape_fun(fun,points,i)
     gc=matmul(fun,coord)
     tensor(2,i,iel)=gc(2)*gama
     tensor(1,i,iel)=epk0*tensor(2,i,iel)
     tensor(4,i,iel)=tensor(1,i,iel)
     tensor(3,i,iel)=0.0
   end do gauss_pts_0
 end do elements_0
 write(11,'(a)')"Global coordinates "
 do k=1,nn
   write(11,'(a,i5,a,2e12.4)')"Node",k,"       ",g_coord(:,k)
 end do
 write(11,'(a)')"Global node numbers "
 do k=1,nels
   write(11,'(a,i5,a,8i5)')"Element ",k,"       ",g_num(:,k)
 end do
 tot_d=0.0
 ntote=0
! ----------------------  excavate a layer ---------------------------------
 read(10,*)layers,incs
 layer_number: do ii=1,layers
   write(*,*)"Excavation number",ii
   write(11,'(//,a,i3)')"Excavation number",ii
```

```
!------------------- read elements to be removed ----------------------------
!------------------- program computes excavated nodes ------------------------
  read(10,*)noexe
  ntote=ntote+noexe
  allocate(exele(noexe))
  read(10,*)exele
  totex(ntote-noexe+1:ntote)=exele
  solid(exele)=0
!
  do i=1,noexe
    do k=1,nod
      nodex=0
      ncheck=g_num(k,exele(i))
      do iel=1,nels
        modex=0
        do jj=1,ntote
          if(iel==totex(jj))then
            modex=1
            exit
          end if
        end do
        if(modex==1)cycle
        do jj=1,nod
          if(ncheck==g_num(jj,iel))then
            nodex=1
            exit
          end if
        end do
        if(nodex==1)exit
      end do
      if(nodex==0)nf(:,ncheck)=0
    end do
  end do
!
  lnf=nf
  call formnf(lnf)
  neq=maxval(lnf)
  nband=0
!--------- recalculate the number of freedoms neq and half-bandwidth nband ----
  elements_1: do iel=1,nels
    num=g_num(:,iel)
    call num_to_g(num,lnf,g)
    if(nband < bandwidth(g))nband=bandwidth(g)
  end do elements_1
  write(11,'(/,3(a,i5))')                                               &
  "There are ",neq, " freedoms and nband is",nband," in step",ii
  allocate(kb(neq,nband+1),exc_loads(0:neq),bdylds(0:neq),oldis(0:neq),  &
          loads(0:neq))
  kb=0.0
```

```
   exc_loads=0.0
!---------------------- calculate excavation load ---------------------------
   elements_2: do iel=1,noexe
     iq=exele(iel)
     gama=prop(4,etype(iq))
     bload=0.0
     eld=0.0
     num=g_num(:,iq)
     call num_to_g(num,lnf,g)
     coord=transpose(g_coord(:,num))
     gauss_pts_2: do i=1,nip
       call shape_fun(fun,points,i)
       call shape_der(der,points,i)
       jac=matmul(der,coord)
       det=determinant(jac)
       call invert(jac)
       deriv=matmul(jac,der)
       call beemat(bee,deriv)
       s=tensor(:,i,iq)
       eload=matmul(s,bee)
       bload=bload+eload*det*weights(i)
       eld(2:ndof:2)=eld(2:ndof:2)+fun(:)*det*weights(i)
     end do gauss_pts_2
     exc_loads(g)=exc_loads(g)+eld*gama+bload
   end do elements_2
   exc_loads(0)=0.0
!------------------ element stiffness integration and assembly-----------------
   dt=1.e20
   elements_3: do iel=1,nels
     if(solid(iel)==0)then
       e=0.0
     else
       phi=prop(1,etype(iel))
       e=prop(5,etype(iel))
       v=prop(6,etype(iel))
       snph=sin(phi*pi/180.)
       ddt=(4.*(1.+v)*(1.-2.*v))/(e*(1.-2.*v+snph*snph))
       if(ddt<dt)dt=ddt
     end if
     km=0.0
     eld=0.0
     call deemat(dee,e,v)
     num=g_num(:,iel)
     call num_to_g(num,lnf,g)
     coord=transpose(g_coord(:,num))
     gauss_pts_3: do i=1,nip
       call shape_der(der,points,i)
       jac=matmul(der,coord)
       det=determinant(jac)
```

```
      call invert(jac)
      deriv=matmul(jac,der)
      call beemat(bee,deriv)
      km=km+matmul(matmul(transpose(bee),dee),bee)*det*weights(i)
    end do gauss_pts_3
    call formkb(kb,km,g)
  end do elements_3
!--------------- factorise l.h.s. ----------------------------------------
  call cholin(kb)
!------------------ factor excavation load by incs----------------------------
  exc_loads=exc_loads/incs
!------------ apply excavation loads incrementally ---------------------------
  load_incs: do iy=1,incs
    write(*,*)"Increment no",iy
    iters=0
    oldis=0.0
    bdylds=0.0
    evpt=0.0
!-------------------------   iteration loop ---------------------------------
    its: do
      iters=iters+1
      write(*,*)"iteration",iters
      loads=exc_loads+bdylds
      call chobac(kb,loads)
!-------------------------   check convergence ------------------------------
      call checon(loads,oldis,tol,converged)
      if(iters==1)converged=.false.
      if(converged.or.iters==limit)then
        bdylds=0.0
        do iq=1,nn
        do i=1,nodof
          if(lnf(i,iq)/=0)tot_d(i,iq)=tot_d(i,iq)+loads(lnf(i,iq))
        end do
        end do
      end if
!------------------------ go round the Gauss Points --------------------------
      elements_4: do iel=1,nels
        phi=prop(1,etype(iel))
        c=prop(2,etype(iel))
        psi=prop(3,etype(iel))
        e=prop(5,etype(iel))
        v=prop(6,etype(iel))
        if(solid(iel)==0)e=0.0
        bload=0.0
        call deemat(dee,e,v)
        num=g_num(:,iel)
        call num_to_g(num,lnf,g)
        coord=transpose(g_coord(:,num))
        eld=loads(g)
```

```
        gauss_pts_4: do i=1,nip
          call shape_der(der,points,i)
          jac=matmul(der,coord)
          det=determinant(jac)
          call invert(jac)
          deriv=matmul(jac,der)
          call beemat(bee,deriv)
          eps=matmul(bee,eld)
          eps=eps-evpt(:,i,iel)
          s=tensor(:,i,iel)+matmul(dee,eps)
!---------------------- air element stresses are zero ------------------------
          if(solid(iel)==0)s=0.0
          call invar(s,sigm,dsbar,lode_theta)
!-------------------- check whether yield is violated -----------------------
          call mocouf(phi,c,sigm,dsbar,lode_theta,f)
          if(converged.or.iters==limit)then
            devp=s
          else
            if(f>=0.0)then
              call mocouq(psi,dsbar,lode_theta,dq1,dq2,dq3)
              call formm(s,m1,m2,m3)
              flow=f*(m1*dq1+m2*dq2+m3*dq3)
              erate=matmul(flow,s)
              evp=erate*dt
              evpt(:,i,iel)=evpt(:,i,iel)+evp
              devp=matmul(dee,evp)
            end if
          end if
          if(f>=0.0.or.(converged.or.iters==limit))then
            eload=matmul(devp,bee)
            bload=bload+eload*det*weights(i)
          end if
!--------------- if appropriate update the Gauss point stresses ---------------
          if(converged.or.iters==limit)tensor(:,i,iel)=s
        end do gauss_pts_4
!------------------- compute the total bodyloads vector --------------------
        bdylds(g)=bdylds(g)+bload
        bdylds(0)=0.0
      end do elements_4
      if(converged.or.iters==limit)exit
    end do its
    write(11,'(a,i3,a,i5,a)')"Increment",iy," took",iters," iterations to converge"
    if(iy==incs.or.iters==limit)then
      write(11,'(a)')"The displacements are :"
      do i=1,nouts
        write(11,'(i5,2e12.4)')no(i),tot_d(:,no(i))
      end do
    exit
    end if
```

```
    end do load_incs
    loads(lnf(1,:))=tot_d(1,:)
    loads(lnf(2,:))=tot_d(2,:)
    call mesh_exc(g_coord,g_num,totex,ntote,12)
    call dis_exc(loads,lnf,0.1,g_coord,g_num,totex,ntote,13)
    call vec_exc(loads,lnf,0.1,0.1,g_coord,g_num,totex,ntote,14)
    if(iters==limit)exit
    deallocate(kb,exc_loads,bdylds,oldis,loads,exele)
  end do layer_number
stop
end program p610_1
```

Graphics subroutines

The graphics subroutines mentioned above require the following arguments:

`subroutine mesh_exc(g_coord,g_num,totex,ntote,ips)`

Subroutine that creates the PostScript output file `*.msh` of the undeformed mesh following excavation.

`call dis_exc(loads,lnf,ratmax,g_coord,g_num,totex,ntote,ips)`

Subroutine that creates the PostScript output file `*.dis` of the deformed mesh following excavation.

`call vec_exc(loads,lnf,ratmax,cutoff,g_coord,g_num,totex,ntote,ips)`

Subroutine that creates the PostScript output file `*.vec` of the nodal displacment vectors following excavation.

Table 3.2. Graphics routine arguments

g_coord(2,nn)	nodal coordinates
g_num(8,nels)	element node numbers
ntote	number of excavated elements
totex(ntote)	excavated element numbers
ips	output channel number
loads(neq)	nodal displacements
lnf(2,nn)	active nodal freedom array
ratmax	maximum nodal displacement ÷ maximum mesh dimension
cutoff	shortest arrow length ÷ maximum nodal displacement

Results File p610_1.res

```
The total number of elements is     16
The total possible number of equations is:    96
Excavation number  1

There are     86 freedoms and nband is    29 in step     1
Increment  1 took     2 iterations to converge
Increment  2 took     2 iterations to converge
Increment  3 took     2 iterations to converge
Increment  4 took     2 iterations to converge
Increment  5 took     2 iterations to converge
The displacements are :
   29  0.7636E-05 -0.5876E-04
   61  0.0000E+00  0.1223E-03


Excavation number  2

There are     76 freedoms and nband is    29 in step     2
Increment  1 took     2 iterations to converge
Increment  2 took     2 iterations to converge
Increment  3 took     4 iterations to converge
Increment  4 took     6 iterations to converge
Increment  5 took    31 iterations to converge
The displacements are :
   29  0.8717E-04 -0.3952E-03
   61  0.0000E+00  0.2324E-03
```

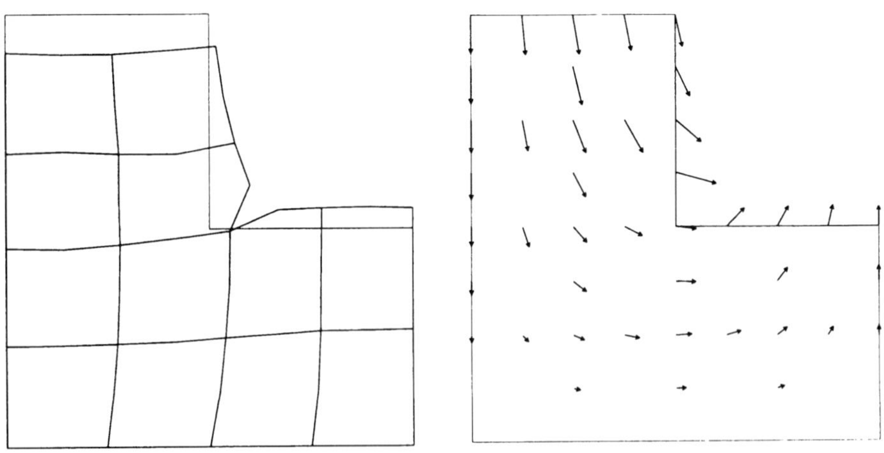

Figure 31. Deformed mesh and displacement vectors following second excavation in vertical cut analysis

Even in this rather simple problem, it is clear that the vertical displacement at the crest of the excavation (node 29) increases quite rapidly (from -0.5876E-04 to -0.3952E-03) following the second excavation of elements 10 and 14. It can also be seen that the algorithm requires a sudden increase in the number of iterations (from 6 to 31) for convergence when the fifth increment of the second excavation is applied.

The critical height of a vertical cut in undrained clay (e.g. Terzaghi and Peck 1967) is given by:

$$H_{cr} \approx \frac{c_u}{0.26\ \gamma} \tag{14}$$

which gives $H_{cr} \approx 1.92$m, hence the finite element analysis agrees quite closely with the classical solution even with a rather crude mesh. Figure 31 shows the deformed mesh and nodal displacement vectors corresponding to the final increment of the second excavation step. The mechanism of failure clearly indicates side wall bulging combined with base heave.

3.3 Excavation in frictional soil

In this example (Calderon 2000), an excavation with a slope of 37° to the horizontal is created in a deep layer of homogeneous soil with properties, $\phi^{'} = 22^o$, $c^{'} = 40\text{kN/m}^2$ and $\gamma = 18\text{kN/m}^3$. Bishop's method indicates that the slope will reach a factor of safety of unity when the depth of the excavation reaches about 63.5m. Using Program `p610_1.f90` Figure 32 shows the computed vertical displacement at the crest of the slope as a function of excavation depth. Following some heave in the initial stages due to unloading, a very sudden increase in crest displacement is apparent following the 13th excavation indicating that failure would occur with an excavated slope height in the range $60\text{m} < H < 65\text{m}$. Figure 33 shows the deformed mesh plot corresponging to the final excavation step.

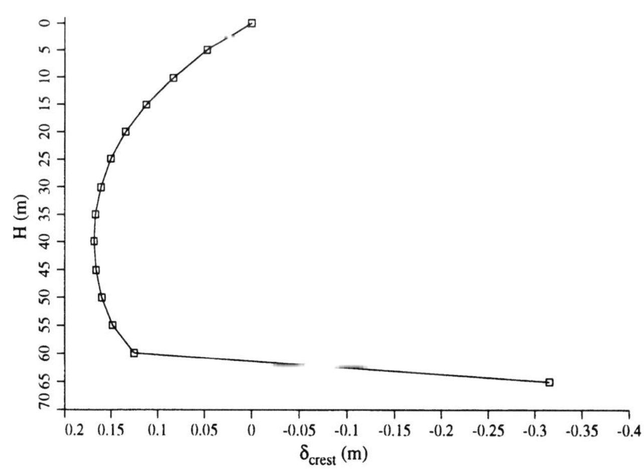

Figure 32. Crest deformation as a function of excavation depth

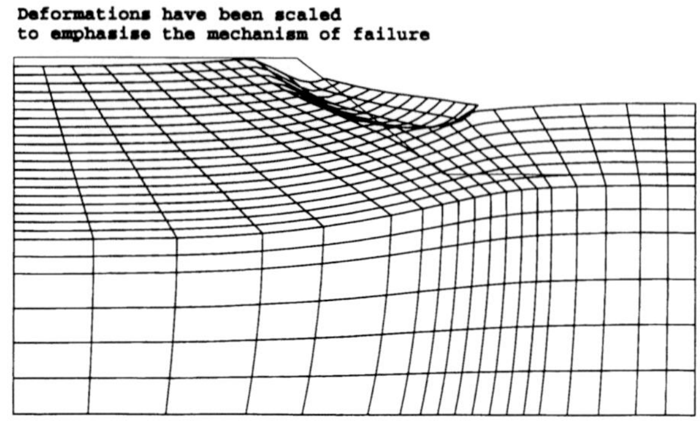

Figure 33. Deformed mesh at excavated slope height of 65m

3.4 Excavation of a tunnel

In this example an analysis of an underground excavation is demonstrated using Program **p610_1.f90**. A 5m high horizontal "tunnel" is constructed in clay with an undrained shear strength of $c_u = 100\text{kN/m}^2$ and total unit weight of $\gamma = 20\text{kN/m}^3$. The centerline of the excavation is at a depth of 5.5m and the in-situ total stresses in the ground prior to excavation have been assumed hydrostatic. Only half the problem is analysed due to symmetry. The excavation is gradually enlarged laterally and the vertical deflection of the ground surface above the center of the excavation recorded. The roof eventually caves when the width of the tunnel reaches 10m as shown in Figures 34 and 35.

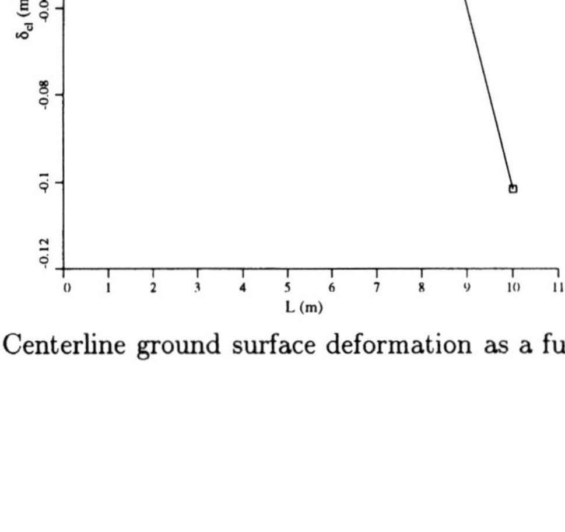

Figure 34. Centerline ground surface deformation as a function of tunnel width

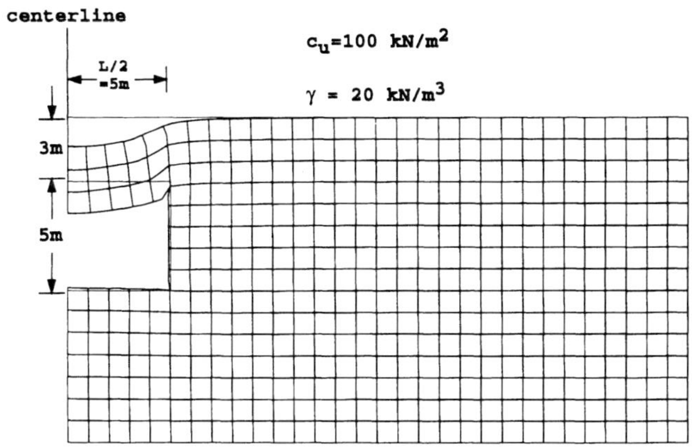

Figure 35. Deformed mesh when tunnel width reaches $L = 10$m

4 Bearing capacity of layered soils

Bearing capacity analysis is a more severe test of an elasto-plastic finite element code than slope stability analysis, due to the more confined nature of the problem. In part, this is due to the sensitivity of the computed values to the volume change tendency of the soil as defined by the plastic potential function or, in the case of "simple" models, the dilation angle ψ.

The volume change issue is not addressed in this section, because in the examples that follow, the soils that are modelled are "undrained clays", thus an associated Tresca failure criterion has been used implying no volume change during yielding.

4.1 A two-layer clay problem

The problem to be considered in this section is the bearing capacity of a rigid smooth strip footing resting on a two-layer clay soil foundation. A typical finite element mesh is shown in Figure 36. This problem has interested researchers for many years, initially using upper-bound limit analysis approaches (Button 1953, Brown and Meyerhof 1969, Chen 1975), and subsequently using numerical approaches (Griffiths 1982). Very recently, the problem has been revisited by Merifield *et al* 1999 using both upper *and* lower bound approaches and these solutions will form the basis of the comparisons with finite element results given in this section.

For the special case of $c_{u1} = c_{u2} = c_u$, the "Prandtl problem" is retrieved in which the ultimate bearing capacity is given by:

$$q_{ult} = (2 + \pi)c_u \tag{15}$$

If $c_{u1}/c_{u2} < 1$ then the footing is resting on soft clays overlying older, stiffer glacial tills. Alternatively, if $c_{u1}/c_{u2} > 1$, then the footing is resting on a strong surface crust

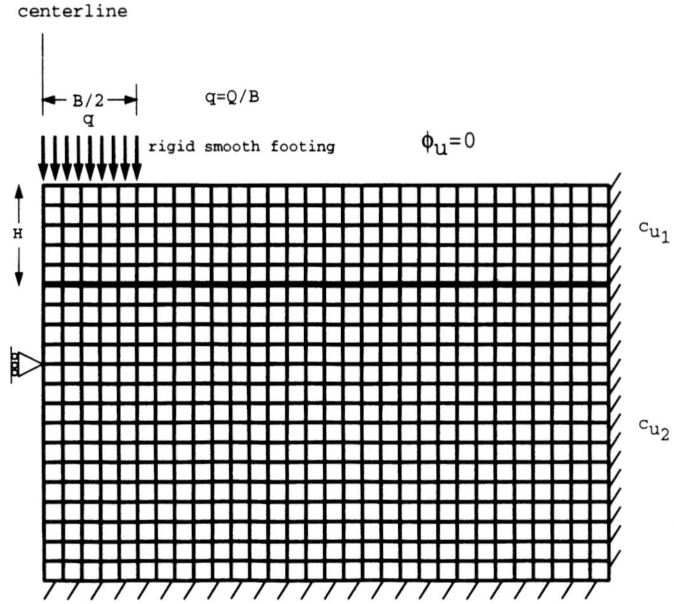

Figure 36. Mesh and configuration of two-layer bearing capacity problem

overlying weaker soils. The stronger crust may have been caused by dessication or weathering. One of the chief questions of interest in layered problems like this is, "How thick must the upper layer be before the influence of the lower layer becomes insignificant?"

Results have indicated that considerably more interaction occurs between the layers if the upper layer is the stronger of the two ($c_{u1}/c_{u2} > 1$). Brown and Meyerhof (1969) recorded a reduction in bearing capacity in such problems up to a depth ratio of $H/B \approx 2.5$, providing $c_{u2} < 0.2c_{u1}$. In the case of a weaker upper layer ($c_{u1}/c_{u2} < 1$), the lower soil strength was found to be immaterial for $H/B > 0.7$ and simply acted as a firm base. This seems consistent with the theoretical Prandtl mechanism which extends to a depth of $H/B = 1/\sqrt{2} = 0.707$.

4.2 Comparison of finite element and limit analysis

The program used in this study is a slightly modified version of Program 6.8 (2-d plane strain, 8-node quadrilaterals, viscoplasticity reduced integration) in the text by Smith and Griffiths (1998). The modifications mainly relate to the organisation of output, and the inclusion of graphics routines for plotting displacement vectors.

In all the analyses (Goss and Griffiths 2001), a vertical downward displacement δ_v, was applied incrementally to the footing. Nodal reactions beneath the footing were back-figured after each increment from the converged stress field (Woodward and Griffiths 1998), by assembly of the nodal forces in the elements directly beneath the footing given by:

$$Q = \sum \int_{V^e} \mathbf{B}^T \sigma dV^e \tag{16}$$

Convergence after each increment was defined as having occurred when the nodal displacements from one iteration to the next were changing by less that 0.01%. Bearing capacity failure was deemed to have occurred when the nodal reactions levelled out to within a tolerance of 0.1%.

The shear strength of each layer of undrained clay was governed by Tresca's failure criterion defined by the dimensionless function:

$$F = \frac{(\sigma_1 - \sigma_3)}{2c_u} - 1 \tag{17}$$

Positive values of F (see Section 1.3) generated within the mesh were considered "illegal" and redistributed to neighbouring regions that still had reserves of strength.

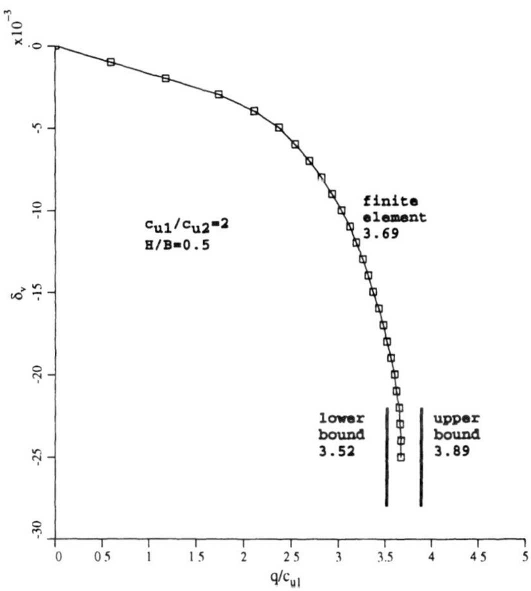

Figure 37. Plot of q/c_{u1} vs. δ_v for $c_{u1}/c_{u2} = 2$, $H/B = 0.5$

A typical plot of q/c_{u1} vs. δ_v is shown in Figure 37 for the case of $c_{u1}/c_{u2} = 2$ and $H/B = 0.5$. The finite element result is compared directly with the upper and lower bound results from Merifield *et al* 1999.

Defining a modified bearing capacity factor with respect to the strength of the upper soil as follows:

$$N_c^* = q_u/c_{u1} \tag{18}$$

it appears that the finite element result of $N_c^* \approx 3.69$ lies between the lower and upper bound solutions of $N_c^* \approx 3.52$ and $N_c^* \approx 3.89$ respectively.

A more complete set of finite element computed N_c^* values for a range of parameters is shown in Tables 4.1 and 4.2, together with upper and lower bound solutions and F_{max}, the largest value of the failure function as given by equation (17) at any point within the mesh at convergence. It should be noted that "perfect convergence" would imply

$F_{max} = 0$ at all Gauss points. A graphical solution to selected results in the table is given in Figure 38.

Table 4.1. Computed values of N_c^*, $H/B < 1$

H/B	c_{u1}/c_{u2}	Lower bound	FE	Upper bound	F_{max}
0.2	0.20	5.44	5.81	5.89	0.0254
	0.25	5.44	5.79	5.89	0.0201
	0.33	5.44	5.78	5.89	0.0154
	0.40	5.44	5.77	5.89	0.0118
	0.50	5.44	5.76	5.89	0.0076
	0.57	5.44	5.75	5.89	0.0062
	0.66	5.42	5.75	5.89	0.0052
	0.80	5.30	5.63	5.71	0.0031
	1.00	4.86	5.11	5.32	0.0014
	1.25	4.06	4.34	4.57	0.0024
	1.50	3.57	3.80	4.02	0.0017
	1.75	3.19	3.40	3.59	0.0018
	2.00	2.90	3.08	3.24	0.0015
	2.50	2.46	2.61	2.77	0.0012
	3.00	2.15	2.28	2.44	0.0008
	3.50	1.93	2.03	2.19	0.0006
	4.00	1.75	1.82	2.00	0.0006
	5.00	1.48	1.51	1.73	0.0006
0.5	0.20	4.86	5.14	5.31	0.0110
	0.25	4.86	5.14	5.31	0.0093
	0.33	4.86	5.14	5.31	0.0070
	0.40	4.86	5.14	5.31	0.0059
	0.50	4.86	5.14	5.31	0.0040
	0.57	4.86	5.14	5.31	0.0034
	0.66	4.86	5.14	5.31	0.0029
	0.80	4.86	5.14	5.31	0.0021
	1.00	4.86	5.11	5.32	0.0015
	1.25	4.42	4.66	4.94	0.0019
	1.50	4.07	4.27	4.48	0.0024
	1.75	3.77	3.95	4.16	0.0024
	2.00	3.52	3.69	3.89	0.0019
	2.50	3.13	3.27	3.47	0.0017
	3.00	2.84	2.96	3.16	0.0015
	3.50	2.62	2.71	2.93	0.0013
	4.00	2.44	2.50	2.74	0.0011
	5.00	2.16	2.15	2.44	0.0008

Table 4.2. Computed values of N_c^*, $H/B \geq 1$

H/B	c_{u1}/c_{u2}	Lower bound	FE	Upper bound	F_{max}
1.0	0.20	4.94	5.11	5.32	0.0074
	0.25	4.94	5.11	5.30	0.0058
	0.33	4.94	5.11	5.30	0.0044
	0.40	4.94	5.11	5.30	0.0037
	0.50	4.94	5.11	5.30	0.0026
	0.57	4.94	5.11	5.30	0.0023
	0.66	4.94	5.11	5.30	0.0020
	0.80	4.94	5.11	5.30	0.0016
	1.00	4.94	5.11	5.30	0.0014
	1.25	4.87	5.11	5.30	0.0011
	1.50	4.77	4.97	5.18	0.0010
	1.75	4.60	4.78	5.00	0.0009
	2.00	4.44	4.61	4.82	0.0010
	2.50	4.14	4.33	4.50	0.0010
	3.00	3.89	4.12	4.24	0.0007
	3.50	3.69	3.95	4.02	0.0006
	4.00	3.46	3.81	3.83	0.0006
	5.00	3.10	3.58	3.54	0.0005
1.5	0.20	4.94	5.11	5.30	0.0070
	0.25	4.94	5.11	5.30	0.0058
	0.33	4.94	5.11	5.30	0.0041
	0.40	4.94	5.11	5.30	0.0034
	0.50	4.94	5.11	5.30	0.0029
	0.57	4.94	5.11	5.30	0.0025
	0.66	4.94	5.11	5.30	0.0021
	0.80	4.94	5.11	5.30	0.0017
	1.00	4.94	5.11	5.32	0.0013
	1.25	4.87	5.11	5.27	0.0024
	1.50	4.87	5.11	5.31	0.0019
	1.75	4.87	5.11	5.31	0.0017
	2.00	4.87	5.11	5.31	0.0014
	2.50	4.84	5.07	5.32	0.0014
	3.00	4.69	4.94	5.15	0.0014
	3.50	4.46	4.79	4.98	0.0012
	4.00	4.24	4.69	4.84	0.0011
	5.00	3.89	4.50	4.56	0.0009

The upper and lower bounds bracket the displacement finite element results, the one exception being when $H/B = 1$ and $c_{u1}/c_{u2} = 5$, where the upper bound solution appears to drift slightly below the finite element result.

The reason for this discrepancy is unclear, however the Tables indicate consistently small values of F_{max}, indicating a high level of convergence and accuracy in the finite element results.

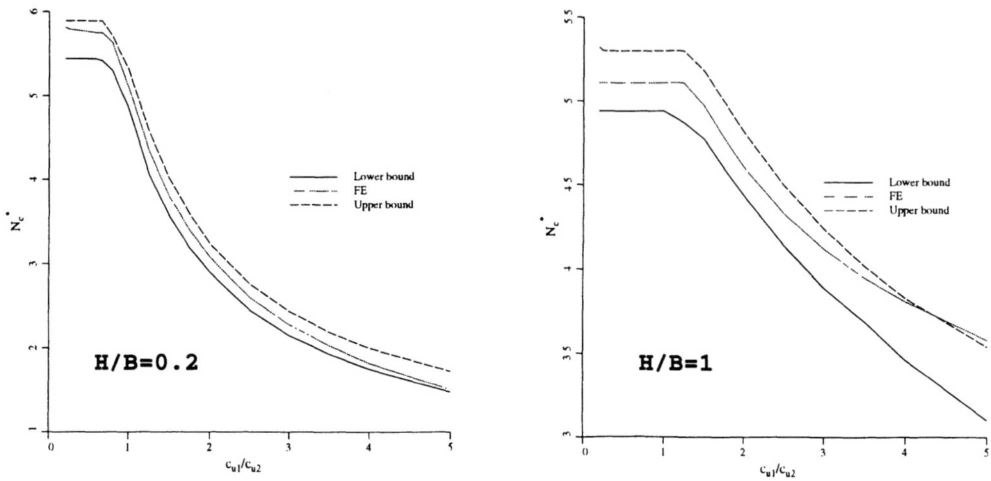

Figure 38. Comparison of finite element with lower and upper bound solutions

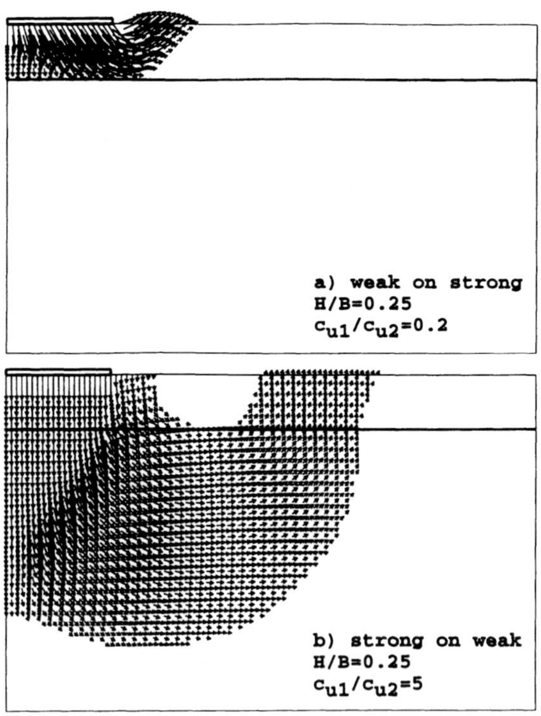

Figure 39. Typical vector plots at failure. a) weak on strong, b) strong on weak.

Figure 39 shows typical nodal displacement patterns from the displacement finite element analyses at failure for the cases of "weak on strong" and "strong on weak". The contrasting nature of the failure mechanisms in each case is clearly indicated.

In summary, while limit analysis can be useful for providing validation checks, the elasto-plastic finite element method is a more powerful and practically useful method for computing bearing capacity. It has been shown that in a single analysis, the method can be relied upon to give robust and accurate solutions to a wide range of geotechnical "failure" problems.

5 A Probabilistic Approach to Bearing Capacity using Finite Elements

The final section of this report extends the elasto-plastic finite element method described above to investigate the bearing capacity of clays with spatially randomly varying shear strength. The methodology is essentially the same as that described in Section 2 for slope stability. The objective of the investigation is to determine the extent to which variance and spatial correlation of the soil's undrained shear strength impacts on the statistics of the bearing capacity. Throughout this section, bearing capacity results are expressed in terms of the bearing capacity factor, N_c, in relation to the *mean* undrained strength. For low coefficients of variation ($C.O.V._{c_u}$) of shear strength, the expected value of the bearing capacity factor tends to the Prandtl solution of $N_c = 5.14$. For higher values of ($C.O.V._{c_u}$) however, the expected value of the bearing capacity factor falls quite steeply. As in the slope stability studies, the spatial correlation length is shown to be an important parameter which should not be ignored. The results of Monte-Carlo simulations on this nonlinear problem are presented in the form of histograms, which enable the interpretation to be expressed in a probabilistic context. Results obtained in this study help to explain why bearing capacity calculations require relatively high factors of safety compared to other branches of geotechnical design.

5.1 Introduction

This section presents results obtained using a program developed by the authors (Griffiths and Fenton 2001) which merges nonlinear elasto-plastic finite element analysis such as described in the previous section, with random field theory. The program computes the bearing capacity of a smooth rigid strip footing (plane strain) at the surface of a undrained clay soil with a shear strength c_u ($\phi_u = 0$) defined by a spatially varying random field.

Rather than deal with the actual bearing capacity, this study focuses on the dimensionless bearing capacity factor N_c defined:

$$N_c = q_f / \mu_{c_u} \tag{19}$$

where q_f is the bearing capacity and μ_{c_u} is the mean undrained shear strength of the soil beneath the footing. For a homogeneous soil with a constant undrained shear strength, N_c is given by the Prandtl solution (eqn. 15) and equals $2 + \pi$ or 5.14. For soils with a variable shear strength, the bearing capacity factor will be defined with respect to the *mean* undrained shear strength.

In this study, the variability of the undrained shear strength is lognormal, and defined by the same three statistical parameters shown in Table 2.1. In the parametric studies that follow, the mean strength (μ_{c_u}) has been held constant at 100 kN/m^2, while the coefficient of variation ($C.O.V._{c_u}$) and spatial correlation length ($\Theta_{c_u} = \theta_{\ln c_u}/B$), non-dimensionalised with respect to the footing width B, are varied systematically.

For each set of assumed statistical properties given by $C.O.V._{c_u}$ and Θ_{c_u}, Monte-Carlo simulations have been performed. These typically involve 1000 repetitions or "realizations" of the shear strength random field and the subsequent finite element analysis of bearing capacity. This means that each realization, while having the same underlying statistics, leads to a quite different spatial pattern of shear strength values beneath the footing. Each realization, therefore, leads to a different value of the bearing capacity and, after normalization by the *mean* undrained shear strength, a different value of the bearing capacity factor,

$$N_{c_i} = q_{f_i}/\mu_{c_u}, \qquad i = 1, 2, ..., n_{sim} \tag{20}$$

IN this study, $n_{sim} = 1000$, and once the bearing capacity factors from all the realisations have been accumulated, they in turn can be subjected to statistical analysis. Estimated (sample) mean bearing capacities will a have standard error ($\pm$ one standard deviation) equal to the sample standard deviation times $1/\sqrt{n} = 1/\sqrt{1000} = 0.032$, or about 3% of the sample standard deviation. Similarly, the sample variance will have standard error of about 0.04 times the sample variance. This means that estimated quantities will generally be within about 5% of the true quantities, statistically speaking.

Of particular interest in the present study, is the *probability* that the actual bearing capacity factor, N_c, as defined in equation (20), will be *less* than the Prandtl value of 5.14 that would be obtained assuming a homogeneous soil with the undrained shear strength everywhere equal to the mean value μ_{c_u}.

5.2 Review of the lognormal distribution

As in the Slope Stability analyses of Section 2, a lognormal distribution for the undrained shear strength c_u has been adopted, meaning that $\ln c_u$ is normally distributed. If the mean and standard deviation of the undrained shear strength are μ_{c_u} and σ_{c_u} respectively, then the standard deviation and mean of the underlying normal distribution of $\ln c_u$ are given by:

$$\sigma_{\ln c_u} = \sqrt{\ln\left\{1 + \left(\frac{\sigma_{c_u}}{\mu_{c_u}}\right)^2\right\}} \tag{21}$$

$$\mu_{\ln c_u} = \ln \mu_{c_u} - \frac{1}{2}\sigma^2_{\ln c_u} \tag{22}$$

and the probability density function of the lognormal distribution by:

$$f(c_u) = \frac{1}{c_u\, \sigma_{\ln c_u} \sqrt{2\pi}} \exp\left\{-\frac{1}{2}\left(\frac{\ln c_u - \mu_{\ln c_u}}{\sigma_{\ln c_u}}\right)^2\right\} \tag{23}$$

The properties of the lognormal distribution can therefore be summarized as:

$$\mu_{c_u} = \exp\left(\mu_{\ln c_u} + \frac{1}{2}\sigma^2_{\ln c_u}\right) \tag{24}$$

$$\sigma_{c_u} = \mu_{c_u}\sqrt{\exp(\sigma^2_{\ln c_u}) - 1} \tag{25}$$

$$\text{Median} = \exp(\mu_{\ln c_u}) \tag{26}$$

$$\text{Mode} = \exp(\mu_{\ln c_u} - \sigma^2_{\ln c_u}) \tag{27}$$

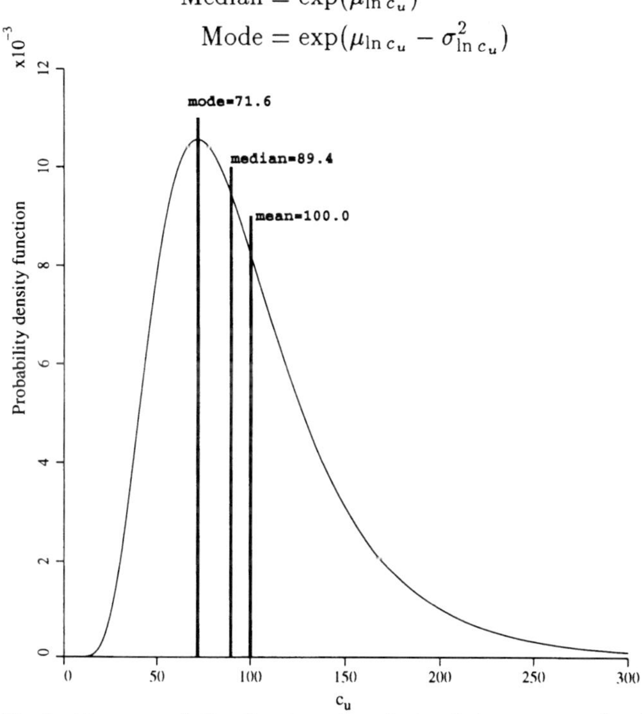

Figure 40. Typical lognormal distribution of undrained shear strength with a mean of 100 and standard deviation of 50 ($C.O.V._{c_u} = 0.5$). All units are in kN/m^2.

A typical lognormal distribution based on equation (23) with mean $\mu_{c_u} = 100$ kN/m^2 and standard deviation $\sigma_{c_u} = 50$ kN/m^2 ($C.O.V._{c_u} = 0.5$) is shown in Figure 40. From equations (21) and (22) it is easily shown that the underlying "normal" statistics are given by $\sigma_{\ln c_u} = 0.472$ and $\mu_{\ln c_u} = 4.494$. Highlighted also on the figure are the median and mode of the distribution, which can be shown from equations (26) and (27) to equal, respectively, 89.4 kN/m^2 and 71.6 kN/m^2. The skewed nature of the lognormal distribution always results in the mode, median and mean being in the sequence indicated. In a lognormal distribution, the median is always smaller than the mean, and this will have implications for the probabilistic interpretation of the bearing capacity results described later in this section.

Use of the lognormal distribution, as opposed to the more familiar normal distribution, or even some other more complex distribution, is based on the following arguments: Firstly, there is a lack of exhaustive field data that would be necessary to conclusively support one kind of distribution over another. However, there is some evidence from the field to support the lognormal distribution for some soil properties (see e.g. Hoeksema and Kitanidis 1985, and Sudicky 1986). Use of the lognormal distribution is also based on the simplicity and familiarity of its two-parameters description. Secondly, and perhaps most importantly from a physical standpoint, the lognormal distribution is strictly non-negative, unlike the normal distribution, and so there is no possibility of generating properties with meaningless negative values, particularly in the extremes of the distribution (which may be important from a reliability standpoint). It might also be noted that a lognormal distribution looks quite similar to a normal distribution for low values of the *C.O.V.*.

Lee *et al* (1983) comment that the "normal or lognormal distributions are adequate for the large majority of geotechnical data" however Harr (1987) finds the unbounded nature of the upper end of the lognormal distribution objectionable. The potential for the lognormal distribution to generate very high property values (albeit with a low probability) is not considered a serious flaw, especially in a study involving the shear strength of heterogeneous soil that is spatially distributed (what is the shear strength of a point that happens to fall inside a boulder of granite?). It is certainly possible that a soil deposit will contain occasional inclusions of very strongly cemented material.

5.3 Brief description of FE method used

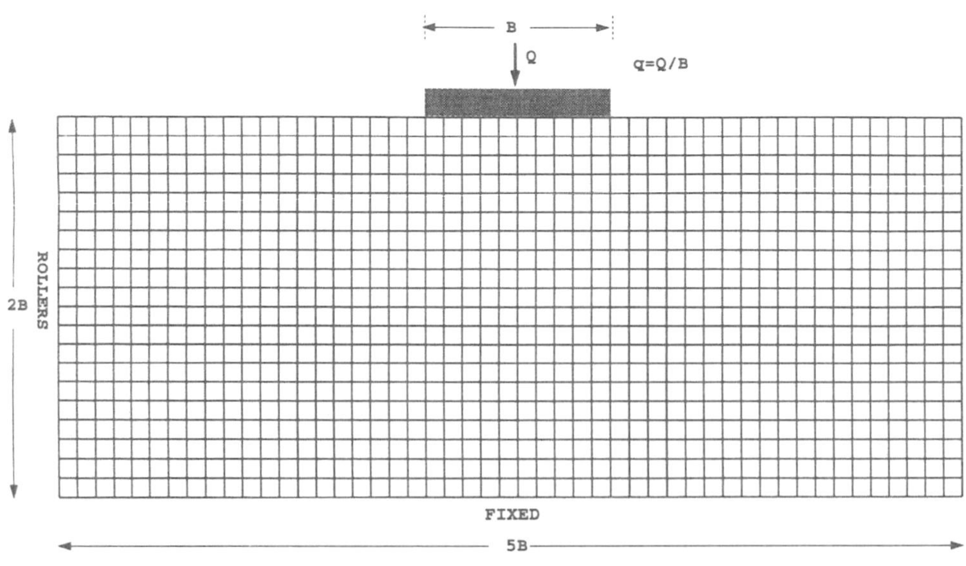

Figure 41. Mesh used in probabilistic bearing capacity analyses (units in m).

A typical mesh is shown in Figure 41 consisting of 1000 elements, with 50 columns and 20 rows. Each element is square, and the strip footing has a width of 10 elements. The bearing capacity analyses use essentially the same program as described in the previous section for a two-layer soil incorporating a Tresca failure criterion.

The finite element model incorporates three parameters; Young's modulus (E), Poisson's ratio (ν) and the undrained shear strength (c_u). The methodology allows for random distributions of all three parameters, however in the present study, E and ν are held constant while c_u is randomized.

At the ith realisation of the Monte-Carlo process, the footing is incrementally displaced vertically (δ_v) into the soil and the sum of the nodal reactions (Q_i) back-figured from the converged stress state. When the sum of the nodal reactions levels out to within a quite strict tolerance (see Section 4.2), "failure" is said to have occurred and the sum of the nodal reactions divided by the footing area is the "bearing capacity" ($q_{f_i} = Q_{f_i}/B$) of that particular realisation.

5.4 Brief description of the random field model

The undrained shear strength is obtained through the transformation

$$c_{u_i} = \exp\{\mu_{\ln c_u} + \sigma_{\ln c_u} g_i\} \tag{28}$$

in which c_{u_i} is the undrained shear strength assigned to the ith element, g_i is the local average of a standard Gaussian random field, g, over the domain of the ith element, and $\mu_{\ln c_u}$ and $\sigma_{\ln c_u}$ are the mean and standard deviation of the logarithm of c_u (obtained from the "point" mean and standard deviation μ_{c_u} and σ_{c_u} after local averaging).

The LAS technique (Fenton 1990, Fenton and Vanmarcke 1990) generates realizations of the local averages g_i which are derived from the random field g having zero mean, unit variance, and a spatial correlation length, $\theta_{\ln c_u}$. As the spatial correlation length tends to infinity, g_i becomes equal to g_j for all elements i and j – that is the field of shear strengths tends to become uniform on each realization. At the other extreme, as the spatial correlation length tends to zero, g_i and g_j become independent for all $i \neq j$ – the soil's undrained shear strength changes rapidly from point to point. In the present study, a Markovian spatial correlation function was used, of the form:

$$\rho(|\tau|) = \exp\left\{-\frac{2}{\theta_{\ln c_u}}|\tau|\right\} \tag{29}$$

where ρ is the correlation coefficient between the logarithm of the undrained strength values at any two point separated by a distance τ in a random field with spatial correlation length $\theta_{\ln c_u}$.

In the two-dimensional analyses presented in this paper, the spatial correlation lengths in the vertical and horizontal directions are taken to be equal (isotropic) for simplicity. Fenton (1999) examined CPT data in relation to random field modeling, however the actual spatial correlation structure of soil deposits is not usually well known, especially in the horizontal direction (see e.g. DeGroot and Baecher 1993, de Marsily 1985, Asaoka and Grivas 1982). In this paper therefore, a parametric approach has been employed to study the influence of $\theta_{\ln c_u}$.

The plane strain model used herein implies that the out-of-plane spatial correlation length is infinite, thus soil properties are constant in this direction. This is clearly a deficiency. However, previous studies by the authors (Griffiths and Fenton 1997) involving seepage through two- and three-dimensional random fields have indicated that the difference may not be very great. The role of the third dimension is an area of ongoing research by the authors.

A local averaging process has been included in the formulation to take full account of the level of mesh discretization, and the size of the finite elements onto which the random field is to be mapped. Local averaging preserves the mean, but reduces the standard deviation of the underlying normal field to a "target" value . The amount by which the standard deviation is reduced, depends on the size of the elements and the nature of the spatial correlation function governing the field. More specifically, there is a function called the "variance function", which can be derived from the correlation function, which governs the rate at which the standard deviation drops as the averaging domain grows larger. The interested reader is referred to Vanmarcke (1984) for a detailed description of this formulation.

Although the mean of the underlying Gaussian field is unaltered by local averaging, equations (24) and (25) indicate that since both the mean *and* standard deviation of the lognormal field are functions of $\sigma_{\ln c_u}$ they will *both* be reduced by the local averaging process. Thus the coarser the mesh, the greater the reduction in the "target" statistics from their nominal "point" values. This local averaging approach is fully implemented in this study, and removes any "mesh effects" that might otherwise be present. It might also be commented that this approach is quite consistent with the philosophy of the finite element method in which finer meshes resolve the finer variations in the stress and material property fields.

5.5 Parametric studies

Analyses were performed using the mesh of Figure 41 with the input parameters in the following ranges:

$$\begin{aligned} 0.125 \leq\ & \Theta_{c_u} < \infty \\ 0.125 \leq\ & C.O.V._{c_u} \leq 4 \end{aligned} \tag{30}$$

To indicate the nature of the different solutions obtained at each realisation of the Monte-Carlo process, load/deformation results for 10 typical realisations of the footing analysis are shown in Figure 42 for the case when $\Theta_{c_u} = 1$ and $C.O.V._{c_u} = 1$. The average stress q under the footing has been non-dimensionalised by dividing it by the mean undrained shear strength μ_{c_u}. The reader should bear in mind the Prandtl solution of 5.14 when viewing this figure. It is clear that a majority of the curves flatten out at bearing capacity values below the Prandtl solution. This trend will be confirmed in all the results shown in this paper.

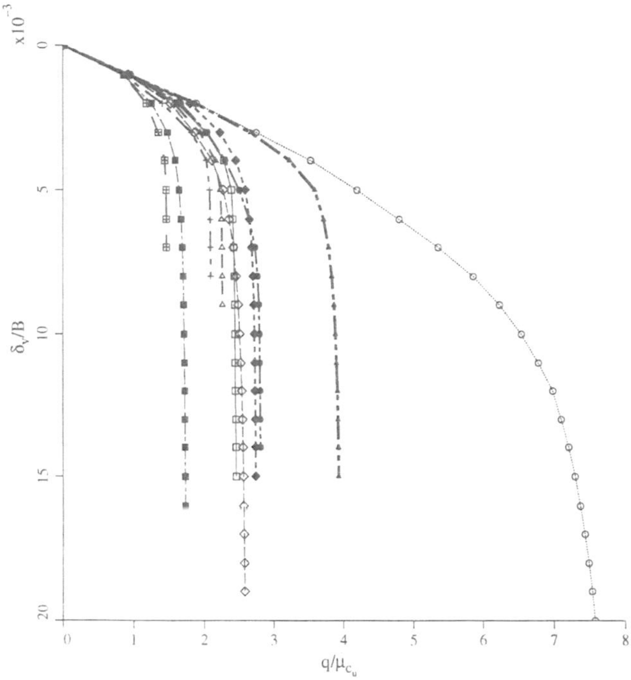

Figure 42. Typical load/deformation curves corresponding to different realizations in the bearing capacity analysis of an undrained clay with $\Theta_{c_u} = 1$ and $C.O.V._{c_u} = 1$.

Figure 43 shows a typical deformed mesh at failure with a superimposed greyscale corresponding to $\Theta_{c_u} = 1$, in which lighter regions indicate stronger soil and darker regions indicate weaker soil. In this case the dark zones and the light zones are roughly the width of the footing itself, and it appears that the weak (dark) region near the ground surface to the right of the footing has triggered a quite non-symmetric failure mechanism. The shape of the non-symmetric mechanism is emphasised further by the plot of displacement vectors for the same realisation shown in Figure 44.

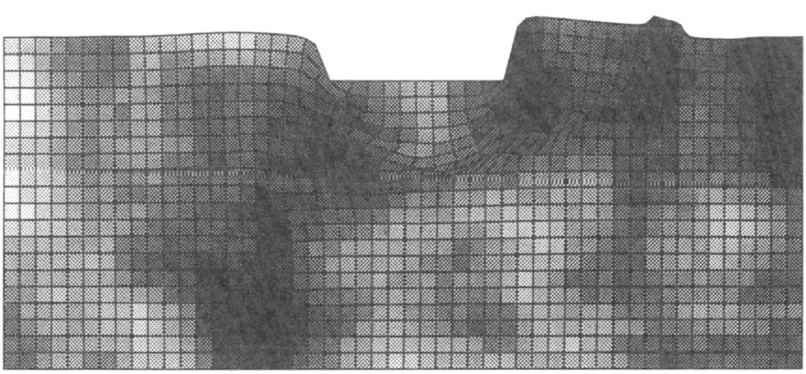

Figure 43. Typical deformed mesh and greyscale at failure with $\Theta_{c_u} = 1$. The darker regions indicate weaker soil.

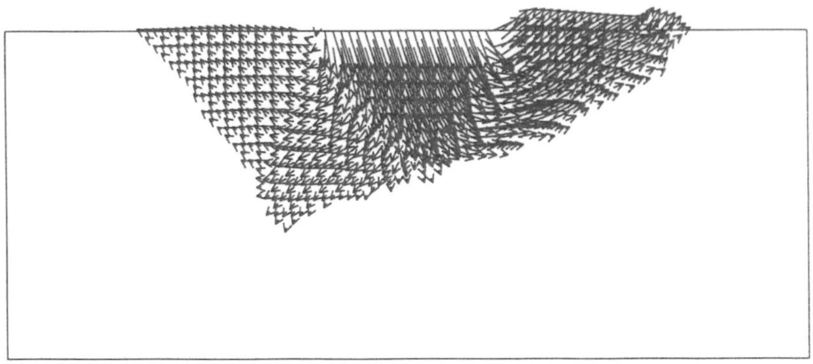

Figure 44. Displacement vectors at failure for the same case shown in Figure 43. The non-symmetric shape of the failure mechanism is clearly visible.

For each combination of Θ_{c_u} and $C.O.V._{c_u}$, $n_{sim} = 1000$ realisations of the Monte-Carlo process were performed, and the estimated mean (m_{N_c}) and standard deviation (s_{N_c}) of the resulting 1000 bearing capacity factors from equation (20) computed.

Figure 45a shows how the estimated mean bearing capacity factor m_{N_c} varies with Θ_{c_u} and $C.O.V._{c_u}$. The plot confirms that for low values of $C.O.V._{c_u}$, m_{N_c} tends to the deterministic Prandtl value of 5.14. For higher values of $C.O.V._{c_u}$ however, the mean bearing capacity factor falls steeply, especially for lower values of Θ_{c_u}. For example, in a highly variable case where $\Theta_{c_u} = 0.5$ and $C.O.V._{c_u} = 4$, the predicted m_{N_c} value is less than unity – over five times smaller than the Prandtl value! For the recommended upper limit of $C.O.V._{c_u} = 0.5$ suggested by Lee *et al* (1983) and others, the m_{N_c} value is closer to 4, corresponding to a more modest reduction of 20%.

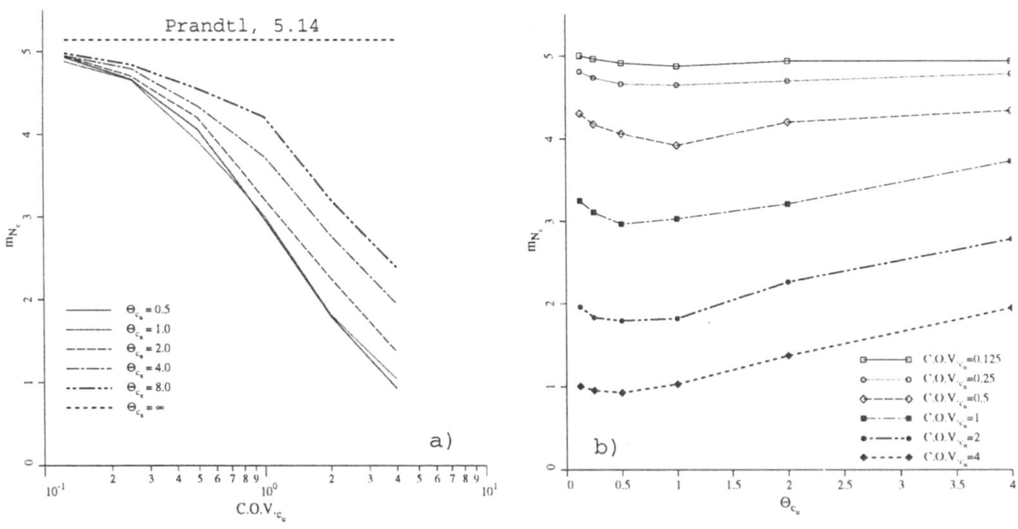

Figure 45. Estimated mean bearing capacity factor m_{N_c} as a function of undrained shear strength statistics, Θ_{c_u} and $C.O.V._{c_u}$.

What this implies from a design standpoint, is that the bearing capacity of a heterogeneous soil, will on average be *less* than the Prandtl solution that would be predicted assuming the soil is homogeneous with its strength given by the mean value. The influence of Θ_{c_u} is also pronounced, with the greatest reduction from the Prandtl solution being observed with values around $\Theta_{c_u} \approx 0.5$. As the value of Θ_{c_u} is reduced further towards zero, there is evidence of a gradual increase in the value of m_{N_c} as shown in Figure 45b. From a theoretical point of view, it could be speculated that as Θ_{c_u} becomes vanishingly small, the mean bearing capacity factor will continue to increase towards the deterministic Prandtl solution of 5.14. The explanation lies in the fact that as the spatial correlation length decreases, the weakest path becomes increasingly tortuous and its length correspondingly longer. As a result, the weakest path starts to look for shorter routes cutting through higher strength material. In the limit, as $\Theta_{c_u} \rightarrow 0$, it is expected that the optimum failure path will be the same as in a uniform material with strength equal to the mean value, hence returning to the deterministic Prandtl solution.

Also included on Figure 45a is a horizontal line corresponding to the analytical solution that would be obtained for $\Theta_{c_u} = \infty$. This hypothetical case implies that each realisation of the Monte-Carlo process involves an essentially homogeneous soil, albeit with strength varying only from one realisation to the next. In this case, the distribution of q_f will be statistically similar to the underlying distribution of c_u but magnified by 5.14. The mean bearing capacity will therefore be given by:

$$\mu_{q_f} = 5.14\mu_{c_u} \tag{31}$$

hence $\mu_{N_c} = 5.14$ for all $C.O.V._{c_u}$.

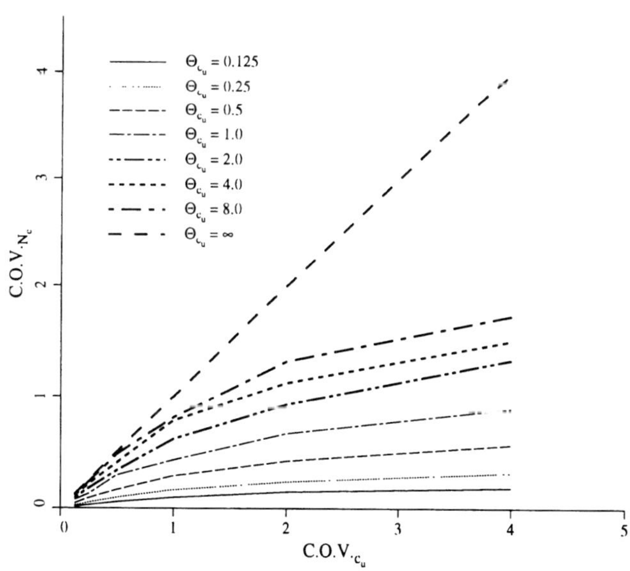

Figure 46. Estimated coefficient of variation of the bearing capacity factor $C.O.V._{N_c} = s_{N_c}/m_{N_c}$ as a function of undrained shear strength statistics, Θ_{c_u} and $C.O.V._{c_u}$.

Figure 46 shows the influence of Θ_{c_u} and $C.O.V._{c_u}$ on the estimated coefficient of variation of the bearing capacity factor, $C.O.V._{N_c} = s_{N_c}/m_{N_c}$. The plots indicate that $C.O.V._{N_c}$ is positively correlated with both $C.O.V._{c_u}$ and Θ_{c_u}. This figure also indicates that the correlation length, Θ_{c_u}, has a significant influence on $C.O.V._{N_c}$. For small correlation lengths, $C.O.V._{N_c}$ is small and rather insensitive to $C.O.V._{c_u}$, however for higher correlation lengths, $C.O.V._{N_c}$ increases quite consistently until it reaches the limiting maximum value corresponding to $\Theta_{c_u} = \infty$, defined by the straight line where $C.O.V._{N_c} = C.O.V._{c_u}$.

5.6 Probabilistic interpretation

Following Monte-Carlo simulations for each parametric combination of input parameters (Θ_{c_u} and $C.O.V._{c_u}$), the suite of computed bearing capacity factor values from equation (20) was plotted in the form of a histogram, and a "best-fit" lognormal distribution superimposed. An example of such a plot is shown in Figure 47 for the case where $\Theta_{c_u} = 2$ and $C.O.V._{c_u} = 1$.

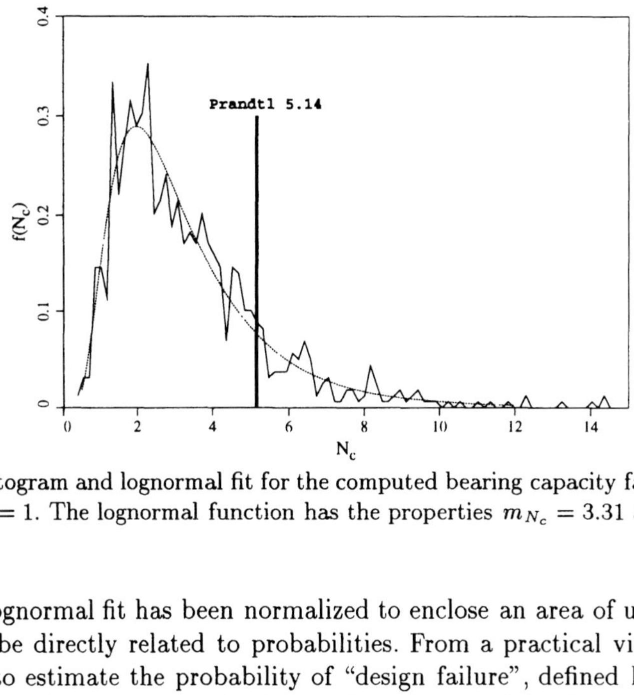

Figure 47. Histogram and lognormal fit for the computed bearing capacity factors when $\Theta_{c_u} = 2$ and $C.O.V._{c_u} = 1$. The lognormal function has the properties $m_{N_c} = 3.31$ and $s_{N_c} = 2.08$.

Since the lognormal fit has been normalized to enclose an area of unity, areas under the curve can be directly related to probabilities. From a practical viewpoint it would be of interest to estimate the probability of "design failure", defined here as occurring when the computed bearing capacity is less than the Prandtl value based on the mean strength, i.e.

$$\text{“Design failure” if } q_f < 5.14\mu_{c_u} \tag{32}$$

Let this probability be $p(N_c < 5.14)$, hence from the properties of the underlying normal distribution we get:

$$p(N_c < 5.14) = \Phi\left(\frac{\ln 5.14 - m_{\ln N_c}}{s_{\ln N_c}}\right) \tag{33}$$

where Φ is the cumulative normal function.

For the particular case shown in Figure 47, the fitted lognormal distribution has the properties $m_{N_c} = 3.31$ and $s_{N_c} = 2.08$, hence from equations (21) and (22) the underlying normal distribution is defined by $m_{\ln N_c} = 1.03$ and $s_{\ln N_c} = 0.58$. Equation (33) therefore, gives $p(N_c < 5.14) = 0.85$, indicating an 85% probability that the actual bearing capacity will be less than the Prandtl value.

Figure 48 gives a summary of $p(N_c < 5.14)$ for a range of values of Θ_{c_u} and $C.O.V._{c_u}$. The figure indicates a wide spread of probability values with respect to Θ_{c_u}, with the highest probabilities corresponding to the lowest values of Θ_{c_u}. For example, a soil with $C.O.V._{c_u} = 0.5$, exhibits a range of $0.59 < p(N_c < 5.14) < 0.95$, with the low and high values corresponding respectively to $\Theta_{c_u} = \infty$ and $\Theta_{c_u} = 0.5$.

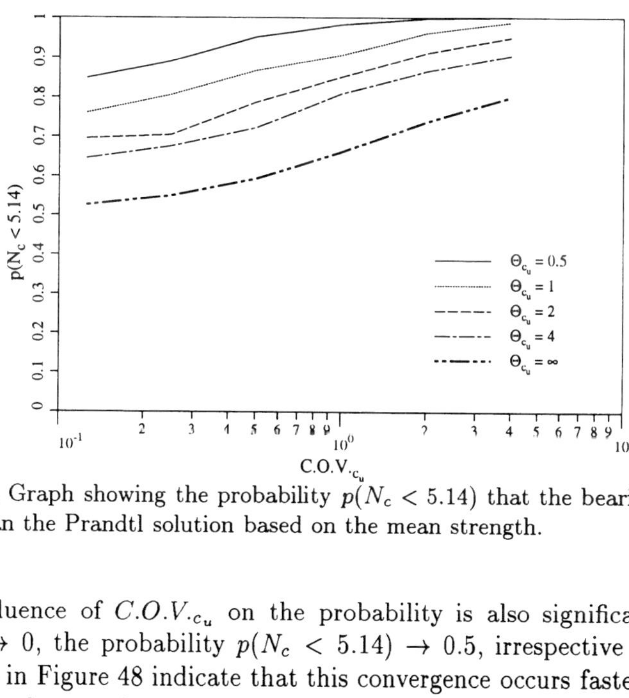

Figure 48. Graph showing the probability $p(N_c < 5.14)$ that the bearing capacity factor will be lower than the Prandtl solution based on the mean strength.

The influence of $C.O.V._{c_u}$ on the probability is also significant. Theoretically, as $C.O.V._{c_u} \to 0$, the probability $p(N_c < 5.14) \to 0.5$, irrespective of the value of Θ_{c_u}. The results in Figure 48 indicate that this convergence occurs faster for higher values of Θ_{c_u} than for lower values. It would appear that low values of Θ_{c_u} permit such widely scattered weak elements, that the probability of the actual bearing capacity lying below the Prandtl value remains high, even for low $C.O.V._{c_u}$ values. This general trend is to be expected however, because for low $C.O.V._{c_u}$ values, the distribution of bearing capacity factors becomes "bunched up" and "centered" on 5.14, giving an almost equal chance of the computed bearing capacity factor lying on either side of the Prandtl solution.

As $C.O.V._{c_u}$ is increased, the probability $p(N_c < 5.14)$ also increases. For example, when $\Theta_{c_u} = 0.5$ and $C.O.V._{c_u} = 0.5$, $p(N_c < 5.14) = 0.95$ indicating a 95% probability that the actual bearing capacity will be lower than the Prandtl solution.

The result corresponding to the limiting case of $\Theta_{c_u} = \infty$ is also indicated in Figure 48. As discussed previously, the distribution of q_f in this case is statistically similar to the underlying distribution of c_u and the required probability $p(N_c < 5.14)$, simply equals the area under the probability density function to the left of the mean. For a lognormal distribution, this probability is always greater than 0.5 and given by:

$$p(N_c < 5.14) = \Phi(0.5\, \sigma_{\ln c_u}) \tag{34}$$

thus from equation (21),

$$p(N_c < 5.14) = \Phi(0.5 \sqrt{\ln(1 + C.O.V._{c_u}^2)}\,) \tag{35}$$

Figure 48 indicates that the expected bearing capacity of a strip footing on an undrained clay with variable shear strength defined by a lognormal distribution, will *always* be lower than the Prandtl value based on the mean strength. It could be argued, however, that this interpretation gives an over-pessimistic impression of the role of soil strength variability by not taking account of the variance of the computed bearing capacity distribution. Even an essentially deterministic analysis with a very small shear strength variance would suggest a > 50% probability that the bearing capacity would lie below the Prandtl value.

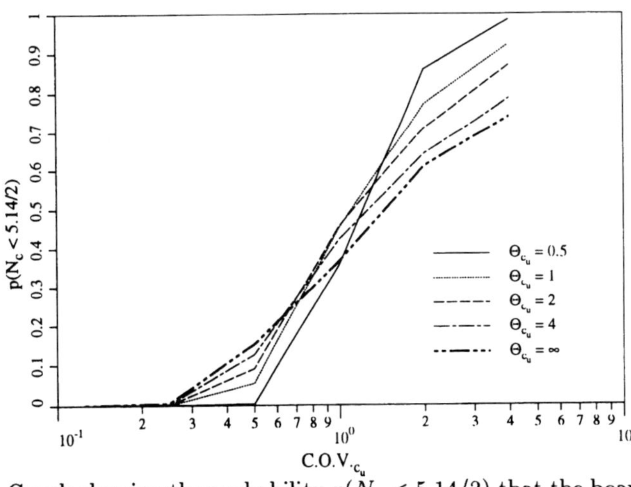

Figure 49. Graph showing the probability $p(N_c < 5.14/2)$ that the bearing capacity factor will be lower than the Prandtl solution based on the mean strength incorporating a Factor of Safety $F = 2$.

In order to remove this anomaly, the results have been reinterpreted in Figures 49 and 50 to compare the computed bearing capacity factor with the Prandtl solution after it has been reduced by a factor F. The factor F is equivalent to a "Factor of Safety" applied to the deterministic bearing capacity based on mean strength. The probability of "design failure" as measured by $p(N_c < 5.14/F)$ is now greatly reduced, giving a more reassuring result from a design viewpoint. For example, from Figure 49 in which $F = 2$,

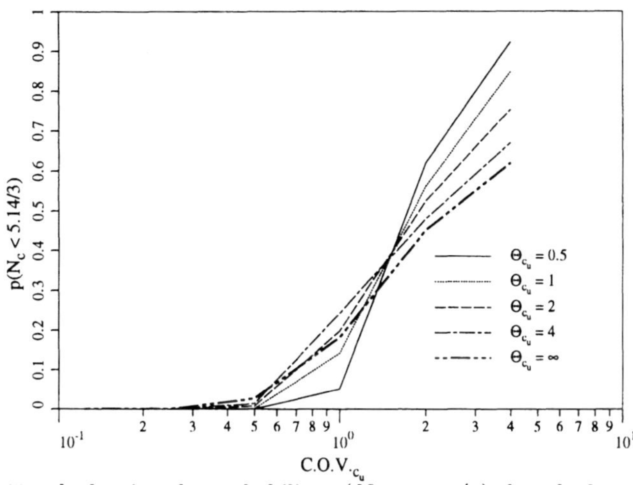

Figure 50. Graph showing the probability $p(N_c < 5.14/3)$ that the bearing capacity factor will be lower than the Prandtl solution based on the mean strength incorporating a Factor of Safety $F = 3$.

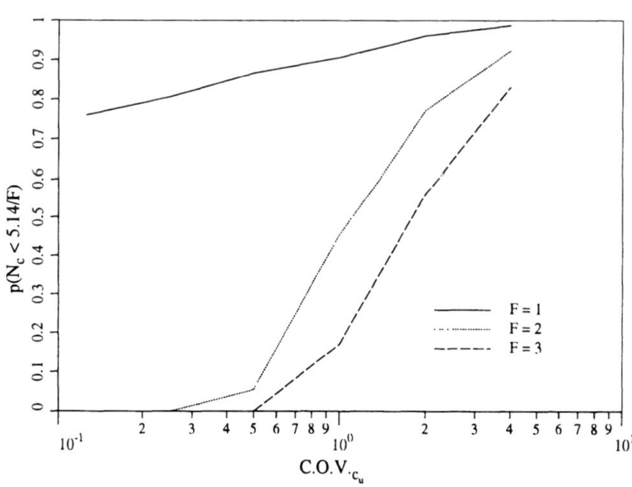

Figure 51. Graph showing the probability $p(N_c < 5.14/F)$ that the bearing capacity factor will be lower than the Prandtl solution based on the mean strength for three different Factors of Safety F for a soil with $\Theta_{c_u} = 1$.

the probability of "design failure" for a soil with $\Theta_{c_u} = 1$ and $C.O.V._{c_u} = 0.5$, is about 6%. This probability is essentially reduced to zero for the same soil, by increasing the factor to $F = 3$, as shown in Figure 50.

Figure 51 shows directly how F affects the probability of design failure for a range of $C.O.V._{c_u}$ in a soil where the correlation length is held constant at $\Theta_{c_u} = 1$. The results indicates that quite high factors of safety are required to reduce the probability of "design failure" to "acceptable" levels. Figures 50 and 51 both suggest that for a soil with $C.O.V._{c_u} = 0.5$ and $\Theta_{c_u} = 1$, a Factor of Safety of at least 3 is needed to essentially eliminate all probability of design failure. This is consistent with geotechnical engineering practice, where a factor of safety of at least 3 (e.g. Lambe and Whitman 1969) is considered necessary to protect against general shear failure.

A further interpretation of the probability of design failure is shown in Figures 52 where a direct comparison is given between the probability of "design failure" $p(N_c < 5.14/F)$ and the Factor of Safety F for a range of Θ_{c_u} and $C.O.V._{c_u}$ values.

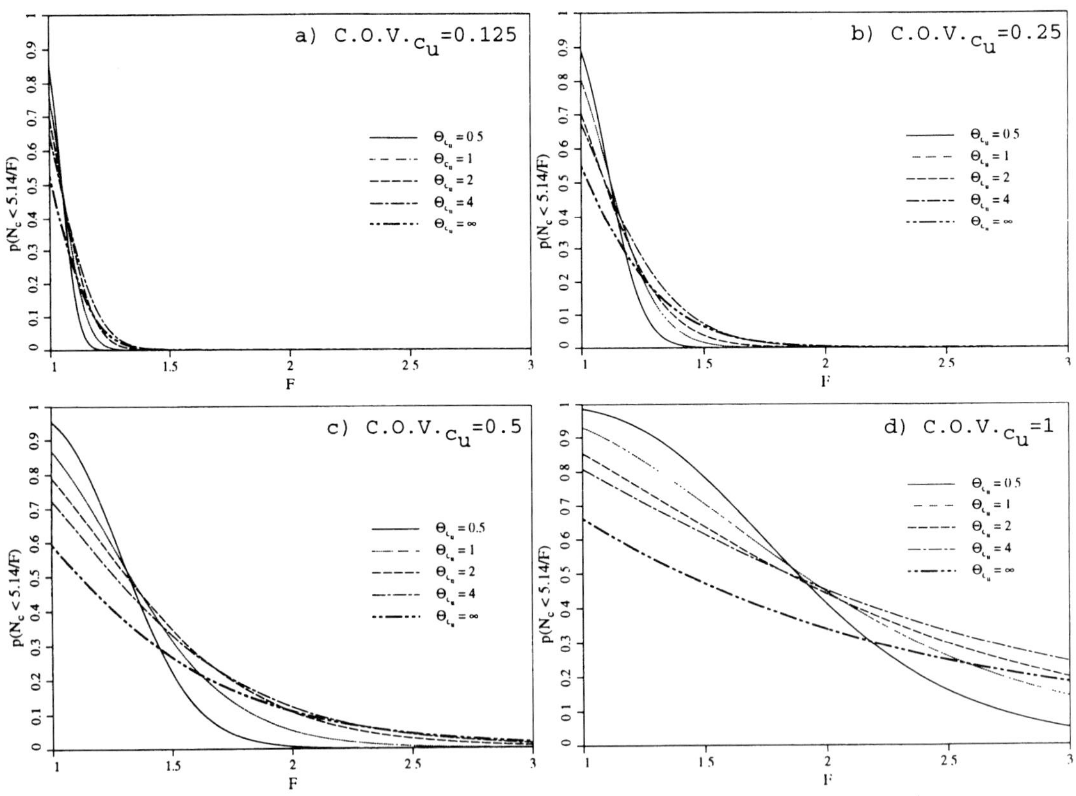

Figure 52. Graphs showing the relationship between $p(N_c < 5.14/F)$ and F for a soil with a) $C.O.V._{c_u} = 0.125$, b) $C.O.V._{c_u} = 0.25$, c) $C.O.V._{c_u} = 0.5$ and d) $C.O.V._{c_u} = 1$.

If the goal is to virtually eliminate any possibility of "design failure" involving a bearing capacity calculation based on the mean strength, Figure 52a indicates that for a soil with $C.O.V._{c_u} = 0.125$ (a value at the lower end of the recommended range of Lee *et al* (1983) and others), a Factor of Safety of $F = 1.5$ would be needed. For an intermediate value of $C.O.V._{c_u} = 0.25$, the required Factor of Safety becomes $F = 2$ (Figure 52b), and for a soil with $C.O.V._{c_u} = 0.5$ (a value at the upper end of the recommended range), the required Factor of Safety increases further to $F = 3$ as shown in Figure 52c. In the case of $C.O.V._{c_u} = 1$ (Figure 52d) - a value that might be considered exceptionally high for most real soils, the need for even higher Factors of Safety is indicated.

An important observation from Figure 52 is that the Correlation Length Θ_{c_u} becomes increasingly relevant to the probabilistic interpretation of the bearing capacity problem as $C.O.V._{c_u}$ gets larger. This is clear from the way the curves are bunched together when $C.O.V._{c_u} = 0.125$ (Figure 52a) while quite divergent when $C.O.V._{c_u} = 1$ (Figure 52d).

In all of Figure 52, the "crossing-over" of the lines corresponding to different Θ_{c_u} values implies that high values of Θ_{c_u} are beneficial to design at low values of F by giving lower probabilities of "design failure", but may be a liability at higher values of F. The explanation lies in the fact that smaller correlation lengths lead to smaller values of $C.O.V._{N_c}$ as shown in Figure 46. Increasing F will therefore result in a steeper fall in the probability of "design failure" as the factored bearing capacity factor rapidly passes through the "bunched up" distribution.

In addition to the expected trend which shows $p(N_c < 5.14/F)$ decreasing as F increases for all Θ_{c_u}, the curves also confirm that a Factor of Safety of 3 is able to reduce the probability of "design failure" to negligible levels for all soils in the recommended range of $0.1 < C.O.V._{c_u} < 0.5$. These results may help explain in a probabilistic context, why Factors of Safety used in bearing capacity calculations are typically much higher than those used in other limit state calculations in geotechnical engineering, e.g. slope stability, earth pressures.

6 Conculding Remarks

This section has shown that soil strength heterogeneity in the form of a spatially varying lognormal distribution can significantly reduce the mean bearing capacity of a strip footing on undrained clay. The following more specific conclusions can be made:

1. As the variance of soil strength increases, the mean bearing capacity decreases. A minimum mean bearing capacity was observed for correlation lengths of approximately one half of the footing width. For still smaller correlation lengths, a modest increase in the mean bearing capacity was detected. It could be speculated that as Θ_{c_u} becomes vanishingly small, the mean bearing capacity factor will continue to increase towards the deterministic Prandtl solution of 5.14. The explanation may lie in the fact that with no spatial correlation, there are no preferred paths of weaker material to attract the mechanism, and the material response is "homogeneous", yielding an essentially deterministic symmetric mechanism at failure.

2. The coefficient of variation of the bearing capacity was observed to be positively correlated with both the coefficient of variation of the soil strength and its spatial correlation length.
3. Results have been presented in a probabilistic context, to determine the probability of "design failure", defined as the probability that the actual bearing capacity would be lower than a factored deterministic prediction of bearing capacity using Prandtl's formula based on the mean strength of the soil.
4. By investigating the role of a Factor of Safety applied to the Prandtl solution, it was observed that a value of $F = 3$ would essentially eliminate any possibility of "design failure" for soils with a strength variability within the recommended range.
5. The influence of correlation length on the probabilistic interpretation of the bearing capacity problem was shown to be significant, especially for soils with higher values of $C.O.V._{c_u}$.

7 Acknowledgements

The writer wishes to acknowledge the support of NSF Grant No. CMS 98-77189.

References

L.W. Abramson, T.S. Lee, S. Sharma, and G.M. Boyce. *Slope stability and stabilisation methods.* John Wiley and Sons, 1995.

A. Asaoka and D.A. Grivas. Spatial variability of the undrained strength of clays. *J Geotech Eng, ASCE*, 108(5):743–756, 1982.

A.W. Bishop. The use of the slip circle in the stability analysis of slopes. *Géotechnique*, 5(1):7–17, 1955.

A.W. Bishop and N.R. Morgenstern. Stability coefficients for earth slopes. *Géotechnique*, 10:129–150, 1960.

R.I. Borja, S.R. Lee, and R.B. Seed. Numerical simulation of excavation in elasto-plastic soils. *Int J Numer Anal Methods Geomech*, 13(3):231–249, 1989.

J.D. Brown and G.G. Meyerhof. Experimental study of bearing capacity in layered clays. In *Proc 7th Int Conf Soil Mech Found Eng*, pages 45–51. 1969.

P.T. Brown and J.R. Booker. Finite element analysis of excavation. *Comput Geotech*, 1:207–220, 1985.

S.J. Button. The bearing capacity of footings on a two-layer cohesive sub-soil. In *Proc 3rd Int Conf Soil Mech Found Eng*, pages 332–335. 1953.

A.R. Calderon. The application of back-analysis and numerical modeling to design a large pushback in a deep open pit mine. Master's thesis, Department of Mining Engineering, Colorado School of Mines, 2000.

W.F. Chen. *Limit analysis and soil plasticity.* Elsevier, Amsterdam, 1975.

R.N. Chowdhury. Discussion on stability analysis of embankments and slopes. *J Geotech Eng, ASCE*, 107:691–693, 1981.

R.W. Clough and R.J. Woodward. Analysis of embankment stresses and deformations. *J Soil Mech Found Div, ASCE*, 93:529–549, 1967.

B.F. Cousins. Stability charts for simple earth slopes. *J Geotech Eng, ASCE*, 104(2):267–279, 1978.

G. de Marsily. Spatial variability of properties in porous media: A stochastic approach. In J. Bear and M.Y. Corapcioglu, editors, *Advances in Transport Phenomena in Porous Media*, pages 719–769. NATO Advanced Study Institute on Fundamentals of Transport Phenomena in Porous Media, Dordrecht, Boston, MA, 1985.

D.J. DeGroot and G.B. Baecher. Estimating autocovariance of in-situ soil properties. *J Geotech Eng, ASCE*, 119(1):147–166, 1993.

J.M. Duncan. State of the Art: Limit equilibrium and finite-element analysis of slopes. *J Geotech Geoenv Eng, ASCE*, 122(7):577–596, 1996.

J.M. Duncan. Factors of safety and reliability in geotechnical engineering. *J Geotech Geoenv Eng, ASCE*, 126(4):307–316, 2000.

J.M. Duncan and P. Dunlop. Slopes in stiff fissured clays and soils. *J Soil Mech Found Div, ASCE*, 95(SM5):467–492, 1969.

J.M. Duncan and K.S. Wong. STABR: A computer program for slope stability analysis with circular slip surfaces. Technical report, Virginia Tech, 1985. Department of Civil Engineering.

W. Fellenius. Calculation of the stability of earth dams. In *Proceedings of the 2nd Congress on Large Dams, Washington, D.C.*, volume 4. U.S. Government Printing Office, 1936.

G.A. Fenton. *Simulation and analysis of random fields*. PhD thesis, Department of Civil Engineering and Operations Research, Princeton University, 1990.

G.A. Fenton. Random field modeling of CPT data. *J Geotech Geoenv Eng, ASCE*, 125(6):486–498, 1999.

G.A. Fenton and E.H. Vanmarcke. Simulation of random fields via local average subdivision. *J Eng Mech, ASCE*, 116(8):1733–1749, 1990.

C.M. Goss and D.V. Griffiths. Discussion on rigorous plasticity solutions for the bearing capacity of two-layered clay. *Géotechnique*, 2001. To appear.

D.V. Griffiths. *Finite element analyses of walls, footings and slopes*. PhD thesis, Department of Engineering, University of Manchester, 1980.

D.V. Griffiths. Computation of bearing capacity factors using finite elements. *Géotechnique*, 32(3):195–202, 1982.

D.V. Griffiths. Computation of bearing capacity on layered soils. In Z. Eisenstein, editor, *Proc 4th Int Conf Numer Methods Geomech*, pages 163–170. A.A. Balkema, Rotterdam, 1982.

D.V. Griffiths. Computation of collapse loads in geomechanics by finite elements. *Ing Arch*, 59:237–244, 1989.

D.V. Griffiths. Failure criterion interpretation based on Mohr-Coulomb friction. *J Geotech Eng, ASCE*, 116(GT6):986–999, 1990.

D.V. Griffiths and G.A. Fenton. Three-dimensional seepage through a spatially random soil. *J Geotech Eng, ASCE*, 123(2):153–160, 1997.

D.V. Griffiths and G.A. Fenton. Influence of soil strength spatial variability on the stability of an undrained clay slope by finite elements. In *Slope Stability 2000, Proceeding of GeoDenver 2000*, pages 184–193. 2000.

D.V. Griffiths and G.A. Fenton. Bearing capacity of spatially random soil: The Prandtl problem revisited. *Géotechnique*, 2001. To appear.

D.V. Griffiths and D.J. Kidger. Enhanced visualization of failure mechanisms in finite elements. *Comput Struct*, 56(2):265–269, 1995.

D.V. Griffiths and P.A. Lane. Slope stability analysis by finite elements. *Géotechnique*, 49(3):387–403, 1999.

D.V. Griffiths, F. Molenkamp, and I.M. Smith. Computer implementation of a double hardening model for sand. In P.A. Vermeer and H.J. Luger, editors, *Proc. IUTAM Symp. on Deformation and Failure of Granular Materials*, pages 213–222. A.A. Balkema, Rotterdam, 1982.

M.E. Harr. *Reliability based design in civil engineering.* McGraw Hill, London, New York, 1987.

M.A. Hicks and R Boughrarou. Finite element analysis of the Nelerk underwater berm failures. *Géotechnique*, 48(2):169–185, 1998.

M.A. Hicks and S.W. Wong. Static liquefaction of loose slopes. In G. Swoboda, editor, *Proc 6th Int Conf Numer Methods Geomech*, pages 1361–1368. A.A. Balkema, Rotterdam, 1988.

R.J. Hoeksema and P.K. Kitanidis. Analysis of the spatial structure of properties of selected aquifers. *Water Resour Res*, 21(4):563–572, 1985.

N. Janbu. Slope stability computations. In *Soil Mech. and Found. Engrg. Rep.* Technical University of Norway, Trondheim, Norway, 1968.

F.H. Kulhawy, M.J.S. Roth, and M.D. Grigoriu. Some statistical evaluations of geotechnical properties. In *Proc. ICASP6, 6th Int. Conf. Appl. Stats. Prob. Civ. Eng.* 1991.

T.W. Lambe and F. Silva. The ordinary method of slices revisited. *Geotechnical News*, 13(3):49–53, 1995.

T.W. Lambe and R.V. Whitman. *Soil Mechanics.* John Wiley and Sons, Chichester, New York, 1969.

P.A. Lane and D.V. Griffiths. Asessment of stability of slopes under drawdown conditions. *J Geotech Geoenv Eng, ASCE*, 126(5):443–450, 2000.

J.B. Lechman. The progression of failure in earth slopes by finite element analysis. Master's thesis, Division of Engineering, Colorado School of Mines, 2000.

I.K. Lee, W. White, and O.G. Ingles. *Geotechnical Engineering.* Pitman, London, 1983.

B. Loret and J.H. Prevost. Dynamic strain localisation in fluid-saturated porous media. *J Eng Mech, ASCE*, 117(4):907, 1991.

J. Lowe and L. Karafiath. Stability of earth dams upon drawdown. In *Proc. of the 1st Pan-Am. Conf. on Soil Mech. and Found. Engrg.*, pages 537–552. Mexico, 1960.

T. Matsui and K-C. San. Finite element slope stability analysis by shear strength reduction technique. *Soils Found*, 32(1):59–70, 1992.

R.S. Merifield, S.W. Sloan, and H.S. Yu. Rigorous plasticity solutions for the bearing capacity of two-layered clays. *Géotechnique*, 49(4):471–490, 1999.

F. Molenkamp. Elasto-plastic double hardening model Monot. Technical report, Delft Geotechnics, 1981.

N.R. Morgenstern. Stability charts for earth slopes during rapid drawdown. *Géotechnique*, 13:121–131, 1963.

N.R. Morgenstern and V.E. Price. The analysis of the stability of general slip surfaces. *Géotechnique*, 15(1):79–93, 1965.

D.J. Naylor and G.N. Pande. *Finite elements in geotechnical engineering.* Pineridge Press, Swansea, U.K., 1981.

G.M. Paice. *Finite element analysis of stochastic soils.* PhD thesis, University of Manchester, U.K., 1997.

G.M. Paice and D.V. Griffiths. Reliability of an undrained clay slope formed from spatially random soil. In J-X. Yuan, editor, *IACMAG 97*, pages 1205–1209. A.A. Balkema, Rotterdam, 1997.

P. Perzyna. Fundamental problems in viscoplasticity. *Advances in Applied Mechanics*, 9:243–377, 1966.

D.M. Potts, G.T. Dounias, and P.R. Vaughan. Finite element analysis of progressive failure of Carsington embankment. *Géotechnique*, 40(1):79–102, 1990.

S. Sarma. Stability analysis of embankments and slopes. *J Geotech Eng, ASCE*, 105:1511–1524, 1979.

I.M. Smith and D.V. Griffiths. *Programming the Finite Element Method.* John Wiley and Sons, Chichester, New York, 3rd edition, 1998.

I.M. Smith and D.K.H. Ho. Influence of construction technique on performance of braced excavation in marine clay. *Int J Numer Anal Methods Geomech*, 16, 1992.

I.M. Smith and R. Hobbs. Finite element analysis of centrifuged and built-up slopes. *Géotechnique*, 24(4):531–559, 1974.

N. Snitbhan and W.F. Chen. Elastic-plastic large deformation analysis of soil slopes. *Comput Struct*, 9:567–577, 1976.

E. Spencer. A method of analysis of the stability of embankments assuming parallel interslice forces. *Géotechnique*, 17(1):11–26, 1967.

E.A. Sudicky. A natural gradient experiment on solute transport in a sand aquifer: Spatial variability of hydraulic conductivity and its role in the dispersion process. *Water Resour Res*, 22(13):2069–2083, 1986.

D.W. Taylor. Stability of earth slopes. *J. Boston Soc. Civ. Eng.*, 24:197–246, 1937.

K. Terzaghi and R.B. Peck. *Soil Mechanics in Engineering Practice*, page 237. John Wiley and Sons, New York, 1967.

R. Torres and T. Coffman. Liquefaction analysis for CAS: Willow Creek Dam. Technical Report WL-8312-2, U.S. Department of the Interior, Bureau of Reclamation, 1997.

E.H. Vanmarcke. *Random fields: Analysis and synthesis*. The MIT Press, Cambridge, Mass., 1984.

R.V. Whitman and W.A. Bailey. Use of computers for slope stability analysis. *J Soil Mech Found Div, ASCE*, 93(SM4):475–498, 1967.

F. S. Wong. Uncertainties in FE modeling of slope stability. *Comput Struct*, 19:777–791, 1984.

P.K. Woodward and D.V. Griffiths. Observations on the computation of the bearing capacity factor N_γ by finite elements. *Géotechnique*, 48(1):137–141, 1998.

O.C. Zienkiewicz and I.C. Cormeau. Viscoplasticity, plasticity and creep in elastic solids. a unified approach. *Int J Numer Methods Eng*, 8:821–845, 1974.

O.C. Zienkiewicz, M. Huang, and M. Pastor. Localization problems in plasticity using finite elements with adaptive remeshing. *Int J Numer Anal Methods Geomech*, 19(2):127–148, 1995.

O.C. Zienkiewicz, C. Humpheson, and R.W. Lewis. Associated and non-associated viscoplasticity and plasticity in soil mechanics. *Géotechnique*, 25:671–689, 1975.

O.C. Zienkiewicz and R.L. Taylor. *The Finite Element Method*, volume 1. McGraw Hill, London, New York, 4th edition, 1989.

Electro-Chemo-Mechanical Couplings in Soils

Benjamin Loret[1], Alessandro Gajo[2] and Tomasz Hueckel[3]

[1] Laboratoire Sols, Solides, Structures, B.P. 53X, 38041 Grenoble Cedex, France
[2] Dipartimento di Ingegneria Meccanica e Strutturale, Università di Trento, via Mesiano 77, 38050 Trento, Italia
[3] Department of Civil Engineering, Duke University, Durham NC 27708-0287, USA

Abstract. Chemically active saturated clays are considered in a two-phase framework. The solid phase contains clay particles, adsorbed water and a single salt. The fluid phase, or pore water, contains free water and salt. Water, and possibly salt, can transfer between the two phases. In addition, part of both species diffuse through the porous medium. A global understanding of all phenomena, deformation, mass transfer and diffusion, is provided. Emphasis is laid on the chemo-mechanical constitutive equations in an elastic-plastic setting.
The presence of several cations in the clay clusters requires modifications with respect to this framework: then the electrolyte nature of pore water has to be accounted for.

1 Introduction

Swelling is an important property of soils being exploited or fought, or engineered in a variety of circumstances, from laying foundations, to setting hydraulic or contamination barriers, to petroleum drilling. Swelling may cause a non-uniform soil heave, resulting in damage of over-structure, or weakening of bore walls in petroleum drillings. On the other hand, swelling of bentonite is taken advantage of to build barriers to water flow, or contaminant transport in hazardous and nuclear waste disposal technologies.

Swelling of soils is generally understood to be driven by capillary forces and/or by chemical gradients. The former one, termed *matric swelling*, occurs when soil is unsaturated. The latter one, termed *chemical swelling*, is characteristic of so called expansive clays, rich in mineral smectite, and occurs when there exists a positive difference in concentration of certain species in the soil and in the water in contact with the soil. Typically, these species are salts of metals (Na, K, Ca, etc...).

This chapter aims at a mathematical modeling of expansive clays. Saturation of pore space with water is assumed. Swelling of unsaturated and chemically neutral soils is discussed within a three-phase framework similar to the present one in a companion chapter and in Loret and Khalili [2000]. Although the framework to be developed is macroscopic and does not address directly the mechanisms that occur at the molecular scale, it intends to keep track at the larger scale of these mechanisms which are briefly described in Loret, Hueckel and Gajo [2000].

1.1 Previous Mechanical Approaches

The approach to modeling of the mechanical aspects of clay swelling usually depends on the application in hand and specific processes involved.

In petroleum engineering, stability of the wellbore can be improved by circulating a mud that creates an osmotic membrane salt impermeable, at least partially depending on the chemical composition of the mud: if p_m, p_r are the pressures and x_m, x_r the salt molar fractions of the mud applied to the wellbore and of the rock respectively, water flow will cease when the chemical potentials equilibrate, which implies the differences $p_m - p_r$ and $x_m - x_r$ to be of the same sign. For a sufficient difference in salinity, the pore pressure in the rock at equilibrium may be forced to be smaller than its initial value, Charlez et al. [1998]. The so called osmotic effect, also called van't Hoff law, can explain this phenomenon only in a qualitative manner: it is based on the equilibrium of the chemical potentials of water, e.g. Spanner [1964].

Simulations of this phenomenon have been presented by Sherwood [1994]a,b, using a constitutive behaviour referred to as *inert*, Sherwood [1993]: the solid particles constitute the solid phase but the fluid phase contain several species, water and salts, and, at fixed chemical potential, the mass of a species squeezed out is proportional to its molar fraction. The problem of chemically induced hydration and dehydration of rocks has been addressed by Heidug and Wong [1996]: they consider the chemical potentials in equilibrium, i.e. they assume the diffusion processes to occur at a much slower rate than the mass transfer processes. However, the above works do not address separately transport and transfer phenomena, that is, they do not recognize that the changes of mass are due to both a transfer (or reactive) contribution and a diffusive contribution. This distinction is necessary to provide elastic constitutive equations that describe coupling effects in swelling clays, and it will be stressed in the present formulation.

Ma and Hueckel [1992], Hueckel [1992]a,b proposed to treat clays exposed to thermal and chemical loads as a two phase mixture, with adsorbed water being a part of solid phase, and an inter-phase mass transfer to model its adsorption and desorption process. Hueckel [1992]a,b extended Terzaghi's principle of effective stress by postulating an a priori equality of the skeletal and adsorbed water isotropic partial stresses, and an explicit "chemical" strain additive with mechanical strain, focusing on its impact on plasticity. Karalis [1993] dealing with simultaneous chemical and capillary swelling discussed a number of ad hoc constitutive functions devised to describe both types of swelling. Bennethum and Cushman [1999] and Murad [1999] have addressed the transfer of water into interlamellar space in clays through a two- or three-spatial scale modeling using homogenization schemes. This type of approach entails a substantial number of constitutive assumptions, requiring sophisticated identification procedures. A multi-constituent mixture, involving charged species and thermal effects, has been suggested by Huyghe and Janssen [1999], but this general framework does not account for the special configuration of clays that will be developed here.

In summary, the above approaches differ mainly in a more or less physical attitude toward exchange of mass between free and adsorbed water and in the role of ions in the process.

1.2 A Model for Chemo-Mechanical Coupling

Saturated clay is viewed as a porous deformable continuum consisting of two overlapping phases, each phase containing several species. However, the strictly physical distinction between the phases may not serve best our purpose, because of ambiguous nature of adsorbed water. Adsorbed water has been considered as a third phase by Murad [1999] and endowed with a partial stress, related thermodynamically to the total fluid volume fraction. Here, we will follow a kinematic criterion of phase identification, Hueckel [1992]a and Ma and Hueckel [1992], and attribute the adsorbed water to the solid phase based on the affinity of their velocities.

The heart of the modeling of deformable porous media is the coupling of the deformation of pore space in soil and the concomitant in- or out-flow of pore liquid. In chemically sensitive soils, this coupling is additionally affected by chemical potentials of the species of the solid and fluid phases. Thus, mechanics of the medium, e.g. balance of momentum, is considered at the phase level, whereas chemical processes, i.e. balances of masses, concern the species. The link between the two levels is obtained through energetic considerations. Whether or not the chemical potentials are in equilibrium, water adsorbed in the clay platelets can transfer into free pore water, or conversely, depending on the chemical composition of the clay and pore water phase, and on the mechanical conditions in terms of volume and pressure. The chemo-mechanical elastic-plastic constitutive equations involve the species present in the solid phase (the clay platelets) but treat the fluid phase as a whole. The species of the latter only diffuse through the porous medium, obeying generalized diffusion equations. The transfer of adsorbed water and species between the solid and fluid phases involves a fictitious membrane surrounding the clay platelets, which may be permeable to the chemical species at various degrees.

In Loret, Hueckel and Gajo [2000], the theoretical framework is illustrated by simulations of typical phenomena observed during laboratory experiments: the change of chemical composition of pore water has, due to chemo-mechanical coupling, consequences on the mechanical state of the porous medium. Parameters involved in the model are calibrated. Mechanical, chemical and chemo-mechanical loading and unloading paths are considered. Increase of the salinity of pore water at constant confinement leads to a volume decrease, so-called *chemical consolidation*. Subsequent exposure to a distilled water solution displays *swelling*: however, the latter is smaller than the chemical consolidation so that the chemical loading cycle results in a net contractancy whose amounts increases with the confinement. With respect to previous publications, the key feature of the present analysis is that the model developed is not only qualitative but it is able to simulate a rather extensive number of experimental data which have been recently published.

Although attention is paid mainly to the elastic-plastic constitutive equations, the latter are shown to be embedded in a general formulation where both mass transfer and diffusion processes can be considered in initial and boundary value problems to be treated through the finite element method in a work under development, Gajo, Loret and Hueckel [2000]. Moreover, clays containing essentially a single counterion, namely Na-Montmorillonite, are considered in a first step: in this situation, electroneutrality makes it possible to treat the phases as electrically neutral since anions chloride Cl^- and cations sodium Na^+ move together. In a companion paper, Gajo, Loret and Hueckel [2000], the framework is extended to include the presence of several counterions, namely sodium Na^+ and potassium K^+, by introduction of electrochemical potentials and enforcement of electroneutrality in both solid and fluid phases. An introduction is provided in the last section of this chapter.

Notation: Compact or index tensorial notation will be used throughout this note. Tensor quantities are identified by boldface letters. $\mathbf{I} = (I_{ij})$ is the second order identity tensor (Kronecker delta). Symbols '·' and ':' between tensors of various orders denote their inner product with single and double contraction respectively. tr denotes the trace of a second order tensor, dev its deviatoric part and div is the divergence operator.

2 The Two-Phase Framework

2.1 General Framework

We consider a two-phase porous medium. Each phase is composed of several species:
- the *solid phase* S contains three species

$$\left\{\begin{array}{ll} \text{clay particles} & \text{denoted by the symbol } S \\ \text{adsorbed water} & w \\ \text{salt in adsorbed water} & s \end{array}\right. \tag{2.1}$$

- the *fluid phase* W contains two species

$$\left\{\begin{array}{ll} \text{pore water} & \text{denoted by the symbol } w \\ \text{salt in pore water} & s \end{array}\right. \tag{2.2}$$

The solid phase is surrounded by a *semi-permeable membrane* which is, see Fig. 1, always impermeable to clay particles,

$$m_{SS} = \text{constant}\,, \tag{2.3}$$

- either *perfect*, that is permeable to water only, not to salt. Then the mass of salt in the solid phase is constant,

$$m_{sS} = \text{constant}\,, \tag{2.4}$$

- or *imperfect* in the sense that it is permeable to both salt and water at various degrees.

To highlight these aspects, we shall introduce the following sets of species:

- species in solid phase $S = \{w, s, S\}$ and species in fluid phase $W = \{w, s\}$;
- species that can cross the membrane when it is imperfect, $W^{\leftrightarrow} = S^{\leftrightarrow} = \{w, s\}$;
- species that can cross the membrane when it is perfect, $W^{\leftrightarrow} = S^{\leftrightarrow} = \{w\}$.

The main assumptions which underly the two-phase model follow the *strongly interacting* model of Bataille and Kestin [1977], namely,

(H1) For each species, only the mass balance is required, not the momentum balance;
(H2) Mass balance for each phase is obtained from mass balances of the species it contains. Momentum balances are required for each phase.

(H3) The velocity of each species in the solid phase is that of the phase, $\mathbf{v}_{kS} = \mathbf{v}_S$, $\forall k \in S$.
(H4) In the fluid phase, pressure is assumed to be uniform across all species, $p_{kW} = p_W$, $\forall k \in W$, but each species k is a priori endowed with its own velocity $\mathbf{v}_{kW}$.

In the solid phase, the pressures attributed to the phase and to the species of the phase are not set a priori as equal. This is motivated by the role that adsorbed water, which is one of the solid phase species, may have in transmitting a part of the total stress.

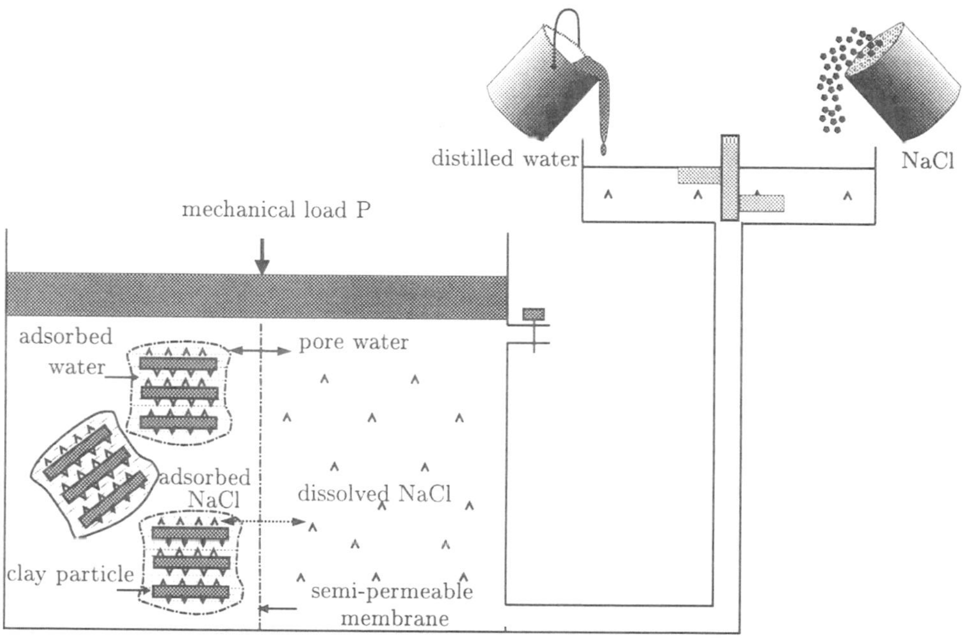

Figure 1 Schematic of a chemo-mechanical experiment. The pore water is in contact with a large recipient whose chemical composition is controlled. In the sample submitted to a variable mechanical load in terms of stress or strain, schematized here by P, the fluid phase (pore water) and the solid phase (clay pockets with adsorbed water and salt) are fictitiously separated by a membrane. Adsorbed water is exchanged with pore water depending on their respective salt contents. Salt may also transfer between the solid and fluid phases depending on the efficiency of the membrane: due to electroneutrality, a cation Na^+ and an anion Cl^- must transfer simultaneously.

2.2 Basic Entities

In an open system that susbstances can enter and leave isothermally, the variation of free energy per unit initial volume, i.e. the work done by the total stress $\boldsymbol{\sigma}$ in the incremental strain $\delta\boldsymbol{\epsilon}$ of the solid phase and by the *chemical potential* μ_{kK} during the addition/subtraction of mass δm_{kK} of the species k of the phase K, is

$$\delta\Psi = \boldsymbol{\sigma} : \delta\boldsymbol{\epsilon} + \sum_{k,K} \mu_{kK}\, \delta m_{kK}\,. \tag{2.5}$$

Summation extends to $k{=}w, s$ and $K{=}W, S$ since the contribution due to the clay particles vanishes due to (2.3). The chemical potentials with unit length^2/time^2 are mass-based[1] and the m_{kK}'s are the mass contents per unit initial volume of the porous medium, with unit mass/volume. Since the volume of the solid phase is also the volume of the porous medium, $\mathrm{tr}\,\boldsymbol{\epsilon}$ is the relative volume change of both these quantities.

The chemical potential of the species k in the fluid phase W is expressed through the classic formula, e.g. Haase [1990], chap. 2-5, Kestin [1968], chap. 21, which includes a purely mechanical contribution involving

- the *intrinsic pressure* of the fluid phase p_W,
- the *intrinsic density of the species* ρ_{kW},

and a chemical contribution which accounts for the *molar fraction* x_{kW} of the species k in phase W, namely in incremental form,

$$\delta\mu_{kW} = \frac{\delta p_W}{\rho_{kW}} + \frac{RT}{m_k^{(M)}}\,\delta\,\mathrm{Ln}\, x_{kW}\,, \quad k = w, s\,. \tag{2.6}$$

As for the species within the solid phase S, the mechanical contribution to the chemical potential of species k involves its intrinsic mean-stress p_{kS},

$$\delta\mu_{kS} = \frac{\delta p_{kS}}{\rho_{kS}} + \frac{RT}{m_k^{(M)}}\,\delta\,\mathrm{Ln}\, x_{kS}\,, \quad k = w, s\,. \tag{2.7}$$

In these formulas, $R = 8.31451\,\mathrm{J/mol/°K}$ is the universal gas constant, T(°K) the absolute temperature and $m_k^{(M)}$ is the molar mass of the species k , e.g. $m_w^{(M)} = 18\,\mathrm{g}$.

The molar fraction x_{kK} of the species k in phase K is defined by the ratio of the mole number N_{kK} of that species within the phase K, namely

$$x_{kK} = \frac{N_{kK}}{\sum_{l\in K} N_{lK}}\,. \tag{2.8}$$

Summation extends to $l{=}w, s$ for $K{=}W$ and to $l{=}S, w, s$ for $K{=}S$. The molar fractions satisfy the constraints,

$$x_{wW} + x_{sW} = 1\,, \quad x_{wS} + x_{sS} + x_{SS} = 1\,. \tag{2.9}$$

Let the initial volume of the porous medium be V_0 and let $V = V(t)$ be its current volume. The current volume of the species k of phase K is denoted by V_{kK} and the current volume of phase K by V_K. Then the *volume fraction* of the species k of phase K is defined as

$$n_{kK} = \frac{V_{kK}}{V}\,, \tag{2.10}$$

[1] Another convention is to use mole-based chemical potentials and measure the m_{kK}'s in moles.

while the volume fraction of phase K is

$$n_K = \frac{V_K}{V} = \sum_{k \in K} n_{kK} \quad \text{with} \quad n_S + n_W = 1\,. \tag{2.11}$$

The ratios of the current volumes to the initial total volume V_0 introduce the *volume contents* v_{kK} for the species k of phase K and v_K for the phase K, namely,

$$v_{kK} = \frac{V_{kK}}{V_0} = n_{kK}\,\frac{V}{V_0}\,, \quad v_K = \frac{V_K}{V_0} = n_K\,\frac{V}{V_0}\,. \tag{2.12}$$

The *mass contents* m_{kK} per unit initial volume V_0, of the species k of phase K and m_K for the phase K, are obtained from the above quantities:

$$m_{kK} = \frac{M_{kK}}{V_0} = \rho_{kK}\, v_{kK} \quad (\text{no summation on } k,\, K)\,, \quad m_K = \frac{M_K}{V_0} = \sum_{k \in K} m_{kK}\,. \tag{2.13}$$

The *apparent* density of the species k in the porous medium, namely ρ^{kK}, and the *apparent* density of the phase K, namely ρ^K, are,

$$\rho^{kK} = n_{kK}\,\rho_{kK} \quad (\text{no summation on } k,\, K)\,, \quad \rho^K = \sum_{k \in K} \rho^{kK}\,. \tag{2.14}$$

Since $N_{kK} = m_{kK} V_0 / m_k^{(M)}$, the molar fractions x_{kK} can be expressed in terms of the mass contents

$$x_{kK} = \frac{m_{kK}/m_k^{(M)}}{\sum_{l \in K} m_{lK}/m_l^{(M)}}\,. \tag{2.15}$$

For the record, let us note the partial derivative,

$$\frac{\partial x_{kK}}{\partial m_{lK}} = \frac{x_{lK}}{m_{lK}}\,(I_{kl} - x_{kK})\,, \quad \forall\, k, l \in K\,. \tag{2.16}$$

Let us denote $v_{kK}^{(M)}$ *the molar volume* of the species k of phase K,

$$m_k^{(M)} = \rho_{kK}\, v_{kK}^{(M)} \quad (\text{no summation on } k,\, K)\,. \tag{2.17}$$

Then the *molar volume* of the fluid phase is

$$v_W^{(M)} = x_{wW}\, v_{wW}^{(M)} + x_{sW}\, v_{sW}^{(M)}\,. \tag{2.18}$$

We will also need the densities referred to the fluid phase rather than to the porous medium, namely

$$\overline{\rho}^{kW} = \frac{\text{mass of species } k \text{ of fluid phase}}{\text{volume of fluid phase}} = \frac{M_{kW}}{V_W} = \frac{\rho^{kW}}{n_W}\,, \quad k = w, s\,. \tag{2.19}$$

From (2.18) and (2.19) follows

$$\overline{\rho}^{kW} = \frac{m_k^{(M)} N_{kW}}{v_W^{(M)} (N_{wW} + N_{sW})} = \frac{m_k^{(M)}}{v_W^{(M)}}\, x_{kW}\,, \quad k = w, s \quad (\text{no summation on } k). \tag{2.20}$$

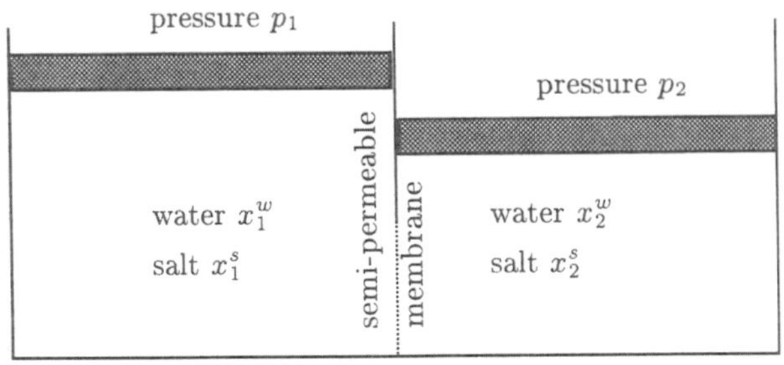

Figure 2 A semi-permeable membrane separates a bucket in two parts [2]. Initially, the chemical content and load is identical on both sides. Addition of salt in side 1 triggers a transfer of water from side 2 to side 1 and the applied load required at equilibrium is larger in side 1.

2.3 The purely osmotic effect or van't Hoff law

Now that we have introduced the notion of chemical potential, we can present the classic osmotic effect. Consider a bucket separated in two parts by a membrane impermeable to salt [2]. Initially, the two sides of the membrane, numbered 1 and 2, have identical chemical content and are submitted to the same external pressure p. At any time, the molar fractions satisfy the identities $x_1^w + x_1^s = 1$ and $x_2^w + x_2^s = 1$.

Since only water can transfer through the membrane, chemo-mechanical equilibrium along the membrane will be phrased in terms of the chemical potential of water only, namely [3]

$$\mu_1^w = v^w\, p_1 + RT\, \ln x_1^w \;\; = \;\; \mu_2^w = v^w\, p_2 + RT\, \ln x_2^w , \tag{2.21}$$

with v^w specific volume of water. The initial concentration of salt is very small, so that:

$$\text{Initial equilibrium :} \quad p_1 = p_2,\; x_1^w = x_2^w \sim 1,\; x_1^s = x_2^s \sim 0. \tag{2.22}$$

Salt is then added in side 1:

$$\text{Add salt :} \quad \Delta x_1^s > 0, \rightarrow \Delta x_1^w = -\Delta x_1^s < 0. \tag{2.23}$$

The transfer of water through the membrane is assumed to be governed by the following law defined by a constant $k > 0$ (motivation for and generalization of this law will be addressed be later),

$$\begin{aligned} \text{Law of transfer :} \quad \dot{m}_1^w &= k(\Delta\mu_2^w - \Delta\mu_1^w) \\ &= k\,(\,0\, - (-RT\,\Delta x_1^s)\,) \quad \text{instantaneously.} \end{aligned} \tag{2.24}$$

[2] Unlike the usual presentation which uses a thin U-shaped tube, gravity is neglected here and the mechanical loads are applied by external agents. In the usual presentation, the external load on both sides of the tube is the atmospheric pressure and the difference of pressures gives rise to different elevations of water on the two sides of the tube.

[3] In this section, we use mole-based chemical potentials following the usual convention of textbooks of physical chemistry.

Consequently, we have
Rule 1: *flow is directed towards the region of higher salt concentration.*

The final equilibrium, corresponding to a vanishing flow through the membrane, occurs when

$$\mu_1^w \sim v^w p_1 - RT x_1^s = \mu_2^w \sim v^w p_2 - RT x_2^s \tag{2.25}$$

This result is encapsulated as
Rule 2 or van't Hoff law: *pressure is higher where salt concentration is higher, namely*

$$v^w (p_1 - p_2) = RT(x_1^s - x_2^s) > 0. \tag{2.26}$$

Hence, osmotic effect generates a counterflow to usual Darcy flow.

2.4 The Gibbs-Duhem Relation

For a fluid phase, the Gibbs-Duhem relation provides the fluid pressure p_W in terms of the chemical potentials of the species μ_{kW}, namely Haase [1990], chap. 1-13,

$$\delta p_W = \overline{\rho}^{wW} \, \delta\mu_{wW} + \overline{\rho}^{sW} \, \delta\mu_{sW} \, . \tag{2.27}$$

This relation is easy to retrieve from the definitions above by forming a linear combination of the chemical potentials (2.6) that eliminates the chemical contribution via $(2.9)_1$; the result follows by using successively (2.17), (2.18) and (2.20). In order to define the mechanical constitutive equations of the porous medium in Section 4, we will need to isolate the chemical effects in pore water. For that purpose, following Heidug and Wong [1996], we define the chemical energy per current unit volume of the fluid phase through its differential

$$\delta\psi_W = \mu_{wW} \, \delta\overline{\rho}^{wW} + \mu_{sW} \, \delta\overline{\rho}^{sW} \, , \tag{2.28}$$

which, via the Gibbs-Duhem relation (2.27), can be integrated, up to a constant, to

$$\psi_W = \mu_{wW} \, \overline{\rho}^{wW} + \mu_{sW} \, \overline{\rho}^{sW} - p_W \, . \tag{2.29}$$

Using $(2.12)_2$, $(2.13)_1$ and (2.19), one has $\overline{\rho}^{kW} v_W = m_{kW}$, $k=w,s$. Therefore, the chemical energy of the fluid phase per unit initial volume of porous medium is, using (2.12),

$$\Psi_W = \psi_W \, \frac{V_W}{V_0} = \psi_W \, v_W = \mu_{wW} \, m_{wW} + \mu_{sW} \, m_{sW} - p_W \, v_W \, . \tag{2.30}$$

Indeed, due to the Gibbs-Duhem relation (2.27), the differential of Ψ_W, eqn (2.30), simplifies to

$$\delta\Psi_W = \sum_{k\in W} \mu_{kW} \, \delta m_{kW} - p_W \, \delta v_W \, . \tag{2.31}$$

2.5 Incompressibility Constraint

A situation of particular interest arises when all species are incompressible,

$$\delta \rho_{kK} = 0 \,, \quad \forall\, k, K \,. \tag{2.32}$$

Then, there exists a relation between the set of variables $\{\boldsymbol{\epsilon}, v_W, m_{wS}, m_{sS}\}$. With (2.13), the increment of fluid volume content can be expressed as,

$$\delta v_W = \delta \operatorname{tr} \boldsymbol{\epsilon} - \frac{\delta m_{wS}}{\rho_{wS}} - \frac{\delta m_{sS}}{\rho_{sS}} \,. \tag{2.33}$$

3 The Global Picture: Deformation, Mass Transfer and Diffusion

The change of mass of each species is due a priori to both a *transfer*, i.e. a physico-chemical reaction, and a *diffusion*, namely

$$\frac{\delta\, m_{kK}}{\delta t} = \frac{\delta}{\delta t} m_{kK}^{\text{reactive}} + \frac{\delta}{\delta t} m_{kK}^{\text{diffusive}} \,. \tag{3.1}$$

In practice, the changes in the species of the solid phase are purely reactive, namely by transfer, through the membrane, of water and salt between the solid and fluid phases. On the other hand, the species of the fluid phase may also undergo mass changes by exchanges (diffusion) with outside, i.e.

$$\frac{\delta}{\delta t} m_{kK}^{\text{diffusive}} = -\operatorname{div} \mathbf{M}_{kK} \,. \tag{3.2}$$

where $\mathbf{M}_{kK}$ is the mass flux of the species k of phase K through the solid, namely,

$$\mathbf{M}_{kK} = \rho^{kK} \left(\mathbf{v}_{kK} - \mathbf{v}_S \right) . \tag{3.3}$$

Here $\delta\, m_{kK}^{\text{reactive}} / \delta\, t$, abbreviated to $\hat{\rho}^{kK}$ in the sequel, is the *rate of transfer* of mass density towards the species k of phase K. Since this transfer occurs between the species as indicated by Fig. 1, then

$$\hat{\rho}^{wS} + \hat{\rho}^{wW} = 0 \,, \quad \hat{\rho}^{sS} + \hat{\rho}^{sW} = 0 \,. \tag{3.4}$$

In absence of thermal effects, starting from the statements of balance of mass for each species, and of momentum and energy for the phases, e.g. Eringen and Ingram [1965], the Clausius-Duhem inequality for a mixture as a whole can be cast in the following form,

$$\delta D = -\delta \Psi + \boldsymbol{\sigma} : \delta \boldsymbol{\epsilon} - \sum_{k,K} \operatorname{div} \left(\mu_{kK}\, \mathbf{M}_{kK} \right) \delta t \geq 0 \,, \tag{3.5}$$

or, equivalently,

$$\delta D = -\delta \Psi + \boldsymbol{\sigma} : \delta \boldsymbol{\epsilon} + \sum_{k,K} \mu_{kK} \left(\delta m_{kK} - \hat{\rho}^{kK}\, \delta t \right) - \sum_{k,K} \nabla \mu_{kK} \cdot \mathbf{M}_{kK}\, \delta t \geq 0 \,. \tag{3.6}$$

The Clausius-Duhem inequality can be viewed as the sum of three contributions $\delta D = \delta D_1 + \delta D_2 + \delta D_3$ which will be required to be positive individually,

$$\begin{cases} \delta D_1 = -\delta\Psi + \boldsymbol{\sigma} : \delta\boldsymbol{\epsilon} + \sum_{k,K} \mu_{kK}\, \delta m_{kK} \geq 0\,, \\ \delta D_2/\delta t = -(\mu_{wS} - \mu_{wW})\, \hat{\rho}^{wS} - (\mu_{sS} - \mu_{sW})\, \hat{\rho}^{sS} \geq 0\,, \\ \delta D_3/\delta t = -\nabla\mu_{wW} \cdot \mathbf{M}_{wW} - \nabla\mu_{sW} \cdot \mathbf{M}_{sW} \geq 0\,. \end{cases} \tag{3.7}$$

If body force density $\mathbf{b}_{kW}$ and acceleration $d^{kW}\mathbf{v}_{kW}/dt$ are accounted for, $\nabla\mu_{wW}$ should be substituted by $\nabla\mu_{wW} - \mathbf{b}_{kW} + d^{kW}\mathbf{v}_{kW}/dt$ in $\delta D_3/\delta t$. The chemo-hyperelastic behaviour, Section 4, will be constructed in order the first term δD_1 to exactly vanish. Satisfaction of the second and third inequalities leads to a generalized mass transfer law and a generalized diffusion law respectively, as outlined below.

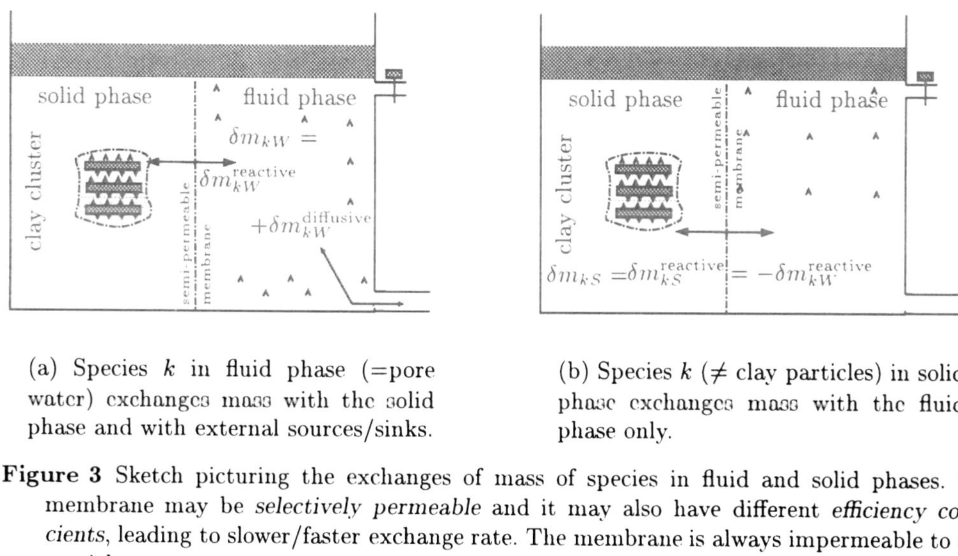

(a) Species k in fluid phase (=pore water) exchanges mass with the solid phase and with external sources/sinks.

(b) Species k ($\neq$ clay particles) in solid phase exchanges mass with the fluid phase only.

Figure 3 Sketch picturing the exchanges of mass of species in fluid and solid phases. The membrane may be *selectively permeable* and it may also have different *efficiency coefficients*, leading to slower/faster exchange rate. The membrane is always impermeable to clay particles.

3.1 Generalized Mass Transfer (or Adsorption/Desorption) Equations

If the semi-permeable membrane around the solid phase is imperfect, see Fig. 1 and Section 2.1, inequality $\delta D_2 \geq 0$ is ensured if one postulates the generalized mass transfer law in the format

$$\begin{cases} \dfrac{\delta m_{wS}}{\delta t} = k_{ww}\,(\mu_{wW} - \mu_{wS}) + k_{ws}\,(\mu_{sW} - \mu_{sS})\,, \\ \dfrac{\delta m_{sS}}{\delta t} = k_{sw}\,(\mu_{wW} - \mu_{wS}) + k_{ss}\,(\mu_{sW} - \mu_{sS})\,, \end{cases} \tag{3.8}$$

where we have used the fact that the species in the solid phase are purely reactive, that is, $\delta m_{kS}/\delta t = \hat{\rho}^{kS}$, $k = w, s$. The transfer matrix $\mathbf{k} = (k_{..})$ is assumed symmetric positive definite,

$$k_{ws} = k_{sw}\,, \quad k_{ww} \geq 0, \quad k_{ww}\, k_{ss} - k_{ws}^2 \geq 0\,. \tag{3.9}$$

On the other hand if the semi-permeable membrane is perfect, eqn $(3.8)_2$ is replaced by eqn (2.4), while eqn $(3.8)_1$ for $\delta m_{wS}/\delta t = \hat{\rho}^{wS}$ becomes

$$\frac{\delta m_{wS}}{\delta t} = k_{ww}\,(\mu_{wW} - \mu_{wS})\,, \qquad k_{ww} \geq 0 \;\; \text{and} \;\; k_{ws} = k_{sw} = k_{ss} = 0\,. \tag{3.10}$$

Moreover, for a perfect semi-permeable membrane, the constitutive chemo-elastic relation, e.g. $(4.6)_2$ below, does not exist for the chemical potential μ_{sS} since $\delta m_{sS} = 0$ in (4.3).

The transfer coefficients $k_{..}$ can be taken in a format that accounts for membrane efficiency, e.g. Staverman and Smit [1975]: in particular, the perfect semi-permeable membrane should be recovered as a limit case of the general situation of an imperfect membrane.

3.2 Generalized Diffusion Equations

Inequality $\delta D_3 \geq 0$ is ensured by appropriate generalization of Darcy's seepage law of water and Fick's diffusion of salt in the water phase through the porous medium.

A generalized diffusion law of pore water and salt through the porous medium can be introduced in the form,

$$\begin{cases} \mathbf{M}_{wW} = -\mathbf{K}_{ww}\, \nabla \mu_{wW} - \mathbf{K}_{ws}\, \nabla \mu_{sW}\,, \\ \mathbf{M}_{sW} = -\mathbf{K}_{sw}\, \nabla \mu_{wW} - \mathbf{K}_{ss}\, \nabla \mu_{sW}\,. \end{cases} \tag{3.11}$$

The inequality $\delta D_3 \geq 0$ is ensured if the symmetric matrices $\mathbf{K}_{ww}$, $\mathbf{K}_{ss}$, $\mathbf{K}_{ws} = \mathbf{K}_{sw}$, are such that $\mathbf{K}_{ss}$ and $\mathbf{K}_{ww} - \mathbf{K}_{ws}\mathbf{K}_{ss}^{-1}\mathbf{K}_{sw}$ are positive definite.

4 Elastic Constitutive Equations

The adsorption and desorption of water (and salt) introduces a chemo-mechanical coupling. On the other hand, the presence of salt in the water phase does not affect directly the mechanical behaviour of the porous medium, it just flows through. Its amount is governed by an equation of mass conservation and a flow equation. Therefore, to develop the chemo-mechanical constitutive equations, we will treat the fluid phase as a whole and, temporarily, ignore its chemical composition. Therefore constitutive equations are developed for the following variables: two stress-strain couples characterizing the mechanical state of each phase, namely $(\boldsymbol{\sigma}, \boldsymbol{\epsilon})$ for the solid phase and (p_W, v_W) for the fluid phase, as many couples chemical potential-mass content as species that can cross the membrane, namely (μ_{kS}, m_{kS}), $k \in S^{\leftrightarrow}$. Incompressibility of the constituents will reduce the number of equations by one.

In view of extension to the elastic-plastic behaviour, the strain-like entities will now be denoted by a superimposed symbol ^{el}.

4.1 Dependent and Independent Variables

When the behaviour is elastic, the energy per unit initial volume of porous medium $\mathcal{W}^{el} = \Psi - \Psi_W$ can be viewed as the elastic energy of the porous medium where the chemical effects in the fluid phase are disregarded. It depends on the restricted set of independent variables $\{\boldsymbol{\epsilon}^{el}, v_W^{el}, m_{wS}^{el}, m_{sS}^{el}\}$; indeed, with the work definition (2.5) and the Gibbs-Duhem relation (2.27), its differential simplifies to

$$\delta\mathcal{W}^{el} = \boldsymbol{\sigma} : \delta\boldsymbol{\epsilon}^{el} + p_W\, \delta v_W^{el} + \mu_{wS}\, \delta m_{wS}^{el} + \mu_{sS}\, \delta m_{sS}^{el} \,. \tag{4.1}$$

We thus have constitutive equations for the dependent variables $\{\boldsymbol{\sigma}, p_W, \mu_{wS}, \mu_{sS}\}$ in terms of the independent variables $\{\boldsymbol{\epsilon}^{el}, v_W^{el}, m_{wS}^{el}, m_{sS}^{el}\}$:

$$\boldsymbol{\sigma} = \frac{\partial\mathcal{W}^{el}}{\partial\boldsymbol{\epsilon}^{el}}\,, \quad p_W = \frac{\partial\mathcal{W}^{el}}{\partial v_W^{el}}\,, \quad \mu_{kS} = \frac{\partial\mathcal{W}^{el}}{\partial m_{kS}^{el}}\,, \quad k = w, s \,. \tag{4.2}$$

Alternative choices in the sets of independent and dependent variables can be postulated by partial or total Legendre transforms of $\mathcal{W}^{el}$.

4.2 Incompressible Constituents

In the sequel, we will restrict the formulation by assuming that all species are incompressible. We shall adopt $\{\boldsymbol{\epsilon}^{el}, m_{wS}^{el}, m_{sS}^{el}\}$ as the set of independent variables and then the increment of elastic fluid volume content is given by (2.33) since we require the incompressibility condition to hold in both the elastic and elastic-plastic regimes. Hence, $\delta\mathcal{W}^{el}$, eqn (4.1), becomes

$$\delta\mathcal{W}^{el} = \bar{\boldsymbol{\sigma}} : \delta\boldsymbol{\epsilon}^{el} + \overline{\mu}_{wS}\, \delta m_{wS}^{el} + \overline{\mu}_{sS}\, \delta m_{sS}^{el} \,, \tag{4.3}$$

where we have introduced Terzaghi's effective stress $\bar{\boldsymbol{\sigma}}$,

$$\bar{\boldsymbol{\sigma}} = \boldsymbol{\sigma} + p_W\, \mathbf{I} \,, \tag{4.4}$$

and the effective chemical potentials $\overline{\mu}_{kK}$,

$$\overline{\mu}_{kK} = \mu_{kK} - \frac{p_W}{\rho_{kK}} \,. \tag{4.5}$$

If we assume the density of any species $k \in S \cap W$ to be one and the same in both solid and fluid phases, i.e. $\rho_{kS} = \rho_{kW}$, equilibrium of the chemical potentials, i.e. $\mu_{kS} = \mu_{kW}$, is equivalent to equilibrium of the effective chemical potentials, i.e. $\overline{\mu}_{kS} = \overline{\mu}_{kW}$. Then the transfer equations (3.8),(3.10), may be phrased identically in terms of the chemical potentials or in terms of effective chemical potentials. Of course, equilibrium makes sense only for species that can cross the membrane.

The constitutive relations take now the form,

$$\bar{\boldsymbol{\sigma}} = \frac{\partial\mathcal{W}^{el}}{\partial\boldsymbol{\epsilon}^{el}}\,, \quad \overline{\mu}_{kS} = \frac{\partial\mathcal{W}^{el}}{\partial m_{kS}^{el}}\,, \quad k = w, s \,. \tag{4.6}$$

Notice that, when there is only a single species present per phase as in the usual poroelasticity, that is, $\delta m^{el}_{wS}=\delta m^{el}_{sS}=0$, the classical relations are retrieved. Indeed, the incremental energy reduces to $\boldsymbol{\sigma} : \delta\boldsymbol{\epsilon}^{el} + p_W\, \delta v^{el}_W$ in the compressible case. In the incompressible case, the incremental energy is equal to $\bar{\boldsymbol{\sigma}} : \delta\boldsymbol{\epsilon}^{el}$ since then the incompressibility condition is $\delta v^{el}_W = \delta \mathrm{tr}\, \boldsymbol{\epsilon}^{el}$ and the remaining constitutive equations $(4.6)_1$ specify the effective stress only, that is the pore pressure p_W does not enter explicitly the constitutive equations: in a boundary value problem, it is defined by boundary conditions.

5 Elastic-Plastic Constitutive Equations

5.1 Experimental Observations

The present analysis is devoted to clays which contain essentially one cation, like Ponza bentonite studied by Di Maio [1996] which is a 98% Na-Montmorillonite. The presence of several cations requires to account explicitly for electroneutrality and will be the subject of the next section. The main features typical of Na-Montmorillonite that emanates from the oedometric tests of Di Maio [1996] and that we intend to model are the following.

5.1.1 Mechanical loading

The specimen is in contact with a large reservoir of constant chemical composition and at atmospheric pressure, so that $p_W \sim 0$. The load is continuously varied, sufficiently slowly however in such a way that chemical equilibrium can be established at the end of each load increment. Therefore, one may assume the chemical potentials of species that can transfer to be equal, at the end of each load increment, to their known counterparts in pore water and in the water contained in the reservoir. Experiments show that the $e - \mathrm{Ln}\,\overline{p}$ curves are nearly straight and converging to a narrow interval void ratio, Fig. 4. However, the slopes of the loading and unloading curves decrease as the Na-content of the pore water increases. This trend holds whatever this content, that is from zero Na-content (distilled water) to saturated solutions (that is at 20^0K and under atmospheric pressure, 6.15 moles/liter of Na). The model considers that all the loading (resp. unloading) curves meet at a common point $\overline{p}_\lambda$ (resp. $\overline{p}_\kappa$). In addition, for the simplifie model, $\overline{p}_\kappa=\overline{p}_\lambda$.

5.1.2 Chemical loading: chemical consolidation and swelling

Under constant mechanical conditions, a chemical loading consists in varying the Na-content of the pore water. When the latter increases, the void ratio decreases, and this decrease rate is especially large at low salt content: in fact most of the volume change occurs for Na-concentrations between 0 and 1 mole/liter. If the Na-content of pore water is decreased, either immediately at constant stress, or after a mechanical loading path, the volume of the sample increases. Notice that these volume changes are in qualitative agreement with the osmotic effect: increase of salt leads to a decrease of pore water pressure, and this in turn leads to water desorption. Alternatively, one may say that water desorption/adsorption to clay surfaces occurs to equilibrate the salt contents in pore water and clay pockets.

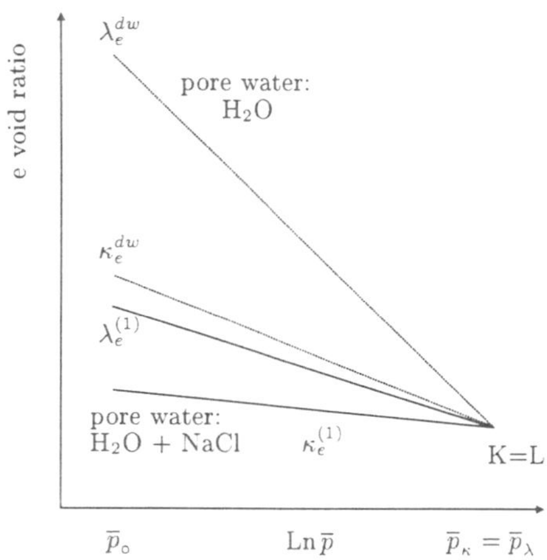

(a) A mechanical loading cycle on a sample in contact with distilled water and on another sample in contact with a saline solution.

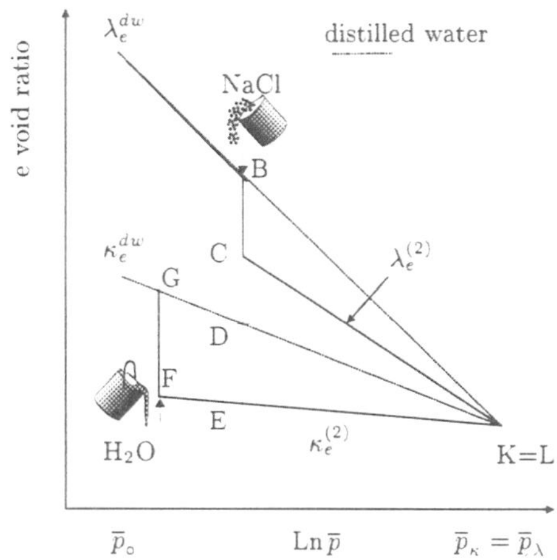

(b) Idealized mixed mechanical and chemical cycle. The purely mechanical test with distilled water as a pore fluid is shown for comparison.

Figure 4 In the simplified modeling, a sample keeps approximatively its chemical composition during a purely mechanical loading cycle and behaves like a Cam-Clay material. Chemical loading and unloading occurs in an elastic regime: after a chemical cycle, the final point G is back on the initial curve, corresponding here to a sample in contact with distilled water. Increase of salt fraction up to solution saturation BC leads to *chemical consolidation* while later re-exposure to distilled water FG displays *swelling*.

5.1.3 Chemical softening and preconsolidation: a paradox

Mechanical loading at constant chemical composition corresponds clearly to an elastic-plastic behaviour while mechanical unloading can be conjectured to be purely elastic, see Fig. 8 of Di Maio [1996]. The situation is more complex for chemical loadings. At relatively low stresses, a chemical loading cycle seems to be practically reversible while the amount of plastic contractancy increases with the applied stress, Fig. 7 of Di Maio [1996].

Also, when chemical consolidation performed at a constant low stress is followed by mechanical loading, the slope is the elastic-plastic slope right from the beginning of the stress increase, so that there seems to be almost no chemical preconsolidation. The data available do not allow to unveil if this description is still correct or not at larger stresses.

The above interpretation may have to account for the large void ratio of Ponza bentonite. Other tests on Bisaccia clay at smaller void ratios, Fig. 7 of Di Maio and Fenelli [1997], show in fact some chemical preconsolidation. This state of affairs, chemical softening leading to plasticity and chemical consolidation, seems somehow paradoxical. The model developed here is able to provide an explanation of these phenomena. In fact, we shall proceed with two levels of sophistication:

- a first conceptual model for which chemical loading and unloading are purely elastic, Fig. 4.
- a more complete model that accounts for chemical softening and preconsolidation. In this model, chemical consolidation has first an elastic-plastic stage, possibly followed by a purely elastic stage, the relative size of the former increasing with the stress level, Fig. 5 (a).

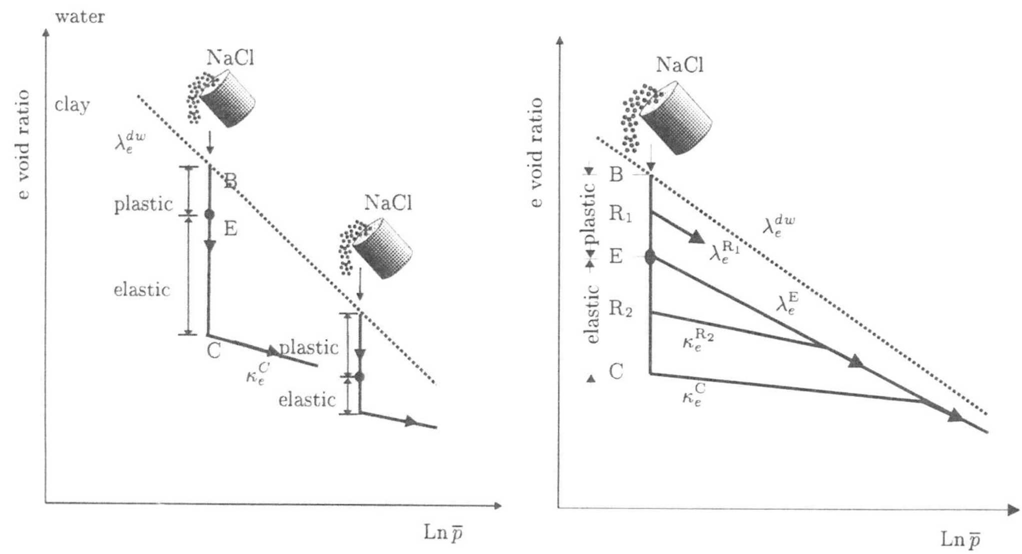

(a) During chemical loading, the relative part of the plastic stage increases with the level of the effective mean-stress. The chemical loading path BC is enlarged on Fig.5 (b).

(b) Mechanical reloading from R_1 in the plastic stage displays a plastic behaviour. Mechanical reloading from R_2 in the elastic stage displays an elastic behaviour: it reveals *chemical consolidation*.

Figure 5 Chemical loading (increase of salt concentration) occurs first plastically and later elastically. Mechanical reloading shows *preconsolidation* in the second case only.

5.2 Incremental Elastic-Plastic Relations

5.2.1 General case: compressible constituents and imperfect membrane

The *generalized strains* are endowed with an elastic part (superimposed symbol ^{el}) and a plastic part (superimposed symbol ^{pl}), namely

$$\boldsymbol{\epsilon} = \boldsymbol{\epsilon}^{el} + \boldsymbol{\epsilon}^{pl}, \quad v_W = v_W^{el} + v_W^{pl}, \quad m_{kS} = m_{kS}^{el} + m_{kS}^{pl}, \; k = w, s. \tag{5.1}$$

The plastic incremental flow relations are motivated by the dissipation inequality $(3.7)_1$. In fact, let us first substitute $\delta\Psi^{el}$ by $\delta\Psi_W^{el} + \delta\mathcal{W}^{el}$ using (2.31) and (4.1), and second let us use the strain decomposition (5.1). The resulting dissipation δD_1,

$$\delta D_1 = \boldsymbol{\sigma} : \delta \boldsymbol{\epsilon}^{pl} + p_W \, \delta v_W^{pl} + \sum_{k \in S^{\leftrightarrow}} \mu_{kS} \, \delta m_{kS}^{pl} \geq 0 \,, \tag{5.2}$$

motivates the generalized normality flow rule,

$$\delta \boldsymbol{\epsilon}^{pl} = \delta \Lambda \, \frac{\partial g}{\partial \boldsymbol{\sigma}} \,, \quad \delta v_W^{pl} = \delta \Lambda \, \frac{\partial g}{\partial p_W} \,, \quad \delta \, m_{kS}^{pl} = \delta \Lambda \, \frac{\partial g}{\partial \mu_{kS}} \,, k = w, s, \tag{5.3}$$

where $\delta \Lambda \geq 0$ is the plastic multiplier and $g = g(\boldsymbol{\sigma}, p_W, \mu_{wS}, \mu_{sS})$ the plastic potential.

5.2.2 Incompressible constituents and salt-permeable membrane

Henceforth, the analysis is restricted to incompressible constituents and to a membrane permeable to water only. The incompressibility constraint (2.33), that holds in both the elastic and elastic-plastic regimes, provides the increment of plastic volume change of the fluid phase,

$$\delta v_W^{pl} = \delta \, \mathrm{tr} \, \boldsymbol{\epsilon}^{pl} - \frac{\delta m_{wS}^{pl}}{\rho_{wS}} - \frac{\delta m_{sS}^{pl}}{\rho_{sS}} \,. \tag{5.4}$$

The dissipation inequality (5.2) becomes

$$\delta D_1 = \bar{\boldsymbol{\sigma}} : \delta \boldsymbol{\epsilon}^{pl} + \overline{\mu}_{wS} \, \delta m_{wS}^{pl} = -\overline{p} \, \mathrm{tr} \, \delta \boldsymbol{\epsilon}^{pl} + \mathbf{s} : \mathrm{dev} \, \delta \boldsymbol{\epsilon}^{pl} + \overline{\mu}_{wS} \, \delta m_{wS}^{pl} \geq 0 \,, \tag{5.5}$$

where $\mathbf{s} = \mathrm{dev} \, \boldsymbol{\sigma}$ is the stress deviator. If, for simplicity, one restricts the analysis to stress paths with constant Lode angles, that expression motivates the generalized normality flow rule

$$\mathrm{tr} \, \delta \boldsymbol{\epsilon}^{pl} = -\delta \Lambda \, \frac{\partial g}{\partial \overline{p}} \,, \quad \mathrm{dev} \, \delta \boldsymbol{\epsilon}^{pl} = \delta \Lambda \, \frac{\partial g}{\partial q} \frac{3}{2} \frac{\mathbf{s}}{q} \; (\Rightarrow \delta \epsilon_q^{pl} = \delta \Lambda \, \frac{\partial g}{\partial q}), \quad \delta \, m_{wS}^{pl} = \delta \Lambda \, \frac{\partial g}{\partial \overline{\mu}_{wS}} . \tag{5.6}$$

Notice that in general the plastic increment of volume change of the fluid phase (5.4) has no reason to vanish: in fact, it has been introduced specifically in order the incompressibility condition be satisfied. Its existence in the compressible case allows a smooth transition between compressible and incompressible materials.

The plastic potential g and consequently the yield function f depend on the three arguments $\overline{p}$, q, $\overline{\mu}_{wS}$, to which we add $\mathrm{tr} \, \boldsymbol{\epsilon}^{pl}$, which will allow hardening or softening. When there is plastic loading, that is the stress point is on the yield surface and stays there, $f = 0$ and $\delta f = 0$, the incremental constitutive equations become,

$$\begin{bmatrix} -\delta \overline{p} \\ \\ \delta q \\ \\ \delta \overline{\mu}_{wS} \end{bmatrix} = \begin{bmatrix} B - \frac{1}{H} g_p \, f_p & -\frac{1}{H} g_p \, f_q & B_w - \frac{1}{H} g_p \, f_w \\ -\frac{1}{H} g_q \, f_p & 3G - \frac{1}{H} g_q \, f_q & -\frac{1}{H} g_q \, f_w \\ B_w - \frac{1}{H} g_w \, f_p & -\frac{1}{H} g_w \, f_q & \beta - \frac{1}{H} g_w \, f_w \end{bmatrix} \begin{bmatrix} \delta \mathrm{tr} \, \epsilon \\ \\ \delta \epsilon_q \\ \\ \delta m_{wS} \end{bmatrix} . \tag{5.7}$$

The general expressions of the coefficients entering in (5.7), namely B, B_w, β, f_p, f_q, f_w, g_p, g_q, g_w, of the plastic modulus $H > 0$ and of the hardening modulus h are reported in

Appendix C of Loret, Hueckel and Gajo [2000], together with their specializations when the yield function and plastic potential are of the Modified Cam-Clay type, namely

$$f = f(\overline{p}, q, \overline{\mu}_{wS}, \mathrm{tr}\, \boldsymbol{\epsilon}^{pl}) = \frac{q^2}{M^2\, \overline{p}} + \overline{p} - p_c\,, \tag{5.8}$$

with $M = M(\overline{\mu}_{wS})$ and $p_c = p_c(\overline{\mu}_{wS}, \mathrm{tr}\, \boldsymbol{\epsilon}^{pl})$. Notice that the major symmetry of the elastic-plastic incremental relations holds iff the flow rule is associative, namely $f = g$.

The calibrations of typical interpolations functions for $\lambda(\overline{\mu}_{wS})$ and $M = M(\overline{\mu}_{wS})$ are provided in Appendix D of Loret, Hueckel and Gajo [2000], with data corresponding to distilled and salt-saturated pore water.

Remark 5.1: The simplified model without chemical preconsolidation.
When the difference $\lambda - \kappa$ [4] is constant and the effective mean-stresses $\overline{p}_\kappa$ and $\overline{p}_\lambda$ are equal, the consolidation stress p_c does not depend on the chemical effects. Chemical loadings and unloadings are then elastic, Fig. 4. Furthermore, the elastic-plastic incremental response to isotropic paths simplifies, since then f and g do not depend on $\overline{\mu}_{wS}$: $\delta q = 0$, $\delta \epsilon_q = 0$, and, for the associative Modified Cam-Clay (5.8),

$$\begin{bmatrix} -\delta \overline{p} \\ \\ \delta \overline{\mu}_{wS} \end{bmatrix} = \frac{1}{\lambda} \begin{bmatrix} \overline{p} & B_w\, \kappa \\ \\ B_w\, \kappa & \beta\, \lambda - \dfrac{B_w^2}{B}\, (\lambda - \kappa) \end{bmatrix} \begin{bmatrix} \delta \mathrm{tr}\, \boldsymbol{\epsilon} \\ \\ \delta m_{wS} \end{bmatrix}, \tag{5.9}$$

where β, B_w and B are elastic coefficients.

6 Electro-Chemo-Mechanical Couplings

When several cations are present in the pore water and in the clay clusters, it is necessary to account explicitly for the electrical potential which appears as a Lagrange multiplier associated to the constraint referred to as electroneutrality. In fact, there are two such constraints, one for each phase and only the difference of potentials is involved in the equations. In the following, we consider the presence of two cations only.

6.1 The Two-Phase Framework

6.1.1 General framework

We consider a two-phase porous medium. Each phase is composed of several species:
- the *solid phase* S contains five species

$$\begin{cases} \text{clay particles} & \text{denoted by the symbol } S \\ \text{adsorbed water} & w \\ \text{ions sodium, potassium, chloride} & Na^+,\ K^+,\ Cl^- \end{cases} \tag{6.1}$$

[4] The compressibility coefficients κ and λ refer to the volumetric strains while κ_e and λ_e refer to the void ratio e, e.g. $\kappa_e = (1 + e_o)\, \kappa$, with e_o initial void ratio.

- the *fluid phase* W contains four species

$$\begin{cases} \text{pore water} & \text{denoted by the symbol } w \\ \text{ions sodium, potassium, chloride} & \mathrm{Na}^+,\ \mathrm{K}^+,\ \mathrm{Cl}^- \end{cases} \tag{6.2}$$

The clay clusters are surrounded by a fictitious membrane which is impermeable to clay particles only. Then the mass of clay in the solid phase, obtained by aggregation of clay clusters, is constant, eqn (2.3). We shall introduce several sets of species with specific properties, namely:

- species in solid phase $S = \{w, \mathrm{Na}^+, \mathrm{K}^+, \mathrm{Cl}^-, S\}$;
- species in fluid phase $W = \{w, \mathrm{Na}^+, \mathrm{K}^+, \mathrm{Cl}^-\}$;
- species that can cross the membrane, $S^{\leftrightarrow} = W^{\leftrightarrow} = W = \{w, \mathrm{Na}^+, \mathrm{K}^+, \mathrm{Cl}^-\}$;
- cations in the solid phase $S^+ = \{\mathrm{Na}^+, \mathrm{K}^+\}$.

Like for the single cation analysis, the main assumptions which underly the two-phase model follow mainly the *strongly interacting* model of Bataille and Kestin [1977], namely,

(H1) For each species, only the mass balance is required, not the momentum balance;
(H2) Mass balance for each phase is obtained from mass balances of the species it contains. Momentum balances are required for each phase.
(H3) The velocity of any species in the solid phase is that of the latter, $\mathbf{v}_{kS} = \mathbf{v}_S$, $\forall k \in W$.
(H4) In the fluid phase, pressure is assumed to be uniform across all species, $p_{kW} = p_W$, $\forall k \in W$, but each species k is a priori endowed with its own velocity $\mathbf{v}_{kW}$.
(H5) Electroneutrality is required in both phases. In the solid phase, negatively charged clay particles require the presence of the cations.

In the solid phase, the pressures attributed to the phase and to the species of the phase are not set a priori as equal. This is motivated by the role that adsorbed water, which is one of the solid phase species, may have in transmitting a part of the total stress.

6.1.2 Basic entities

The incremental work done by the total stress $\boldsymbol{\sigma}$ in the incremental strain $\delta\boldsymbol{\epsilon}$ of the solid phase and by the *electro-chemical potentials* μ^{ec}_{kK} during the addition/subtraction of mass δm_{kK} in/from the species k of the phase K is

$$\delta\Psi = \boldsymbol{\sigma} : \delta\boldsymbol{\epsilon} + \sum_{k,K} \mu^{ec}_{kK}\, \delta m_{kK}\,. \tag{6.3}$$

Summation extends to $k \in S^{\leftrightarrow} = W^{\leftrightarrow} = W$ and $K{=}S, W$ since the contribution due to the clay particles vanishes due to (2.3). The electro-chemical potentials with unit length^2/time^2 are mass-based. In the fluid phase, for charged species in presence of the electrical field ϕ^W, the electro-chemical potential μ^{ec}_{kW} involves, in addition to the chemical potential, an electrical contribution namely in incremental form,

$$\delta\mu^{ec}_{kW} = \frac{\delta p_W}{\rho_{kW}} + \frac{RT}{m^{(M)}_k}\,\delta\,\mathrm{Ln}\,x_{kW} + \xi_k\,\delta\phi^W\,, \quad k \in W\,. \tag{6.4}$$

Similarly for species in the solid phase,

$$\delta \mu_{kS}^{ec} = \frac{\delta p_{kS}}{\rho_{kS}} + \frac{RT}{m_k^{(M)}} \, \delta \operatorname{Ln} x_{kS} + \xi_k \, \delta \phi^S \,, \quad k \in S \,. \tag{6.5}$$

The constant ξ_k involves the valence ζ_k, the molar mass $m_k^{(M)}$ and Faraday's equivalent charge $\mathcal{F} = 96487$ Coulomb/mol,

$$\xi_k = \zeta_k \, \frac{\mathcal{F}}{m_k^{(M)}} \,, \quad k \in W, S \,. \tag{6.6}$$

By convention, $\zeta_w = 0$, and so $\xi_w = 0$.

Two additional phase entities of physical importance in electrolytes need to be defined, namely the *electrical current* and the *electrical current density*. The electrical current I_{eK} in phase K is defined as

$$I_{eK} = \frac{\mathcal{F}}{V} \sum_{k \in K} \zeta_k \, N_{kK} = \frac{V_0}{V} \sum_{k \in K} \xi_k \, m_{kK} \,. \tag{6.7}$$

The equality of the two first expressions is due to the relation $N_{kK} = V_0 \, m_{kK}/m_k^{(M)}$ and to the definition (6.6). Satisfaction of *electroneutrality* in phase K can be expressed in various forms, e.g.,

$$\sum_{k \in K} \zeta_k \, N_{kK} = 0 \quad \Leftrightarrow \quad \sum_{k \in K} \zeta_k \, x_{kK} = 0 \quad \Leftrightarrow \quad I_{eK} = 0 \,. \tag{6.8}$$

The electroneutrality condition restricts the minimal admissible values of the molar fractions of the cations, especially when the pore solution is distilled water. Then at equilibrium, the molar fraction of adsorbed water overweights the other molar fractions. However, electroneutrality implies

$$\zeta_{\mathrm{Na}^+} \, x_{\mathrm{Na}^+ S} + \zeta_{\mathrm{K}^+} \, x_{\mathrm{K}^+ S} = X \equiv -\zeta_S \, x_{SS} - \zeta_{\mathrm{Cl}^-} \, x_{\mathrm{Cl}^- S} > 0 \,. \tag{6.9}$$

Therefore, in the plane $(x_{\mathrm{Na}^+ S}, \, x_{\mathrm{K}^+ S})$, the triangle defined by the points $(0,0)$, $(\zeta_{\mathrm{Na}^+}/X, 0)$ and $(0, \zeta_{\mathrm{K}^+}/X)$ is inaccessible.

The electrical current density $\mathbf{I}_{eK}$ in phase K can be defined in several forms,

$$\mathbf{I}_{eK} = \frac{\mathcal{F}}{V} \sum_{k \in K} \zeta_k \, N_{kK} \, \mathbf{v}_{kK} = \mathcal{F} \frac{V_0}{V} \sum_{k \in K} \zeta_k \frac{m_{kK}}{m_k^{(M)}} \, (\mathbf{v}_{kK} - \mathbf{v}_S) = \sum_{k \in K} \xi_k \, \mathbf{M}_{kK} \,. \tag{6.10}$$

The equality of the two first expressions, which uses the relation $N_{kK} = V_0 \, m_{kK}/m_k^{(M)}$, holds only when electroneutrality holds in the solid phase and the last expression uses the definitions (6.6), (2.13), (2.14) and (3.3) and the flux of mass of species (k, K) through the solid, $\mathbf{M}_{kK} = \rho^{kK} \, (\mathbf{v}_{kK} - \mathbf{v}_S)$, eqn (3.3). On comparing $(6.8)_3$ and $(6.10)_2$, a uniform velocity for all species of a phase satisfying electroneutrality is seen to be a sufficient condition for the electrical current density to vanish in that phase. So, as a consequence of assumptions (H3)-(H5), $\mathbf{I}_{eS} = \mathbf{0}$ but a priori $\mathbf{I}_{eW} \neq \mathbf{0}$. However, one can show

$$\operatorname{div} \mathbf{I}_{eW} = 0 \,. \tag{6.11}$$

6.2 The Global Picture: Deformation, Mass Transfer, Diffusion and Electroneutrality

The adsorption and desorption of water and ionic species introduces electro-chemo-mechanical couplings. On the other hand, the presence of ions in the water phase does not affect directly the mechanical behaviour of the porous medium, it just flows through. Their amount are governed by an equation of mass conservation and a flow equation. Therefore, to develop the electro-chemo-mechanical constitutive equations, we will treat the fluid phase as a whole and, temporarily ignore its chemical composition. Therefore constitutive equations are developed for the following variables: two stress-strain couples characterizing the mechanical state of each phase, namely $(\boldsymbol{\sigma}, \boldsymbol{\epsilon})$ for the solid phase and (p_W, v_W) for the fluid phase, as many couples electro-chemical potential-mass content as species that can cross the membrane, namely $(\mu^{ec}_{kS}, m_{kS}), k \in S^{\leftrightarrow}$. Incompressibility of the constituents will reduce the number of unknowns and equations by one. The electroneutrality condition furnishes the additional equation for the electrical field ϕ^S.

In absence of thermal effects, starting from the statements of balance of mass for each species, and of momentum and energy for the phases, e.g. Eringen and Ingram [1965], the Clausius-Duhem inequality for a mixture as a whole can be cast in the following form,

$$\delta D = -\delta\Psi + \boldsymbol{\sigma} : \delta\boldsymbol{\epsilon} - \sum_{k,K} \operatorname{div}\,(\mu^{ec}_{kK}\,\mathbf{M}_{kK})\,\delta t \geq 0\,. \tag{6.12}$$

Using (3.1), (3.2), δD may be advantageously rewritten in a form that highlights mechanical, mass transfer and diffusion contributions, namely,

$$\delta D = -\delta\Psi + \boldsymbol{\sigma} : \delta\boldsymbol{\epsilon} + \sum_{k,K} \mu^{ec}_{kK}\,(\delta m_{kK} - \hat{\rho}^{kK}\,\delta t) - \sum_{k,K} \nabla\mu^{ec}_{kK} \cdot \mathbf{M}_{kK}\,\delta t \geq 0\,. \tag{6.13}$$

Consequently, the Clausius-Duhem inequality can be viewed as the sum of three contributions $\delta D = \delta D_1 + \delta D_2 + \delta D_3$ which will be required to be positive individually,

$$\left\{\begin{aligned} \delta D_1 &= -\delta\Psi + \boldsymbol{\sigma} : \delta\boldsymbol{\epsilon} + \sum_{k,K} \mu^{ec}_{kK}\,\delta m_{kK} \geq 0\,,\\ \delta D_2/\delta t &= -\sum_{k\in S^{\leftrightarrow}} (\mu^{ec}_{kS} - \mu^{ec}_{kW})\,\hat{\rho}^{kS} \geq 0\,,\\ \delta D_3/\delta t &= -\sum_{k\in W} \nabla\mu^{ec}_{kW} \cdot \mathbf{M}_{kW} \geq 0\,. \end{aligned}\right. \tag{6.14}$$

The chemo-hyperelastic behaviour will be constructed in order the first term δD_1 to exactly vanish. Observe that due to electroneutrality the electro-chemical potentials in δD_1 can be replaced by the chemical potentials: consequently, the constitutive relations for $\boldsymbol{\sigma}$ and $\boldsymbol{\epsilon}$ do not depend on the electrical field. Satisfaction of the second and third inequalities leads to generalized mass transfer equations and generalized diffusion equations respectively, as outlined in Loret, Hueckel and Gajo [2000] in the simpler context of non-electrolyte solutions. Mass transfer equations can readily be read from δD_2 while diffusion equations require some rewriting of δD_3 that highlights specific conjugate flux-force couples. If not neglected, body force and acceleration terms would appear in δD_3.

References

Bataille, J. and J. Kestin (1977). On the structuring of thermodynamic fluxes: a direct implementation of the dissipation inequality. *Int. J. Eng. Science*, **17**, 563-572.

Bennethum, L. S. and J.H.Cushman (1999). Coupled solvent and heat transport of a mixture of swelling porous particles and fluids: single time-scale problem. *Transport in Porous Media*, **36**, 211-244.

Charlez, Ph., V. Pradet, R. Pollard, A. Onaisi and M. Grégoire (1998). How to manage wellbore stability in the Vicking Graben tertiary shales by using mud systems environmentally friendly ? Offshore Technology Conference, Houston, 4-7 May 1998.

Di Maio, C. (1996). Exposure of bentonites to salt solution: osmotic and mechanical effects. *Géotechnique*, **46**(4), 695-707.

Di Maio, C. and G. Fenelli (1997). Influenza delle interazioni chimico-fisiche sulla deformabilità di alcuni terreni argillosi. *Rivista Italiana di Geotecnica*, **1**, 695-707.

Eringen, A.C. and J.D. Ingram (1965). A continuum theory for chemically reacting media -I. *Int. J. Eng. Science*, **3**, 197-212.

Gajo, A., Loret, B. and T. Hueckel (2000). Electro-chemo-mechanical couplings in saturated porous media: elastic-plastic behaviour of heteroinic expansive clays. submitted.

Haase, R. (1990). *Thermodynamics of Irreversible Processes.* Dover Publications, New York.

Heidug, W.K. and S.-W. Wong (1996). Hydration swelling of water-absorbing rocks: a constitutive model. *Int. J. Num. Anal. Meth. Geomechanics*, **20**, 402-430.

Hueckel, T. (1992)a. Water-Mineral Interaction in Hygro-Mechanics of Clays Exposed to Environmental Loads: a Mixture Approach. *Canadian Geotechnical J.*, **29**, 1071-1086.

Hueckel, T. (1992)b. On Effective Stress Concepts and Deformation in Clays Subjected to Environmental Loads. *Canadian Geotechnical J.*, **29**, 1120-1125.

Huyghe, J.M. and J.D. Janssen (1999). Thermo-chemo-electro-mechanical formulation of saturated charged porous solids. *Transport in Porous Media*, **34**(1-3), 129-141.

Karalis, T.K. (1993). Water flow in non-saturated swelling soil. *Int. J. Eng. Science*, **31**(5), 751-774.

Kestin, J. (1968). *A Course in Thermodynamics.* Blaisdell Publishing Co., Waltham, Massachusetts.

Loret, B. and N. Khalili (2000). A three-phase model for unsaturated soils. *Int. J. Num. Anal. Meth. Geomechanics*, **24**, 893-927.

Loret, B., Hueckel, T. and A. Gajo (2000). Chemo-mechanical coupling in saturated porous media: elastic-plastic behaviour of homoionic expansive clays. submitted.

Ma, C. M. and T. Hueckel (1992). Effects of Inter-phase Mass Transfer in Heated Clays: a Mixture Theory. *Int. J. Eng. Science*, **30**(11), 1567-1582.

Murad, M.A. (1999). Thermo-mechanical model for hydration swelling in smectitic clays. *Int. J. Num. Anal. Meth. Geomechanics*, **27**(7), Part I: 673-696, Part II: 697-720.

Sherwood, J.D. (1993). Biot poroelasticity of a chemically active shale. *Proc. R. Soc. Lond.*, **440**, 365-377.

Sherwood, J.D. (1994)a. Swelling of shale around a cylindrical wellbore. *Proc. R. Soc. Lond.*, **444**, 161-184.

Sherwood, J.D. (1994)b. A model of hindered solute transport in a poroelastic shale. *Proc. R. Soc. Lond.*, **445**, 679-692.

Spanner, D.C. (1964). *Introduction to thermodynamics.* Academic Press, New York.

Staverman, A.J. and J.A.M. Smit (1975). Thermodynamics of Irreversible Processes. Membrane Theory. in *Physical Chemistry: enriching topics from Colloid and Surface Science*, H. van Olphen and K.J. Mysels eds, Theorex, La Jolla, Ca.

Thermo-Mechanical Potentials for Unsaturated Soils

Benjamin Loret[1] and Nasser Khalili[2]

[1] Laboratoire Sols, Solides, Structures, B.P. 53X, 38041 Grenoble Cedex, France
[2] School of Civil and Environmental Engineering,
The University of New South Wales, 2052 Sydney, Australia

Abstract. The behaviour of unsaturated soils is known to depend strongly and in a nonlinear way of suction. Usual elastic formulations store energy during arbitrary paths in the stress and strain spaces. To avoid polluting the interpretation of the energy balance when thermal effects are accounted for, a hyperelastic formulation is systematically developed, first in a purely mechanical analysis, and in a second step, in a thermo-mechanical analysis where each of the three phases is endowed with its own temperature. The secant constitutive matrices are shown to lack symmetry, in contrast with the tangent matrices. For the mechanical analysis, the potentials are obtained in explicit form for the stress-based formulation and for a mixed formulation. The thermo-mechanical potential accounting for the menisci effect is also provided for the formulation where the solid strain, the two fluid pressures and the three phase temperatures are the primary variables.

1 Introduction

Unsaturated soils occupy almost 40 percent of earth's land surface. They are present in man-made structures like roads, airfields, earth dams, and in special formations like swelling clays, collapsing soils and residual soils.

A comprehensive framework to define the mechanical behaviour of unsaturated soils is developed within the theory of mixtures applied to three-phase porous media in Loret and Khalili [2000]. Each of the three phases is endowed with its own strain and stress. Elastic and elastic-plastic constitutive equations are presented. Particular emphasis is laid on the interactions between the phases both in the elastic and plastic regimes. Nevertheless the clear structure of the constitutive equations requires a minimal number of material parameters. Their identification is provided: in particular, it incorporates directly the soil-water characteristic curve. Crucial to the formulation is an appropriate definition of the effective stress. The coupled influence of this effective stress and of suction makes it possible to describe qualitatively many of the characteristic features observed in experiments, e.g. for normally consolidated soils, a plastic behaviour up to air entry followed by an elastic behaviour at increasing suctions, and, upon suction unloading, an elastic behaviour, unless compression is applied in which case plastic collapse may occur.

Capitalising upon the model for unsaturated soils above, a thermo-mechanical model where three phases (solid, liquid and gas) and four constituents (solid particle, water, vapour, dry air) are identified is developed in Khalili and Loret [2000]: this framework can handle phase changes (vaporisation and condensation). In fact, the study of non-isothermal mass flow in porous media and development of strategies for its management and/or containmement in natural formations and engineered materials has emerged as a key research area in modern environmental geomechanics. The goal is to develop predictive capabilities for storage of nuclear wastes, disposal of hazardous wastes and remediation of contaminated sites using air-sparging.

The elastic constitutive equations for unsaturated soils depend strongly, and crucially, on suction, mainly through the effective stress and through the soil-water characteristic curve. Therefore, secant formulations and tangent formulations are not endowed a priori with the same type of symmetry. There exists three types of elastic formulations for purely mechanical analyses, see Loret [1985]. To paths closed in the stress space, *hypoelastic* (rate) formulations generally do not associate closed paths in the strain space. For so called *elastic* formulations, there is a one-to-one relationship between strains and stresses, so that to closed stress paths correspond closed strain paths, and conversely. However, for both hypoelastic and elastic formulations, some energy is stored in the material during a closed path in stress or strain space: this energy is not *dissipated*, it can be recovered, but only if the paths are exactly reversed. Therefore, since the energy balance is one of the ingredients of a thermo-mechanical problem, it is preferable to avoid polluting the analysis by these pitfalls. That state of affairs leads to consider *hyperelastic* formulations, which avoid completely the two problems mentioned above, by assuming the existence of a potential for the stresses or the strains.

In fact, this framework can be shown to apply not only in a purely mechanical analysis, but also in more general contexts involving electro-chemo-mechanical couplings, or as presented below, thermo-mechanical effects in unsaturated soils. Section 2 is devoted to the analysis of purely mechanical aspects of the behaviour of deformable unsaturated porous media. Section 3 presents an extension including thermal couplings, where a priori each phase is endowed with its own temperature. Both elastic and elastic-plastic formulations are considered, and the constitutive equations are given in a format appropriate to immediate use in mixed finite element computations.

This chapter should be understood as a complement to Loret and Khalili [2000] and Khalili and Loret [2000] where the physical aspects of the constitutive equations are motivated. For completeness, we begin by defining basic entities of the theory of porous media: a more complete account can be found in Eringen and Ingram [1965], Biot [1973], Bowen [1976], Hassanizadeh and Gray [1993]. The reader may also consult the proceedings of the CISM course on unsaturated soils, Gens et al. [1995], for alternative models and experimental data reported in Fleureau et al. [1993].

Notation: Compact or index tensorial notation will be used throughout. Tensor quantities are identified by boldface letters. **I** is the second order identity tensor (Kronecker delta). Symbols '·' and ':' between tensors of various orders denote their inner product with single and double contraction respectively. Time derivative following the solid phase of any time-dependent quantity a is denoted either by a superimposed $\dot{a}$ or by $\delta a/\delta t$.

2 Mechanical Behaviour of Unsaturated Soils

2.1 Basic entities

Unsaturated soils are made up of three constituents, a solid and two fluids. One fluid, water, is wetting, the other, air, is not. *Phases* (solid (s), water (w) and air (a)) represent the constituents when viewed as part of the mixture, also referred to as *porous medium*. In the context of the theory of mixtures, are viewed as three independent overlapping continua. The solid phase is also referred to as *solid skeleton*. Each constituent, and each phase, is endowed with its own kinematics, mass and momentum.

Each constituent has a mass M_k and a volume V_k, k=s, w, a, which make up the total mass M=M_s+M_w+M_a and the total volume $V = V_s + V_w + V_a$ of the mixture. At a reference time $t = 0$, V was equal to $V(0)$=$V_\circ$. The solid phase plays a special role and it occupies the volume V of the whole mixture.

At each point of each phase are defined *intrinsic* quantities, labeled by subscripts, and *apparent* or *partial* quantities, labeled by superscripts. For example, the intrinsic mass density is defined as ρ_k=M_k/V_k, whereas the apparent mass density is defined by ρ^k=M_k/V; hence ρ^k=$n^k\, \rho_k$, where n^k is the *volume fraction* of phase k,

$$n^k = \frac{V_k}{V}\,. \tag{2.1}$$

By definition, the volume fractions sum up to one,

$$n^s + n^w + n^a = 1\,. \tag{2.2}$$

The solid phase is endowed with its own (infinitesimal) strain tensor $\boldsymbol{\epsilon}$ defined from the macroscopic displacement field. The *volume content of fluid phase k* per unit reference volume of porous medium is denoted by v^k,

$$v^k \equiv \frac{V_k}{V_\circ} = n^k\, \frac{V}{V_\circ}\,, \tag{2.3}$$

and the *fluid mass content* per unit reference volume of porous medium by m^k,

$$m^k \equiv \frac{M_k}{V_\circ} = \rho_k\, v^k = \rho^k\, \frac{V}{V_\circ}\,. \tag{2.4}$$

The total stress $\boldsymbol{\sigma}$ is the sum of the three *partial* or *apparent* stresses of the phases,

$$\boldsymbol{\sigma} = \boldsymbol{\sigma}^s + \boldsymbol{\sigma}^w + \boldsymbol{\sigma}^a\,. \tag{2.5}$$

The apparent stresses are linked to the *intrinsic* stresses of the associated constituents (if there is a single constituent per phase) through the volume fractions,

$$\boldsymbol{\sigma}^s = n^s \boldsymbol{\sigma}_s\,, \quad \boldsymbol{\sigma}^k = -p^k\,\mathbf{I} = -n^k p_k\,\mathbf{I}\ \text{(no sum over)}\ k, \quad k = w, a. \tag{2.6}$$

Other stresses play central roles in the behaviour of unsaturated soils, namely the *effective stress* $\bar{\boldsymbol{\sigma}}$,

$$\bar{\boldsymbol{\sigma}} \equiv \boldsymbol{\sigma} + \chi_w\, p_w\,\mathbf{I} + \chi_a\, p_a\,\mathbf{I}\,, \tag{2.7}$$

with $\chi_k = \chi_k(s) \in [0,1]$, $k = w, a$, suction dependent functions, Khalili and Khabbaz [1995],[1998], the *net stress* $\boldsymbol{\sigma}^{\text{net}}$,

$$\boldsymbol{\sigma}^{\text{net}} \equiv \boldsymbol{\sigma} - \boldsymbol{\sigma}_a = \boldsymbol{\sigma} + p_a\,\mathbf{I}\,, \tag{2.8}$$

and the *suction* s,

$$s \equiv p_a - p_w\,. \tag{2.9}$$

To make contact with the standard convention in soil mechanics, we shall define mean-stresses with the appropriate sign, for example,

$$p = -\frac{1}{3}\,\text{tr}\,\boldsymbol{\sigma}\,,\quad \bar{p} = -\frac{1}{3}\,\text{tr}\,\bar{\boldsymbol{\sigma}} = p - \chi_w\,p_w - \chi_a\,p_a\,,\quad p^{\text{net}} = -\frac{1}{3}\,\text{tr}\,\boldsymbol{\sigma}^{\text{net}} = p - p_a\,, \tag{2.10}$$

such that, for cohesionless soils, p is positive. Therefore, the decomposition of the stress into a spherical part and a deviatoric part writes

$$\boldsymbol{\sigma} = -p\,\mathbf{I} + \mathbf{s}\,,\quad \bar{\boldsymbol{\sigma}} = -\bar{p}\,\mathbf{I} + \mathbf{s}\,. \tag{2.11}$$

Since the elastic equations depend strongly and in a nonlinear way on suction, it is preferable to work on finite increments of stresses and strains reckoned to some predefined reference state, say at time $t = 0$. For any time-dependent quantity $a = a(t)$, we introduce the operator Δa,

$$\Delta a = a(t) - a(0)\,. \tag{2.12}$$

We assume that the configuration at time $t = 0$ is submitted to atmospheric pressure p_{atm}, and we reckon stresses and strains to this configuration, so that, for these quantities, the operator Δ can be omitted.

2.2 Hyperelastic Equations for Compressible Solid Grains

The elastic constitutive equations for unsaturated soils are known to depend non-linearly on suction. Therefore, it is more general, if not more convenient, to start with secant formulations. Rate (tangent) relations follow by differentiation. We begin by

---

Proposition 2.2.1 Secant Relations with $(\boldsymbol{\sigma}, p_w, p_a)$ as Primary Variables.

The effective stress $\bar{\boldsymbol{\sigma}}$ *is by definition linked by a one-to-one relationship to the strain* $\boldsymbol{\epsilon}$, *namely* [1] [2]

$$\bar{\boldsymbol{\sigma}} \equiv \mathbb{E}^{DS} : \boldsymbol{\epsilon} \qquad [\Leftrightarrow \text{tr}\,\boldsymbol{\epsilon} = -\frac{\bar{p}}{K^{DS}}\,,\ \text{dev}\,\boldsymbol{\epsilon} = \frac{\mathbf{s}}{2\,\mu^{DS}}\,]. \tag{2.13}$$

---

[1] In a first step, this relation will be assumed to be linear, but later in Sect. 2.3, we shall consider logarithmic hyperelasticity.

[2] Here and in the sequel, the soils considered are elastically isotropic and the relation between the deviatoric parts of stress and strain is always assumed to be given $(2.13)_3$, with a constant shear modulus.

The existence of an elastic potential implies elastic secant relations between the primary variables total stress, fluid and air pressures, $(\boldsymbol{\sigma}, p_w, p_a)$*, and the dependent variables strain and changes in fluid volume contents,* $(\boldsymbol{\epsilon}, \Delta v^w, \Delta v^a)$*, of the form:*

$$\begin{bmatrix} -\operatorname{tr}\boldsymbol{\epsilon} \\ \Delta v^w \\ \Delta v^a \end{bmatrix} = \begin{bmatrix} \dfrac{1}{K^{DS}} & -\dfrac{\chi_w}{K^{DS}} & -\dfrac{\chi_a}{K^{DS}} \\ -\dfrac{\psi_w}{K^{DS}} & c_{ww} & c_{wa} \\ -\dfrac{\psi_a}{K^{DS}} & c_{aw} & c_{aa} \end{bmatrix} \begin{bmatrix} p \\ p_w \\ p_a \end{bmatrix}, \tag{2.14}$$

where the suction dependent coefficients $\chi_k(s)$ *and* $\psi_k(s)$ *satisfy the following relations,*

$$\psi_k \equiv \frac{d}{ds}(s\,\chi_k)\,,\quad k = w, a\,;\qquad \psi_w + \psi_a = \chi_w + \chi_a = 1 - \frac{K^{DS}}{K_s}\,. \tag{2.15}$$

Here K_s *is the bulk modulus of the solid grains while* K^{DS} *is the bulk modulus of the drained solid, i.e. of the solid skeleton. The coefficients* $c_{kl}(s)$, $k, l = w, a$, *will be obtained during the analysis. Crucially, the secant relations do not display the major symmetry despite the existence of an elastic potential; in fact, due to the nonlinearities, the matrix format used to cast the secant relations is purely formal. On the other hand, the tangent (rate) relations display this symmetry,* Loret and Khalili [2000].

Before we consider the proof, let us first record a useful formula,

$$\frac{\partial}{\partial p_k}(\chi_w\, p_w + \chi_a\, p_a) = \psi_k\,,\quad k = w, a. \tag{2.16}$$

The proof proceeds in five steps, as follows.
1/ The complementary energy Ψ_C:

Since the chosen independent variables are the stress and pressures, the potential we look for is the complementary energy $\Psi_C(\boldsymbol{\sigma}, p_w, p_a)$, defined through its differential[3]

$$\delta\Psi_C = -\delta\boldsymbol{\sigma} : \boldsymbol{\epsilon} - \delta p_w\, \Delta v^w - \delta p_a\, \Delta v^a\,. \tag{2.17}$$

Its existence implies the constitutive relations,

$$\boldsymbol{\epsilon} = -\frac{\partial \Psi_C}{\partial \boldsymbol{\sigma}}\ \Big[\Leftrightarrow \operatorname{tr}\boldsymbol{\epsilon} = \frac{\partial \Psi_C}{\partial p},\ \operatorname{dev}\boldsymbol{\epsilon} = -\frac{\partial \Psi_C}{\partial \mathbf{s}} - \frac{\mathbf{s}}{2\,\mu^{DS}}\Big],\ \Delta v^w = \frac{\partial \Psi_C}{\partial p_w},\ \Delta v^a = -\frac{\partial \Psi_C}{\partial p_a}. \tag{2.18}$$

For simplicity, the potential $\Psi_C(\boldsymbol{\sigma}, p_w, p_a)$ is isotropic and quadratic in $\boldsymbol{\sigma}$, which justifies the decomposition in the parentheses in (2.18); consequently the tensor of drained moduli $\mathbb{E}^{DS}$ with constant bulk modulus K^{DS} and constant shear modulus μ^{DS} should be isotropic. On the other hand, we wish to leave its suction dependance as arbitrary as possible in

[3] All the potentials considered are measured per unit volume of porous medium.

order to fit experimental data and accommodate analytical expressions proposed in the literature. However, this arbitrariness is restricted by the existence of the effective stress $\bar{\boldsymbol{\sigma}}$ and by

Proposition 2.2.2 Strain Averaging Relation.
The apparent strains of the solid grains and individual fluids sum up to the strain of the solid skeleton,

$$n^s \operatorname{tr} \boldsymbol{\epsilon}_s + \Delta v^w + \Delta v^a = \operatorname{tr} \boldsymbol{\epsilon}\,, \quad n^s \operatorname{dev} \boldsymbol{\epsilon}_s = \operatorname{dev} \boldsymbol{\epsilon}\,. \tag{2.19}$$

2/ Determination of $\Psi_C = \Psi_C(\boldsymbol{\sigma}, \cdot, \cdot)$:
Using (2.7), (2.13) and (2.18) yields

$$\frac{\partial \Psi_C}{\partial p} = -\frac{1}{K^{\mathrm{DS}}} (p - \chi_w\, p_w - \chi_a\, p_a)\,, \tag{2.20}$$

which can be integrated, up to an arbitrary function $\phi_1(p_w, p_a)$, to

$$\Psi_C = -\frac{1}{2}\frac{p^2}{K^{\mathrm{DS}}} + \frac{p}{K^{\mathrm{DS}}} (\chi_w\, p_w + \chi_a\, p_a) + \phi_1(p_w, p_a) - \frac{1}{4\,\mu^{\mathrm{DS}}}\, \mathbf{s} : \mathbf{s}\,. \tag{2.21}$$

3/ Consequences of the strain averaging relation:
The solid grains are assumed to be randomly oriented and positioned in space so that the tensor of the solid constituent with bulk modulus K_s and shear modulus μ_s satisfy a micro-isotropy,

$$\boldsymbol{\sigma}_s = \mathbf{E}_s : \boldsymbol{\epsilon}_s \quad [\Leftrightarrow \quad \frac{1}{3} \operatorname{tr} \boldsymbol{\sigma}_s = K_s \operatorname{tr} \boldsymbol{\epsilon}_s\,, \quad \operatorname{dev} \boldsymbol{\sigma}_s = 2\,\mu_s \operatorname{dev} \boldsymbol{\epsilon}_s]\,. \tag{2.22}$$

Since the intrinsic stress $\boldsymbol{\sigma}_s$ is related to the partial stress $\boldsymbol{\sigma}^s$ by $\boldsymbol{\sigma}^s = n^s\, \boldsymbol{\sigma}_s$, one thus can express $\operatorname{tr} \boldsymbol{\epsilon}_s$ in terms of stress and pressures, namely using (2.5),

$$\Delta\, v^s = n^s \operatorname{tr} \boldsymbol{\epsilon}_s = \frac{1}{K_s} \frac{\operatorname{tr} \boldsymbol{\sigma}^s}{3} = \frac{1}{K_s} (-p + n^w\, p_w + n^a\, p_a)\,. \tag{2.23}$$

Similarly, using $(2.18)_{2,3}$ and (2.21) and the definitions $(2.15)_1$, the changes of fluid volume contents become,

$$-\Delta v^w = \frac{p}{K^{\mathrm{DS}}} \psi_w + \frac{\partial \phi_1}{\partial p_w}\,, \quad -\Delta v^a = \frac{p}{K^{\mathrm{DS}}} \psi_a + \frac{\partial \phi_1}{\partial p_a}\,. \tag{2.24}$$

Calculating $\operatorname{tr} \boldsymbol{\epsilon}$ by $(2.10)_2$, $(2.13)_2$, and inserting (2.23), (2.24) in (2.19) provides a null linear relation in terms of the three arbitrary primary variables. The vanishing of the three factors yield (2.15) and

$$\frac{\partial \phi_1}{\partial p_w} + \frac{\partial \phi_1}{\partial p_a} = \left(\frac{n^w}{K_s} - \frac{\chi_w}{K^{\mathrm{DS}}}\right) p_w + \left(\frac{n^a}{K_s} - \frac{\chi_a}{K^{\mathrm{DS}}}\right) p_a\,. \tag{2.25}$$

Also the equality of the grain and macroscopic shear moduli, μ_s and $\mu^{\rm DS}$ respectively, follows as a consequence of the introduction of $(2.13)_3$, $(2.22)_3$ in $(2.19)_2$.

4/ Soil water characteristic curve:

The soil-water characteristic curve is defined by the volumetric water content w,

$$w \equiv \frac{V_w}{V_s}\,. \tag{2.26}$$

with approximate increment[4], $\Delta w = \Delta v^w/n^s - \Delta v^s\, n^w/(n^s)^2$. Inserting (2.23) and (2.24) in this expression provides $\partial\phi_1/\partial p_w$ and (2.25) yields further $\partial\phi_1/\partial p_a$,

$$\begin{aligned}\frac{\partial\phi_1}{\partial p_w} &= -n^s\,\Delta w - \frac{p}{K^{\rm DS}}\,\psi_w + \frac{n^w}{n^s}\frac{1}{K_s}\,(p - n^w\,p_w - n^a\,p_a)\\ \frac{\partial\phi_1}{\partial p_a} &= \;\;n^s\,\Delta w + \frac{p}{K^{\rm DS}}\,\psi_w - \frac{n^w}{n^a}\frac{1}{K_s}\,(p - a^w\,p_w - a^a\,p_a)\,,\end{aligned} \tag{2.27}$$

where $a^k = n^k(n^s+n^w)/(n^s/K_s) - \chi_k/K^{\rm DS}$, $k = w, a$. Differentiating $(2.27)_1$ wrt p yields $n^s\partial w/\partial p = -\phi_2(s)$, with $\phi_2(s) = \psi_w/K^{\rm DS} - n^w/(n^s K_s)$, and therefore (a): $n^s\,\Delta w = -p\,\phi_2(s) + \phi_3(s,t)$ with $t \equiv p_a + p_w$. Equating the partial crossed derivatives of (2.27) provides another constraint for w, $n^s(\partial w/\partial p_w + \partial w/\partial p_a) = \phi_2(s)$, and therefore $2\,\partial\phi_3/\partial t = \phi_2(s)$. By integration, $\phi_3(s,t) = t\,\phi_2(s)/2 + n^s\,\Delta C(s)$. Substituting in (a) above implies that the change from its reference value of w should vary as a function of $p^{\rm net}$ and s in the following format where $\Delta C = \Delta C(s)$ is an arbitrary function of s,

$$\Delta w = -\frac{p^{\rm net}}{n^s}\left(\frac{\psi_w}{K^{\rm DS}} - \frac{n^w}{n^s\,K_s}\right) + \Delta C(s)\,. \tag{2.28}$$

Backsubstituting Δw in (2.27) and next in (2.24) yields the relations $(2.14)_{2,3}$ with

$$c_{aa} = c_{ww} + \frac{\psi_a - \chi_w}{K^{\rm DS}} - \frac{n^a - n^w}{K_s}\,,\qquad c_{wa} - \frac{\psi_w}{K^{\rm DS}} = c_{aw} - \frac{\chi_w}{K^{\rm DS}} = -c_{ww} - \frac{n^w}{K_s}\,, \tag{2.29}$$

together with c_{ww}. In explicit form,

$$\begin{cases} c_{ww} = \dfrac{1}{n^s\,K_s}\,(n^w)^2 - n^s\dfrac{\Delta C}{\Delta s}\,, & c_{wa} = \dfrac{\psi_w}{K^{\rm DS}} - \dfrac{n^w(1-n^a)}{n^s K_s} + n^s\dfrac{\Delta C}{\Delta s}\,,\\ c_{aw} = \dfrac{\chi_w}{K^{\rm DS}} - \dfrac{n^w(1-n^a)}{n^s K_s} + n^s\dfrac{\Delta C}{\Delta s}\,, & c_{aa} = \dfrac{\psi_a + \chi_a - 1}{K^{\rm DS}} + \dfrac{(1-n^a)^2}{n^s K_s} - n^s\dfrac{\Delta C}{\Delta s}\,.\end{cases} \tag{2.30}$$

5/ Reconstruction of the potential Ψ_C:

To establish the secant relations, only the existence of the potential was required, not its explicit expression. The latter can be obtained, if needed, as follows. Substitution of

[4] Some linearizations are in force from now on. Since the volume change of the solid grains remains always small in comparison with the phase volume changes due to the large value of the bulk modulus K_s, v^s is approximated as n^s which is assumed to be constant; also the coefficients of $1/K_s$ are linearized to their initial values.

Δw, (2.30), into $(2.27)_1$, and integration of the resulting expression yields $\phi_1(p_w, p_a)$ up to an arbitrary function $\phi_5(p_a)$. Insertion of its partial differential $\partial\phi_1/\partial p_a$ in $(2.27)_2$ with Δw taken from (2.30) yields $\partial\phi_5/\partial p_a$ which can be integrated up to a constant, thus defining $\phi_1(p_w, p_a)$. Then $\Psi_C(\boldsymbol{\sigma}, p_w, p_a)$, eqn (2.21), becomes

$$\begin{aligned}\Psi_C = {} & \frac{p^{\rm net}}{K^{\rm DS}}\Big(-\frac{1}{2}(p+p_a)+(\chi_w\,p_w+\chi_a\,p_a)\Big) \\ & -\frac{1}{2\,n^s K_s}\Big(n^w p_w-(1-n^a)p_a\Big)^2+n^s\int_\circ^s \Delta C(s')\,ds'-\frac{1}{4\,\mu^{\rm DS}}\,\mathbf{s}:\mathbf{s}\,.\end{aligned} \tag{2.31}$$

□

In view of finite element applications where the solid displacement and pressures are the primary variables, it is worth recording the elastic relations in mixed form.

Proposition 2.2.3 Mixed Secant Relations with $(\boldsymbol{\epsilon},\, p_w,\, p_a)$ as Primary Variables.
The elastic relations in mixed form where the primary variables are $(\boldsymbol{\epsilon},\, p_w,\, p_a)$, can be cast in the following matrix format from (2.14):

$$\begin{bmatrix} p \\ \Delta v^w \\ \Delta v^a \end{bmatrix} = \begin{bmatrix} -K^{DS} & \chi_w & \chi_a \\ \psi_w & d_{ww} & d_{wa} \\ \psi_a & d_{aw} & d_{aa} \end{bmatrix} \begin{bmatrix} \mathrm{tr}\,\boldsymbol{\epsilon} \\ p_w \\ p_a \end{bmatrix}, \tag{2.32}$$

with

$$d_{kl} = c_{kl} - \frac{\psi_k \chi_l}{K^{DS}}, \quad k, l = w, a\,. \tag{2.33}$$

While these relations are hyperelastic, like their strain-based counterparts, they do not display the major symmetry, at variance with the tangent (rate) relations, eqn (66) of Loret and Khalili [2000] *and eqn (3.33) below.*

The potential here is $\Psi_M(\boldsymbol{\epsilon}, p_w, p_a) = \Psi_C(\boldsymbol{\sigma},\, p_w,\, p_a) + \boldsymbol{\sigma} : \boldsymbol{\epsilon}$, with the complementary energy Ψ_C defined through its differential (2.17), so that,

$$\delta\Psi_M = \delta\boldsymbol{\epsilon} : \boldsymbol{\sigma} - \delta p_w\,\Delta v^w - \delta p_a\,\Delta v^a\,, \tag{2.34}$$

and therefore,

$$\boldsymbol{\sigma} = \frac{\partial\Psi_M}{\partial\boldsymbol{\epsilon}}\;[\Leftrightarrow p = -\frac{\partial\Psi_M}{\partial \mathrm{tr}\,\boldsymbol{\epsilon}},\ \mathbf{s} = \frac{\partial\Psi_M}{\partial \mathrm{dev}\boldsymbol{\epsilon}} = 2\,\mu^{\rm DS}\,\mathbf{s}],\ \Delta v^w = -\frac{\partial\Psi_M}{\partial p_w},\ \Delta v^a = -\frac{\partial\Psi_M}{\partial p_a}\,. \tag{2.35}$$

The explicit form of $\Psi_M(\boldsymbol{\epsilon}, p_w, p_a)$ is,

$$\begin{aligned}\Psi_M = {} & \frac{1}{2}K^{\rm DS}\,\mathrm{tr}^2\boldsymbol{\epsilon} - \mathrm{tr}\,\boldsymbol{\epsilon}\,(\chi_w\,p_w+\chi_a\,p_a) + \frac{1}{2\,K^{\rm DS}}(\chi_w\,p_w+\chi_a\,p_a-p_a)^2 \\ & -\frac{1}{2\,n^s K_s}\Big(n^w p_w-(1-n^a)\,p_a\Big)^2+n^s\int_\circ^s \Delta C(s')\,ds' + \mu^{\rm DS}\,\mathrm{dev}\,\boldsymbol{\epsilon}:\mathrm{dev}\,\boldsymbol{\epsilon}\,.\end{aligned} \tag{2.36}$$

□

To check satisfaction of the strain averaging relation (2.19), one may record the following formulas,

$$d_{ww}+d_{aw} = \frac{1}{K_s}(\chi_w - n^w), \quad d_{wa}+d_{aa} = \frac{1}{K_s}(\chi_a - n^a)\,. \tag{2.37}$$

2.3 Hyperelastic Equations for Incompressible Solid Grains

The case of incompressible solid grains, $K_s = \infty$, is obtained as a limit from eqns (2.15) and (2.29),

$$d_{ww} = d_{aa} = -d_{wa} = -d_{aw} = -n^s \frac{\Delta C}{\Delta s} - \frac{\psi_w \chi_w}{K^{\rm DS}} . \tag{2.38}$$

One can however develop another point of view and start the analysis accounting from the start for incompressibility. Then, the three strain variables are no longer independent, since, by (2.19) and (2.22), there exists a condition of incompressibility of kinematic nature, $\Delta v^w + \Delta v^a = \mathrm{tr}\,\boldsymbol{\epsilon}$. The complementary energy which serves as a stress-based potential admits the following differential, accounting for $\boldsymbol{\sigma} : \boldsymbol{\epsilon} = -p\,\mathrm{tr}\,\boldsymbol{\epsilon} + \mathbf{s} : \mathrm{dev}\,\boldsymbol{\epsilon}$,

$$\delta\Psi_{CI}(p^{\rm net}, s, \mathbf{s}) = -\delta\mathbf{s} : \mathrm{dev}\,\boldsymbol{\epsilon} + \delta p^{\rm net}\,\mathrm{tr}\,\boldsymbol{\epsilon} + \delta s\,\Delta v^w . \tag{2.39}$$

If the above relation is a total differential, one then has the constitutive relations,

$$\mathrm{tr}\,\boldsymbol{\epsilon} = \frac{\partial\Psi_{CI}}{\partial p^{\rm net}}, \quad \mathrm{dev}\,\boldsymbol{\epsilon} = -\frac{\partial\Psi_{CI}}{\partial \mathbf{s}}\Big[= \frac{\mathbf{s}}{2\,\mu^{\rm DS}}\Big], \quad \Delta v^w = \frac{\partial\Psi_{CI}}{\partial s}, \tag{2.40}$$

and Δv^a is given by the incompressibility condition. As a particular case, let us examine logarithmic elasticity.

Proposition 2.3.1 Incompressible Solid Grains: Secant Logarithmic Relations.
Assume there exists an effective elastic law which is isotropic and logarithmic,

$$\mathrm{tr}\,\boldsymbol{\epsilon} = -\kappa\,\mathrm{Ln}\frac{\bar p}{p_\circ}, \quad \mathrm{dev}\,\boldsymbol{\epsilon} = \frac{\mathbf{s}}{2\,\mu^{DS}}, \tag{2.41}$$

with κ constant and $p_\circ$ a reference stress. If it exists, a stress-based elastic potential Ψ_{CI} depends then necessarily on $(p^{\rm net}, s, \mathbf{s})$ and the elastic secant relation for the fluid volume content Δv^w is of the form,

$$\Delta v^w = -\psi_w\,\kappa\,\mathrm{Ln}\frac{\bar p}{p_\circ} + n^s\,\Delta C(s), \tag{2.42}$$

where $\Delta C(s)$ is defined by the soil-water characteristic curve. The mixed form of the secant relations where $(p,\ \Delta v^w,\ \Delta v^a)$ are the dependent variables is

$$p = p_\circ \exp(-\frac{\mathrm{tr}\,\boldsymbol{\epsilon}}{\kappa}) - s\,\chi_w + p_a, \quad \Delta v^w = \psi_w\,\mathrm{tr}\,\boldsymbol{\epsilon} + n^s\,\Delta C(s). \tag{2.43}$$

Before we proceed with the proof, let us note first that, as a consequence of the incompressibility constraint, a single pressure representative is involved directly in the constitutive equations. In a boundary value problem, the other pressure representative should be given at the boundaries. Also, it is useful to record the specializations of some formulas, namely,

$$\psi_w + \psi_a = \chi_w + \chi_a = 1, \quad \bar p = p^{\rm net} + s\,\chi_w . \tag{2.44}$$

Now, from (2.40),(2.41),

$$\Psi_{CI}(p^{\text{net}}, s, \mathbf{s}) = -\kappa\, F(\bar{p}, p_\circ) + n^s \int_\circ^s \Delta C(s')ds' - \frac{1}{4\,\mu^{\text{DS}}}\,\mathbf{s} : \mathbf{s}\,, \tag{2.45}$$

with $F(\bar{p}, p_\circ) \equiv \bar{p}\,(\text{Ln}(\bar{p}/p_\circ)-1)$. The above additive suction dependent function $\Delta C(s)$ is provided by the soil-water characteristic curve. Indeed, now, $\Delta w = \Delta v^w/v^s \sim \Delta v^w/n^s$, and so, the counterpart of (2.28) is

$$\Delta w = -\frac{\psi_w}{n^s}\,\kappa\,\text{Ln}\frac{\bar{p}}{p_\circ} + \Delta C(s)\,. \tag{2.46}$$

The rate relations in mixed form, counterpart of eqn (66) of Loret and Khalili [2000] in the linear and compressible case, are then

$$\begin{bmatrix} \dot{p} \\ \dot{v}^w \\ \dot{v}^a \end{bmatrix} = \begin{bmatrix} -\bar{p}/\kappa & \psi_w & \psi_a \\ \psi_w & a_{ww} & -a_{ww} \\ \psi_a & -a_{ww} & a_{ww} \end{bmatrix} \begin{bmatrix} \text{tr}\,\dot{\boldsymbol{\epsilon}} \\ \dot{p}_w \\ \dot{p}_a \end{bmatrix}, \tag{2.47}$$

with

$$a_{ww} = -\frac{d\psi_w}{ds}\,\text{tr}\,\boldsymbol{\epsilon} - n^s\,\frac{dC}{ds}\,. \tag{2.48}$$

The tangent matrix in (2.47) is symmetric as expected and its determinant is equal to $-a_{ww}$.

2.4 Plastic Equations for Compressible and Incompressible Solid Grains

The structure of the elastic-plastic relations is motivated by the complementary power P_C which provides conjugate triplets, $(\boldsymbol{\sigma}, p_w, p_a)$ for the stress triplet and $(\boldsymbol{\epsilon}, \Delta v^w, \Delta v^a)$ for the strain triplet. The rates of strain-like variables are additively decomposed into an elastic and a plastic contribution, namely

$$\dot{\boldsymbol{\epsilon}} = \dot{\boldsymbol{\epsilon}}_{el} + \dot{\boldsymbol{\epsilon}}_{pl} \qquad \text{and} \qquad \dot{v}^k = \dot{v}^k_{el} + \dot{v}^k_{pl}\,, \quad k = w, a. \tag{2.49}$$

Decomposing P_C into an elastic part and a plastic part, we are left with,

$$\delta P_C^{pl} = -\delta\boldsymbol{\sigma} : \dot{\boldsymbol{\epsilon}}_{pl} - \delta p_w\,\dot{v}^w_{pl} - \delta p_a\,\dot{v}^a_{pl}\,. \tag{2.50}$$

The flow rule emanates from a generalized normality rule applied to the above stress and strain triplets: we assume the existence of a plastic potential $g = g(\boldsymbol{\sigma}, p_w, p_a, \mathcal{H})$ where stresses act in fact only through the effective stress $\bar{\boldsymbol{\sigma}}$ and the suction s and, although an abuse of notation, we note $g = g(\boldsymbol{\sigma}, p_w, p_a, \mathcal{H}) = g(\bar{\boldsymbol{\sigma}}, s, \mathcal{H})$. Here $\mathcal{H}$ represent a set of hardening parameters, of arbitrary tensorial nature, which embody the history of the material. The yield function f is assumed to have the same arguments. Then

$$\dot{\boldsymbol{\epsilon}}_{pl} = \dot{\Lambda}\,\frac{\partial g}{\partial \boldsymbol{\sigma}}\,, \quad \dot{v}^k_{pl} = \dot{\Lambda}\,\frac{\partial g}{\partial p_k}\,, \quad k = w, a, \tag{2.51}$$

with $\dot{\Lambda} \geq 0$ the plastic multiplier. Since we adopt a generalized normality rule, the behaviour would then be associative, and the elastic-plastic stiffness of the underlying drained solid would display the major symmetry property, if the directions of the normals to the yield function $f(\boldsymbol{\sigma}, p_w, p_a)$ and plastic potential $g(\boldsymbol{\sigma}, p_w, p_a)$ in the space $(\boldsymbol{\sigma},\ p_w\ p_a)$ were identical. Using the chain rule, one easily shows, eqn (68) of Loret and Khalili [2000],

$$\dot{v}^w_{pl} = \psi_w \operatorname{tr} \dot{\boldsymbol{\epsilon}}_{pl} - \dot{\Lambda}\, \frac{\partial g}{\partial s}, \qquad \dot{v}^a_{pl} = \psi_a \operatorname{tr} \dot{\boldsymbol{\epsilon}}_{pl} + \dot{\Lambda}\, \frac{\partial g}{\partial s}. \tag{2.52}$$

Let us now reconsider the kinematical constraint that arises when the solid grains are incompressible, namely in rate form $\dot{v}^w + \dot{v}^a = \operatorname{tr} \dot{\boldsymbol{\epsilon}}$. The elastic rates satisfy exactly this constraint and this holds both in the limit of the compressible formulation, as $K_s \to \infty$, and in the direct incompressible formulation (2.47). Therefore, the plastic rates should also satisfy exactly that constraint: (2.52) shows that the constraint is indeed satisfied, since then $\psi_w + \psi_a = 1$.

Observe that if $\partial g/\partial s$ is zero, then the fluid volume contents tend to an asymptotic value when the plastic strain rate vanishes, so as to provide a generalized critical state. Then, the rate elastic-plastic relations in mixed form for the volume contents are defined by exactly the same coefficients as their elastic counterparts: plasticity nevertheless is involved through the strain rate, see eqn (76) of Loret and Khalili [2000] and eqn (3.48) below.

3 Heat and Mass Flow in Unsaturated Soils

3.1 Introduction

The temperatures of the fluid phases T_k, $k = w, a$, are the temperatures averaged over their respective constituents and the temperature of the solid skeleton T_s is assumed to be equal to the average temperature of the solid constituent. The meniscus separating the fluid phases is endowed with a temperature $T^{(m)}$ that will be assumed to be a combination of the phase temperatures. The coefficients χ_w and χ_a which define the effective stress will be assumed to be independent of the temperature. In practice, this may be an approximation as their definition, see Khalili and Khabbaz [1998], involves the air-entry value which may well depend on $T^{(m)}$. Constitutive equations are required for the entropies of each phase. There are now six primary and six dependent variables.

3.2 The global picture: thermo-mechanical, transfer and diffusion equations

For a phase containing several species, the mass properties of the phase are obtained by summing up the contributions of the species. The change of mass content of species K phase k is due a priori to both a *transfer*, i.e a physico-chemical reaction, and a *diffusion*, namely

$$\dot{m}^{Kk} = \dot{m}^{Kk}_{\text{rea}} + \dot{m}^{Kk}_{\text{dif}}\,. \tag{3.1}$$

Here $\dot{m}^{Kk}_{\text{rea}}$ is the rate of transfer of mass density *towards* the species K of phase k. If vaporisation and condensation occur, the fluid phase contains water and dissolved air and the air phase contains dry air and water vapor, see Khalili and Loret [2000]. The reactive changes between the two phases occur by dissolution, vaporisation and condensation and they satisfy the closure relation, for any exchangeable species K,

$$\dot{m}^{Kw}_{\rm rea} + \dot{m}^{Ka}_{\rm rea} = 0\,. \tag{3.2}$$

The exchanges with outside (diffusive changes) are described by

$$\dot{m}^{Kk}_{\rm dif} = -\frac{V}{V_\circ}\,\mathrm{div}\,\mathbf{M}^{Kk}\,, \tag{3.3}$$

where $\mathbf{M}^{Kk}$ is the mass flux of the species K of phase k through the solid, namely, in terms of the absolute velocities of the species and solid phase, $\mathbf{v}_{Kk}$ and $\mathbf{v}_s$ respectively,

$$\mathbf{M}^{Kk} = \rho^{Kk}\,(\mathbf{v}_{Kk} - \mathbf{v}_s)\,. \tag{3.4}$$

In order to simplify the establishment of the constitutive equations, we shall temporarily ignore the species in the phases and consider the latter as homogeneous. There is no difficulty to redefine the reactive and diffusive entities defined above. Then, starting from the statements of balance of mass for each species, and of momentum and energy for the phases, e.g. Eringen and Ingram [1965], the Clausius-Duhem inequality for a mixture as a whole can be cast in the following form,

$$\begin{aligned} \frac{\delta D}{\delta t} = &- \frac{V_\circ}{V}\frac{\delta\Psi}{\delta t} + \boldsymbol{\sigma} : \frac{\delta\epsilon}{\delta t} - \sum_{k=w,a} \mathrm{div}\,(\mu_k\,\mathbf{M}^k) \\ &- \sum_{k=s,w,a} \rho^k\,\eta_k\,\frac{\delta T_k}{\delta t} + \eta_k\,\nabla T_k\cdot\mathbf{M}^k + \nabla T_k\cdot\frac{\mathbf{q}^k}{T_k} + T_k\,e^k_{\rm rea} \geq 0. \end{aligned} \tag{3.5}$$

Here Ψ is the *free energy* per unit initial volume of the porous medium, η_k is the *entropy* per unit mass of phase k, μ_k is the *chemical potential* (also called free enthalpy) per unit mass of the fluid phase k, and $e^k_{\rm rea}$ is the rate of *entropy supply* by the other phases to phase k. Entropy supplies satisfy the closure relation

$$\sum_{k=s,w,a} e^k_{\rm rea} = 0\,. \tag{3.6}$$

Using (2.4), (3.1), (3.3) and (3.6), δD may be advantageously rewritten in a form that highlights thermo-mechanical, transfer and diffusion contributions, namely $\delta D = \delta D_1 + \delta D_2 + \delta D_3$, which will be required to be positive individually,

$$\left\{\begin{aligned} \delta D_1 &= -\delta\Psi + \boldsymbol{\sigma} : \delta\epsilon - \sum_{k=s,w,a} m^k\,\eta_k\,\delta T_k + \sum_{k=w,a} \mu_k\,\delta m^k \geq 0\,, \\ \delta D_2 &= -(\mu_w - \mu_a)\,\delta m^w_{\rm rea} - \sum_{k=w,a} (T_k - T_s)\,e^k_{\rm rea} \geq 0\,, \\ \delta D_3/\delta t &= -\sum_{k=w,a} (\nabla\mu_k + \eta_k\nabla T_k)\cdot\mathbf{M}^k - \sum_{k=s,w,a} \nabla T_k\cdot\frac{\mathbf{q}^k}{T_k} \geq 0\,. \end{aligned}\right. \tag{3.7}$$

In passing from (3.5) to (3.7), the volume ratio $V/V_\circ$ has been dropped in front of Cauchy stress $\boldsymbol{\sigma}$ and of δD_1.

The thermo-hyperelastic behavior will be constructed in order the first term δD_1 to exactly vanish: then the energy dissipation will be due exclusively to transfer of matter and energy between phases, eqn $(3.7)_2$, and diffusion of fluids through the solid skeleton and of conductive heat, eqn $(3.7)_3$. On the other hand, when the behaviour is thermo-elastic-plastic, energy dissipation occurs by thermo-plastic effects that make $(3.7)_1$ strictly positive. Satisfaction of the second and third inequalities leads to generalized transfer equations and generalized diffusion equations respectively, see Khalili and Loret [2000]. If not neglected, body forces and acceleration terms would appear in δD_3. We are concerned here with the thermo-mechanical constitutive behaviour dictated by $(3.7)_1$.

3.3 Thermo-Elastic Equations: Secant Form

The vanishing of δD_1 implies the following formal constitutive relations, where the free energy $\Psi(\boldsymbol{\epsilon}, m^w, m^a, T_s, T_w, T_a)$ serves as a potential,

$$\boldsymbol{\sigma} = \frac{\partial \Psi}{\partial \boldsymbol{\epsilon}}; \quad \mu_k = \frac{\partial \Psi}{\partial m^k}, \; k = w, a; \quad m^k \eta_k = -\frac{\partial \Psi}{\partial T_k}, \; k = s, w, a. \tag{3.8}$$

However, in order to mimic the mechanical analysis and use the same primary variables, we will work instead with the mixed potential Ψ_M,

$$\Psi_M = \Psi - \sum_{k=w,a} \mu_k \, m^k \, . \tag{3.9}$$

Using the differential of Ψ defined by $\delta D_1 = 0$, the differential of Ψ_M is,

$$\delta \Psi_M = \boldsymbol{\sigma} : \delta \boldsymbol{\epsilon} - \sum_{k=w,a} m^k \, \delta \mu_k - \sum_{k=s,w,a} m^k \eta_k \, \delta T_k . \tag{3.10}$$

3.3.1 Influence of Menisci Neglected

For each fluid phase k, the chemical potential is defined through its differential which includes a mechanical contribution and a thermal contribution, namely, e.g. Kestin [1968], Haase [1990],

$$\delta \mu_k = \frac{\delta p_k}{\rho_k} - \eta_k \, \delta T_k, \; k = w, a. \tag{3.11}$$

Insertion of these expressions in (3.10) shows that, to lower order, Ψ_M does not depend any longer on the temperatures of the fluids,

$$\delta \Psi_M = \boldsymbol{\sigma} : \delta \boldsymbol{\epsilon} - \sum_{k=w,a} v^k \, \delta p_k - \rho^s_\circ \, \eta_s \, \delta T_s \, , \tag{3.12}$$

where we have used the fact $m^s = \rho^s_\circ$. The resulting formal constitutive equations are the following, with $\Psi_M(\boldsymbol{\epsilon}, p_w, p_a, TS)$ as a potential,

$$\boldsymbol{\sigma} = \frac{\partial \Psi_M}{\partial \boldsymbol{\epsilon}}; \quad v^k = \frac{m^k}{\rho_k} = -\frac{\partial \Psi_M}{\partial p_k}, \; k = w, a; \quad \rho^s_\circ \, \eta_s = -\frac{\partial \Psi_M}{\partial T_s}. \tag{3.13}$$

The thermal modification to the purely mechanical part of Ψ_M, eqn (2.36), may be guessed in the following form that provides an elastic behaviour, linear with respect to thermal changes,

$$\Psi_M(\epsilon, p_w, p_a, T_s) = \Psi_M{}^{\mathrm{mech}}(\epsilon, p_w, p_a) + \\ + (-K^{\mathrm{DS}}\,\mathrm{tr}\,\epsilon + c_w\,p_w + c_a\,p_a)\,\tau^{\mathrm{DS}}\,\Delta T_s - \frac{C^s}{2\,T_s^{\circ}}\,(\Delta T_s)^2 \,. \tag{3.14}$$

Here C^s,

$$\frac{C^s}{T_s^{\circ}} = \frac{\partial(\rho_{\circ}^s\,\eta_s)}{\partial T_s}\,, \tag{3.15}$$

is the *apparent heat capacity*, at constant strain and fluid pressures, of the solid phase, per unit initial volume of porous medium, i.e. heat capacity times volume fraction n^s. The constitutive equations in explicit form are, from (3.13),

$$\begin{bmatrix} p \\ \Delta v^w \\ \Delta v^a \\ \rho_{\circ}^s\,\Delta\eta_s \end{bmatrix} = \begin{bmatrix} -K^{\mathrm{DS}} & \chi_w & \chi_a & K^{\mathrm{DS}}\tau^{\mathrm{DS}} \\ \psi_w & d_{ww} & d_{wa} & -\tau^{\mathrm{DS}}X_w \\ \psi_a & d_{aw} & d_{aa} & -\tau^{\mathrm{DS}}X_a \\ K^{\mathrm{DS}}\tau^{\mathrm{DS}} & -\tau^{\mathrm{DS}}c_w & -\tau^{\mathrm{DS}}c_a & C^s/T_s^{\circ} \end{bmatrix} \begin{bmatrix} \mathrm{tr}\,\epsilon \\ p_w \\ p_a \\ \Delta T_s \end{bmatrix} \tag{3.16}$$

with $X_k \equiv \partial X/\partial p_k$, $k = w, a$, and $X \equiv c_w\,p_w + c_a\,p_a$. The suction dependent coefficients c_k are obtained using the following

---

Proposition 3.3.1 Homogeneous Deformation under Purely Thermal Loading.

For a purely thermal loading, i.e. $\Delta T_s \neq 0$ but the stresses and pressures are maintained constant, $p = p_a = p_w = 0$, the deformation should be homogeneous throughout the phases: the contributions of each phase is proportional to its volume fraction, $\Delta v^k = n^k\,\mathrm{tr}\,\epsilon$, $k = w, a$.

Under such a loading, one has $\mathrm{tr}\,\epsilon = \tau^{\mathrm{DS}}\Delta T_s$. Comparing with $(3.16)_{2,3}$, this requires $X_k = \psi_k - n^k$, $k = w, a$, with ψ_k defined by (2.15). Integrating X_w considering the volume fractions fixed and using the definition of ψ_w, eqn (2.15), one first finds $X = -s\,\chi_w - n^w p_w + \phi_a(p_a)$. Similarly, integration of X_a gives $X = +s\,\chi_a - n^a p_a + \phi_w(p_w)$. Elimination of the χ's with help of (2.15), shows that $n^k p_k - p_a(1 - K^{\mathrm{DS}}/K_s) + \phi_k(p_k)$ is independent of k. Substituting in any of the expressions of X above, using again (2.15), provides X to within a constant, and then

$$c_k = \chi_k - n^k\,, \quad k = w, a. \tag{3.17}$$

Note that the above proposition does not violate the strain averaging relation (2.19). In fact, since the deformation is homogeneous, the strain in the solid grains is equal to the strain of the solid skeleton, i.e. $\mathrm{tr}\,\boldsymbol{\epsilon}_s = \mathrm{tr}\,\boldsymbol{\epsilon}$, and the strain averaging relation trivially follows from (2.2). □

In addition, the form of the thermal relations (3.16) has tacitly assumed that the cubical coefficient of thermal expansion of the solid skeleton τ^{DS} is equal to the cubical coefficient of thermal expansion of the solid constituent τ_s,

$$\tau^{\mathrm{DS}} = \tau_s \,. \tag{3.18}$$

Indeed, the volumetric strain of the solid constituent is, see eqn (2.23) for the mechanical part,

$$\mathrm{tr}\,\boldsymbol{\epsilon}_s = \frac{1}{K_s}\,\frac{\mathrm{tr}\,\boldsymbol{\sigma}_s}{3} + \tau_s\,\Delta T_s \,. \tag{3.19}$$

Now calculating the strain $\mathrm{tr}\,\boldsymbol{\epsilon}$ and the changes of fluid volume contents from (3.16) in terms of p, p_w, p_a and ΔT_s, the volume change of the solid constituent resulting from the strain averaging relation (2.19) is found in the form (3.19) with τ^{DS} in place of τ_s: this proves (3.18).

Remark 3.3.1 on a nonlinear temperature-dependance.

In the potential Ψ_M, a term of the form $-(1/2T_k^\circ)C^s\,(\Delta T_k)^2$ gives to the entropy and entropy rate the following contributions, $C^s\,\Delta T_k/T_k^\circ$ and $C^s\,\dot{T}_k/T_k^\circ$ respectively, the heat capacity C^s being assumed to be a constant. On the other hand, a term of the form $-C^s\,T_k\,(\mathrm{Ln}(T_k/T_k^\circ)-1)$ gives to the entropy and entropy rate the following contributions, $C^s\,\mathrm{Ln}(T_k/T_k^\circ)$ and $C^s\,\dot{T}_k/T_k$: the secant matrix form then involves the diagonal coefficient $C^s\,\mathrm{Ln}(T_k/T_k^\circ)/\Delta T_k$.

3.3.2 Accounting for the Influence of Menisci

The presence of the meniscus modifies the entropies of the fluids: let $\eta_k + \mu_k^{(m)}$ be the modified entropy of fluid k. The chemical potential of fluid k is then defined through a modified differential,

$$\delta\mu_k = \frac{\delta p_k}{\rho_k} - (\,\eta_k + \mu_k^{(m)}\,)\,\delta T_k, \;\; k = w, a. \tag{3.20}$$

Insertion of these expressions in (3.10) yields

$$\delta\Psi_M = \boldsymbol{\sigma} : \delta\boldsymbol{\epsilon} - \sum_{k=w,a} v^k\,\delta p_k - \rho_\circ^s\,\eta_s\,\delta T_s + \sum_{k=w,a} \rho^k\,\mu_k^{(m)}\,\delta T_k \,, \tag{3.21}$$

from which follow the formal constitutive equations,

$$\boldsymbol{\sigma} = \frac{\partial\Psi_M}{\partial\boldsymbol{\epsilon}}; \quad v^k = -\frac{\partial\Psi_M}{\partial p_k}, \;\; k = w, a; \quad \rho_\circ^s\,\eta_s = -\frac{\partial\Psi_M}{\partial T_s}; \quad \rho^k\,\mu_k^{(m)} = \frac{\partial\Psi_M}{\partial T_k}, \;\; k = w, a. \tag{3.22}$$

The potential Ψ_M now includes the strain-energy due to meniscus, with $\zeta^{(m)}$ coefficient of thermal expansion of the menisci,

$$\begin{aligned}\Psi_M(\boldsymbol{\epsilon}, p_w, p_a, T_s, T_w, T_a) &= \Psi_M{}^{\mathrm{mech}}(\boldsymbol{\epsilon}, p_w, p_a) + \\ &+ (-K^{\mathrm{DS}}\,\mathrm{tr}\,\boldsymbol{\epsilon} + c_w\,p_w + c_a\,p_a)\,\tau^{\mathrm{DS}}\Delta T_s - \frac{C^s}{2T_s^\circ}\,(\Delta T_s)^2 + s\,\zeta^{(m)}\,\Delta T^{(m)}\,,\end{aligned} \tag{3.23}$$

with c_k, $k = w, a$, given by (3.17). The variation of the intrinsic density ρ_k of the fluid k is defined through the *bulk modulus* K_k and *coefficient of thermal expansion* τ_k, namely
5

$$\frac{\Delta\rho_k}{\rho_k} = \frac{\Delta p_k}{K_k} - \tau_k \Delta T_k\,,\ k = w, a. \tag{3.24}$$

So $\partial(1/\rho_k)/\partial T_k$ is equal to τ_k/ρ_k, but also to $-\partial(\eta_k + \mu_k^{(m)})/\partial p_k$ from (3.20) which defines an exact differential. So,

$$\eta_k + \mu_k^{(m)} = \eta_k(p_k^\circ, T_k^\circ) - \tau_k \frac{p_k}{\rho_k} + \frac{C_{pk}}{T_k^\circ}\Delta T_k,\ k = w, a. \tag{3.25}$$

Here C_{pk},

$$\frac{C_{pk}}{T_k^\circ} = \frac{\partial}{\partial T_k}(\eta_k + \mu_k^{(m)}),\ k = w, a, \tag{3.26}$$

is the heat capacity *at constant pressure* per unit mass of the phase k. Alternatively, we will also make use of the *apparent heat capacities* per unit volume of porous medium, namely

$$C_p^k = \rho^k\, C_{pk}\,,\ k = w, a. \tag{3.27}$$

From $(3.22)_4$ and (3.23),

$$\rho^k\,\mu_k^{(m)} = s\,\zeta^{(m)}\frac{\partial T^{(m)}}{\partial T_k},\ k = w, a. \tag{3.28}$$

Therefore, we have now the complete expression of the entropy accounting for the effect of menisci as,

$$\rho^k\,\Delta\eta_k = -n^k\,\tau_k\,p_k + \frac{C_p^k}{T_k^\circ}\Delta T_k - s\,\zeta^{(m)}\frac{\partial T^{(m)}}{\partial T_k},\ k = w, a\,. \tag{3.29}$$

In absence of definite experimental data, the temperature of the menisci is assumed to be a *volume average of the temperatures of the phases*, namely

$$T^{(m)} = \sum_{k=s,w,a} n^k\,T_k\,. \tag{3.30}$$

Therefore the complete thermo-elastic equations in secant form can be cast in the matrix format (3.32). The mechanical coefficients $d_{k\,l}$, $k, l = w, a$, are given by (2.30),(2.33). The constitutive matrix is not symmetric, like its isothermal counterpart (2.32).

Removing the terms in brackets in the 2nd and 3rd lines provides the changes of fluid volume contents v^k rather than fluid mass contents m^k. Indeed using the definitions (2.4)and (2.3) of fluid mass content and fluid volume content their increments (operator $d(\cdot) = \Delta(\cdot)$ below) or their rates (operator $d(\cdot) = d^S(\cdot)/dt$) relate as,

$$\frac{dm^k}{\rho_k} = dv^k + v^k(\frac{dp_k}{K_k} - \tau_k\,dT_k) \simeq dv^k + n^k(\frac{dp_k}{K_k} - \tau_k\,dT_k)\,. \tag{3.31}$$

$$
\begin{bmatrix} p \\ \dfrac{\Delta m^w}{\rho_w} \\ \dfrac{\Delta m^a}{\rho_a} \\ \rho^s_\circ\,\Delta\eta_s \\ \rho^w\,\Delta\eta_w \\ \rho^a\,\Delta\eta_a \end{bmatrix}
=
\left[\begin{array}{c|c|c|c|c|c}
-K^{\rm DS} & \chi_w & \chi_a & K^{\rm DS}\,\overset{\rm DS}{\tau} & 0 & 0 \\ \hline
\psi_w & d_{ww}[+\dfrac{n^w}{K_w}] & d_{wa} & \begin{array}{c}(n^w-\psi_w)\,\overset{\rm DS}{\tau}\\ +n^s\zeta^{(m)}\end{array} & \begin{array}{c}[-n^w\tau_w]\\ +n^w\zeta^{(m)}\end{array} & n^a\zeta^{(m)} \\ \hline
\psi_a & d_{aw} & d_{aa}[+\dfrac{n^a}{K_a}] & \begin{array}{c}(n^a-\psi_a)\,\overset{\rm DS}{\tau}\\ -n^s\zeta^{(m)}\end{array} & -n^w\zeta^{(m)} & \begin{array}{c}[-n^a\tau_a]\\ -n^a\zeta^{(m)}\end{array} \\ \hline
K^{\rm DS}\,\overset{\rm DS}{\tau} & \begin{array}{c}(n^w-\chi_w)\,\overset{\rm DS}{\tau}\\ +n^s\zeta^{(m)}\end{array} & \begin{array}{c}(n^a-\chi_a)\,\overset{\rm DS}{\tau}\\ -n^s\zeta^{(m)}\end{array} & \dfrac{C^s}{T^\circ_s} & 0 & 0 \\ \hline
0 & \begin{array}{c}-n^w\tau_w\\ +n^w\zeta^{(m)}\end{array} & -n^w\zeta^{(m)} & 0 & \dfrac{C^w_p}{T^\circ_w} & 0 \\ \hline
0 & n^a\zeta^{(m)} & \begin{array}{c}-n^a\tau_a\\ -n^a\zeta^{(m)}\end{array} & 0 & 0 & \dfrac{C^a_p}{T^\circ_a}
\end{array}\right]
\begin{bmatrix} \mathrm{tr}\,\epsilon \\ p_w \\ p_a \\ \Delta T_s \\ \Delta T_w \\ \Delta T_a \end{bmatrix}
\tag{3.32}
$$

Secant thermo-elastic matrix: note the lack of major symmetry.
Terms showing to the meniscus effect are highlighted on a lower level.

Note that these relations hold in both incremental and rate forms even for a perfect gas where the bulk modulus K_a and coefficient of thermal expansion τ_a are not constants, see Appendix B.2: this is due to the fact the above coefficients can be viewed both as secant and tangent. The approximation due to the fact the volume content $v^k = n^k\, V/V_\circ$ is replaced by n^k is quite acceptable for small strain analyses [6].

3.4 Thermo-Elastic Equations: Rate Form

In rate form, the constitutive equations phrased in terms of the appropriate primary and dependent variables display the **major symmetry**: in fact the structure of the constitutive equations has been elaborated to satisfy this property. They generalize the purely isothermal formulation of Loret and Khalili [2000], their eqn (66) and Appendix.

[5] For the air phase, these coefficients are not constant, see (B.6).

[6] As a general rule of the so-called updated lagrangian formalism, terms that involved volume changes, or more generally the deformation gradient, are dropped *after* rates have been calculated.

$$
\begin{bmatrix} \dot{p} \\ \dfrac{\dot{m}^w}{\rho_w} \\ \dfrac{\dot{m}^a}{\rho_a} \\ \rho_{\circ}^s\,\dot{\eta}_s \\ \rho^w\dot{\eta}_w \\ \rho^a\dot{\eta}_a \end{bmatrix}
=
\begin{bmatrix}
-K^{\mathrm{DS}} & \psi_w & \psi_a & K^{\mathrm{DS}}\,\tau^{\mathrm{DS}} & 0 & 0 \\
\psi_w & a_{ww}[+\dfrac{n^w}{K_w}] & a_{wa} & (n^w-\psi_w)\,\tau^{\mathrm{DS}} + n^s\zeta^{(m)} & [-n^w\tau_w] + n^w\zeta^{(m)} & n^a\zeta^{(m)} \\
\psi_a & a_{aw} & a_{aa}[+\dfrac{n^a}{K_a}] & (n^a-\psi_a)\,\tau^{\mathrm{DS}} - n^s\zeta^{(m)} & -n^w\zeta^{(m)} & [-n^a\tau_a] - n^a\zeta^{(m)} \\
K^{\mathrm{DS}}\,\tau^{\mathrm{DS}} & (n^w-\psi_w)\,\tau^{\mathrm{DS}} + n^s\zeta^{(m)} & (n^a-\psi_a)\,\tau^{\mathrm{DS}} - n^s\zeta^{(m)} & \dfrac{C^s}{T_s^{\circ}} & 0 & 0 \\
0 & -n^w\tau_w + n^w\zeta^{(m)} & -n^w\zeta^{(m)} & 0 & \dfrac{C_p^w}{T_w^{\circ}} & 0 \\
0 & n^a\zeta^{(m)} & -n^a\tau_a - n^a\zeta^{(m)} & 0 & 0 & \dfrac{C_p^a}{T_a^{\circ}}
\end{bmatrix}
\begin{bmatrix} \mathrm{tr}\,\dot{\epsilon} \\ \dot{p}_w \\ \dot{p}_a \\ \dot{T}_s \\ \dot{T}_w \\ \dot{T}_a \end{bmatrix}
\tag{3.33}
$$

Tangent thermo-elastic matrix: note the major symmetry.
Terms pertaining to the meniscus effect are highlighted on a lower level.

Neglecting the variations of volume fractions, but accounting for the dependence on suction of the coefficients χ_k, ψ_k and d_{kl}, $k,l = w,a$, the elastic coefficients $a_{kl}=a_{lk}$ defining the rate thermo-elastic equations differ only by second-order terms from their elastic counterparts:

$$
\begin{cases}
a_{ww} = -n^s\,\dfrac{dC}{ds} + \dfrac{1}{K^{\mathrm{DS}}}\left(p^{\mathrm{net}}\,\dfrac{d\psi_w}{ds} - \psi_w^2\right) + \dfrac{(n^w)^2}{n^s\,K_s}, \\
a_{\overset{\leftrightarrow}{wa}} = n^s\,\dfrac{dC}{ds} - \dfrac{1}{K^{\mathrm{DS}}}\left(p^{\mathrm{net}}\,\dfrac{d\psi_w}{ds} - \psi_w(1-\psi_a)\right) - \dfrac{n^w(1-n^a)}{n^s\,K_s}, \\
a_{aa} = -n^s\,\dfrac{dC}{ds} + \dfrac{1}{K^{\mathrm{DS}}}\left(p^{\mathrm{net}}\,\dfrac{d\psi_w}{ds} - (1-\psi_a)^2\right) + \dfrac{(1-n^a)^2}{n^s\,K_s},
\end{cases}
\tag{3.34}
$$

where $\Delta C(s)$ has been obtained from the *isothermal* soil-water characteristic curve by (2.28). Removing the terms in brackets in the 2nd and 3rd lines of (3.33) provides the changes of fluid volume contents rather than fluid mass contents.

So far, we have considered constitutive equations for the isotropic parts of stress and strain. In fact, since the materials we consider are isotropic and the shear modulus is constant, the elastic and thermo-elastic constitutive equations for the deviatoric stress and strain decouple from the spherical equations and they are simply $\mathbf{s} = 2\mu^{\rm DS}\,\mathrm{dev}\,\boldsymbol{\epsilon}$. In view of the elastic-plastic rate formulation, one may however recompose these two parts: this operation is facilitated by

---

Proposition 3.4.1 Form-Invariance of the Effective Stress and its Rate.

Due to (2.16), the effective mean-stress associated to the χ's and its rate associated to the ψ's are form-invariant

$$\bar{p} = -\frac{1}{3}\,\mathrm{tr}\,\bar{\boldsymbol{\sigma}} = p - \chi_w\,p_w - \chi_a\,p_a \quad \Rightarrow \quad \dot{\bar{p}} = -\frac{1}{3}\,\mathrm{tr}\,\dot{\bar{\boldsymbol{\sigma}}} = \dot{p} - \psi_w\,\dot{p}_w - \psi_a\,\dot{p}_a\,. \tag{3.35}$$

Therefore the effective thermo-elastic law in secant form,

$$\bar{\boldsymbol{\sigma}} = \mathbb{E}^{DS}\colon \left(\boldsymbol{\epsilon} - \tfrac{1}{3}\,\tau^{DS}\Delta T_s\,\mathbf{I}\right) \Leftrightarrow \boldsymbol{\sigma} = \mathbb{E}^{DS}\colon \left(\boldsymbol{\epsilon} - \tfrac{1}{3}\,\tau^{DS}\Delta T_s\mathbf{I}\right) - \chi_w\,p_w\,\mathbf{I} - \chi_a\,p_a\,\mathbf{I} \tag{3.36}$$

becomes in rate form

$$\dot{\bar{\boldsymbol{\sigma}}} = \mathbb{E}^{DS} : \left(\dot{\boldsymbol{\epsilon}} - \tfrac{1}{3}\,\tau^{DS}\dot{T}_s\,\mathbf{I}\right) \Leftrightarrow \dot{\boldsymbol{\sigma}} = \mathbb{E}^{DS} : \left(\dot{\boldsymbol{\epsilon}} - \tfrac{1}{3}\,\tau^{DS}\dot{T}_s\,\mathbf{I}\right) - \psi_w\,\dot{p}_w\,\mathbf{I} - \psi_a\,\dot{p}_a\,\mathbf{I}. \tag{3.37}$$

These relations, as well as (3.32) and (3.33), apply for linear elasticity where $\mathbb{E}^{\rm DS}$ is the tensor of elastic moduli[7],

$$\mathbb{E}^{\rm DS} = (\,K^{\rm DS} - \tfrac{2}{3}\mu^{\rm DS}\,)\,\mathbf{I}\otimes\mathbf{I} + 2\,\mu^{\rm DS}\,\mathbf{I}\,\overline{\underline{\otimes}}\,\mathbf{I}. \tag{3.38}$$

For logarithmic elasticity, e.g. (2.41), the single modification concerns the bulk modulus: the secant relations use the secant bulk modulus $K^{\rm DS}_{\rm sec}$ while the rate relations use the tangent bulk modulus $K^{\rm DS}_{\rm tan}$,

$$K^{\rm DS}_{\rm sec} = \frac{\bar{p} - p_\circ}{\kappa\,\mathrm{Ln}\dfrac{\bar{p}}{p_\circ}}\,, \qquad K^{\rm DS}_{\rm tan} = \frac{\bar{p}}{\kappa}\,. \tag{3.39}$$

On the other hand, the terms in brackets which have to be removed if one wants to obtain the rates of fluid volume changes are identical to the secant terms in view of the remark accompanying (3.31).

3.5 Thermo-Elastoplastic Equations

The structure of the thermo-elastic-plastic relations is motivated by the complementary power $\mathcal{P}_C$ which provides conjugate triplets, $(\boldsymbol{\sigma}, p_w, p_a, T_s)$ for the stress triplet and $(\boldsymbol{\epsilon}, \Delta v^w, \Delta v^a, \rho^s_\circ\,\Delta\eta_s)$ for the strain triplet. The rates of strain-like variables and solid entropy are additively decomposed into an elastic and a plastic contribution, namely

$$\dot{\boldsymbol{\epsilon}} = \dot{\boldsymbol{\epsilon}}_{el} + \dot{\boldsymbol{\epsilon}}_{pl}\,, \quad \dot{v}^k = \dot{v}^k_{el} + \dot{v}^k_{pl}\,, \quad k = w, a, \quad \text{and} \quad \dot{\eta}_s = \dot{\eta}^{el}_s + \dot{\eta}^{pl}_s\,. \tag{3.40}$$

[7] The dyadic product of two tensors is indicated with '$\otimes$'. The symbol '$\overline{\underline{\otimes}}$' denotes a symmetrized outer product: componentwise in cartesian axes, for any arbitrary symmetric second order tensors $\mathbf{A}$ and $\mathbf{B}$, $(\mathbf{A}\,\overline{\underline{\otimes}}\,\mathbf{B})_{ijkl} = (A_{ik}B_{jl} + A_{il}B_{jk})/2$.

The flow rule emanates from a generalized normality rule applied to the above stress and strain triplets. Decomposing P_C into an elastic part and a plastic part, we have, using (3.12),

$$\delta P_C^{pl} = -\delta\boldsymbol{\sigma} : \dot{\boldsymbol{\epsilon}}_{pl} - \delta p_w\, \dot{v}_{pl}^w - \delta p_a\, \dot{v}_{pl}^a - \delta T_s\, \rho_\circ^s\, \dot{\eta_s}^{pl}\,. \tag{3.41}$$

We assume the existence of a plastic potential $g = g(\boldsymbol{\sigma}, p_w, p_a, T_s, \mathcal{H})$ where, like for the mechanical analysis, stresses act in fact only through the effective stress $\bar{\boldsymbol{\sigma}}$ and the suction s and, although an abuse of notation, we note $g = g(\boldsymbol{\sigma}, p_w, p_a, T_s, \mathcal{H}) = g(\bar{\boldsymbol{\sigma}}, s, T_s, \mathcal{H})$. The yield function f is assumed to have the same arguments. Then

$$\dot{\boldsymbol{\epsilon}}_{pl} = \dot{\Lambda}\,\frac{\partial g}{\partial \boldsymbol{\sigma}}, \quad \dot{v}_{pl}^k = \dot{\Lambda}\,\frac{\partial g}{\partial p_k}, \quad k = w, a, \quad \text{and} \quad \rho_\circ^s\, \dot{\eta_s}^{pl} = \dot{\Lambda}\,\frac{\partial g}{\partial T_s}, \tag{3.42}$$

with $\dot{\Lambda} \geq 0$ the plastic multiplier. The hardening rule is formally

$$\dot{\mathcal{H}} = \dot{\Lambda}\, \mathcal{X}(\bar{\boldsymbol{\sigma}}, s, T_s, \mathcal{H})\,, \tag{3.43}$$

where $\mathcal{X}$ is an appropriate constitutive function, arbitrary in this formal presentation. Since we adopt a generalized normality rule, the behaviour would then be associative, and the elastic-plastic stiffness of the underlying drained solid would display the major symmetry property, to within signs in the first row, if the directions of the normals to the yield function $f(\boldsymbol{\sigma}, p_w, p_a, T_s, \mathcal{H})$ and plastic potential $g(\boldsymbol{\sigma}, p_w, p_a, T_s, \mathcal{H})$ in the space $(\boldsymbol{\sigma}, p_w, p_a, T_s, \mathcal{H})$ were identical.

Inserting $(3.40)_1$ and $(3.42)_1$ into (3.37) provides the plastic multiplier $\dot{\Lambda}$,

$$\dot{\Lambda} = \frac{1}{H}\Big[\mathbf{Q} : \mathbb{E}^{\text{DS}} : \dot{\boldsymbol{\epsilon}} + \frac{\partial f}{\partial s}\dot{s} + (-\mathbf{Q} : \mathbb{E}^{\text{DS}} : (\frac{\tau^{\text{DS}}}{3}\mathbf{I}) + \frac{\partial f}{\partial T_s})\,\dot{T}_s\Big]\,, \tag{3.44}$$

where the plastic modulus H and hardening modulus h are defined as in a purely mechanical theory,

$$H = h + \mathbf{Q} : \mathbb{E}^{\text{DS}} : \mathbf{P}\,, \quad h = -\frac{\partial f}{\partial \mathcal{H}}\mathcal{X}\,. \tag{3.45}$$

Also $\mathbf{P}$ and $\mathbf{Q}$ denote the normals to the plastic potential and yield function in stress-space

$$\mathbf{P} = \frac{\partial g}{\partial \boldsymbol{\sigma}}, \quad \mathbf{Q} = \frac{\partial f}{\partial \boldsymbol{\sigma}}\,. \tag{3.46}$$

The tangent elastic-plastic matrix is shown in eqn (3.48). To simplify the notations, the following conventions have been adopted:

- the terms which retain their elastic values as given by (3.33) are denoted with the superscript ^{el};
- the terms which are modified by plasticity are denoted with the superscript ^{ep} and are given in (A.1);
- the terms which are not modified by plasticity when $\partial g/\partial s = 0$, a constitutive assumption argued in Loret and Khalili [2000], are denoted by the superscript $^{(ep)}$.

The elastic coefficients $\underset{\cdot\cdot}{A}^{el}$ can be read from the tangent elastic matrix (3.33) except some of the first line and first column which should be written in tensorial form, see Proposition 3.4.1,

$$-A_{12}^{el} = A_{21}^{el} = \psi_w\,\mathbf{I}\,, \quad -A_{13}^{el} = A_{31}^{el} = \psi_a\,\mathbf{I}\,. \tag{3.47}$$

$$
\begin{bmatrix} \dot{\sigma} \\ \hline \dfrac{\dot{m}^w}{\rho_w} \\ \hline \dfrac{\dot{m}^a}{\rho_a} \\ \hline \rho_\circ^s \dot{\eta}_s \\ \hline \rho^w \dot{\eta}_w \\ \hline \rho^a \dot{\eta}_a \end{bmatrix}
=
\begin{bmatrix}
A^{ep}_{11} & A^{ep}_{12} & A^{ep}_{13} & A^{ep}_{14} & 0 & 0 \\
A^{(ep)}_{21} & A^{(ep)}_{22} & A^{(ep)}_{23} & A^{(ep)}_{24} & A^{el}_{25} & A^{el}_{26} \\
A^{(ep)}_{31} & A^{(ep)}_{32} & A^{(ep)}_{33} & A^{(ep)}_{34} & A^{el}_{35} & A^{el}_{36} \\
A^{ep}_{41} & A^{ep}_{42} & A^{ep}_{43} & A^{ep}_{44} & 0 & 0 \\
0 & A^{el}_{52} & A^{el}_{53} & 0 & A^{el}_{55} & 0 \\
0 & A^{el}_{62} & A^{el}_{63} & 0 & 0 & A^{el}_{66}
\end{bmatrix}
\begin{bmatrix} \dot{\epsilon} \\ \hline \dot{p}_w \\ \hline \dot{p}_a \\ \hline \dot{T}_s \\ \hline \dot{T}_w \\ \hline \dot{T}_a \end{bmatrix}
\begin{matrix} \\ \leftarrow \text{elastic line if } \dfrac{\partial g}{\partial s} = 0 \\ \leftarrow \quad \text{''} \\ \\ \leftarrow \text{elastic line} \\ \leftarrow \text{elastic line} \end{matrix}
\tag{3.48}
$$

Tangent thermo-elastic-plastic matrix: major symmetry holds if the thermo-plastic flow rule is associative. Some terms are always elastic, some always plastic and other depend on constitutive assumptions, see text.

References

Biot, M.A. (1973). Nonlinear and semilinear rheology of porous solids. *J. of Geophysical Research*, **78**(23), 4924-4937.

Bowen, R.M (1976). Theory of mixtures. *Continuum Physics*, vol. 3, 1-127, A.C. Eringen ed., Academic Press, New York.

Eringen, A.C. and Ingram, J.D. (1965). A continuum theory for chemically reacting media -I. *Int. J. Eng. Science*, **3**, 197-212.

Fleureau, J.M., Kheirbek-Saoud, S., Soemitro, R. and Taibi, S. (1993). Behavior of clayey soils on drying-wetting paths. *Canadian Geotechnical J.*, **30**, 287-296

Gens, A., Jouanna, P. and Schrefler, B.A. (1995). Modern issues in non-saturated soils. CISM Courses and Lectures n°. 357, Springer Verlag, Wien.

Haase, R. (1990). *Thermodynamics of Irreversible Processes.* Dover Publications, New York.

Hassanizadeh, S.M. and Gray, W.G. (1993). Thermodynamic basis of capillarity pressure in porous media. *Water Resources Research*, **29**(10), 3389-3405.

Kestin, J. (1968). *A Course in Thermodynamics.* Blaisdell Publishing Co., Waltham, Massachusetts.

Khalili, N. and Khabbaz, M.H. (1995). On the theory of three-dimensional consolidation in unsaturated soils. *Unsaturated Soils/Sols non saturés*, E.E. Alonso and P. Delage eds, Presses de l'Ecole Nationale des Ponts et Chaussées, Paris, 1995, 745-750.

Khalili, N. and Khabbaz, M.H. (1998). A unique relationship for χ for the determination of the shear strength of unsaturated soils. *Géotechnique*, **48**(2), 1-7.

Khalili, N. and Loret B. (2000). An elasto-plastic model for non-isothermal analysis of flow and deformation in unsaturated porous media: formulation, submitted.

Loret B. (1985). On the choice of elastic parameters for sand. *Int. J. Num. Anal. Meth. Geomechanics*, **9**, 285-292.

Loret B. and Khalili (2000). A three-phase model for unsaturated soils. *Int. J. Num. Anal. Meth. Geomechanics*, **24**, 893-927.

Appendix A. Elastic-plastic coefficients of the mixed formulation in rate form

The coefficients of the elastic-plastic matrix (3.48) which differ from their elastic counterparts in the matrix (3.33) are the following,

$$
\begin{aligned}
&A_{11}^{ep} = \mathbb{E}_{\mathrm{ep}}^{\mathrm{DS}} = \mathbb{E}^{\mathrm{DS}} - \frac{1}{H}\mathbb{E}^{\mathrm{DS}} : \mathbf{P}\otimes\mathbf{Q} : \mathbb{E}^{\mathrm{DS}}; \\
&A_{1i}^{ep} = A_{1i}^{el} + \frac{(-1)^i}{H}(\mathbb{E}^{\mathrm{DS}} : \mathbf{P})\,\frac{\partial f}{\partial s}, \qquad A_{i1}^{(ep)} = A_{i1}^{el} - \frac{(-1)^i}{H}(\mathbf{Q} : \mathbb{E}^{\mathrm{DS}})\,\frac{\partial g}{\partial s},\ i=2,3; \\
&A_{14}^{ep} = -\mathbb{E}_{\mathrm{ep}}^{\mathrm{DS}} : (\frac{\tau^{\mathrm{DS}}}{3}\mathbf{I}) - \frac{1}{H}(\mathbb{E}^{\mathrm{DS}} : \mathbf{P})\frac{\partial f}{\partial T_s}, \qquad A_{41}^{ep} = (\frac{\tau^{\mathrm{DS}}}{3}\mathbf{I}) : \mathbb{E}_{\mathrm{ep}}^{\mathrm{DS}} + \frac{1}{H}(\mathbf{Q} : \mathbb{E}^{\mathrm{DS}})\,\frac{\partial g}{\partial T_s}; \\
&A_{ij}^{(ep)} = A_{ij}^{el} + \frac{-(1)^{i+j}}{H}\frac{\partial f}{\partial T_s}\frac{\partial g}{\partial T_s},\ i,j=2,3; \quad A_{44}^{ep} = A_{44}^{el} + \frac{1}{H}Dg\,Df; \\
&A_{i4}^{(ep)} = A_{i4}^{el} + \frac{(-1)^i}{H}Df\,\frac{\partial g}{\partial s}, \qquad A_{4i}^{ep} = A_{4i}^{el} + \frac{(-1)^i}{H}Dg\,\frac{\partial f}{\partial s},\ i=2,3; \\
&Df = \mathbf{Q} : \mathbb{E}^{\mathrm{DS}} : (\frac{\tau^{\mathrm{DS}}}{3}\mathbf{I}) - \frac{\partial f}{\partial T_s}, \qquad Dg = (\frac{\tau^{\mathrm{DS}}}{3}\mathbf{I}) : \mathbb{E}^{\mathrm{DS}} : \mathbf{P} - \frac{\partial g}{\partial T_s}.
\end{aligned}
\tag{A.1}
$$

Appendix B. Explicit Entropies, chemical potentials and internal energies

Appendix B.1 **Energies for the fluid phase**

The fluid is only slightly compressible, so one can linearize the expansion (3.24) to

$$
\frac{\Delta\rho_w}{\rho_w^\circ} = \frac{\Delta p_w}{K_w} - \tau_w\,\Delta T_w\,, \tag{B.1}
$$

where the bulk modulus K_w and the coefficient of thermal expansion τ_w are considered constant. Then, the change of entropy per unit mass of fluid is

$$
\Delta\eta_w = \eta_w(p_w, T_w) - \eta_w(p_w^\circ, T_w^\circ) = -\frac{\tau_w}{\rho_w^\circ}\,\Delta p_w + C_{pw}\,\frac{\Delta T_w}{T_w^\circ} - \frac{s\,\zeta^{(m)}}{\rho_w}\frac{\partial T^{(m)}}{\partial T_w}. \tag{B.2}
$$

Since the two partial derivatives of μ_w are now known, one can obtain the chemical potential per unit mass of fluid in explicit form,

$$
\begin{aligned}
\mu_w(p_w, T_w) = \mu_w(p_w^\circ, T_w^\circ) &+ \frac{1}{\rho_w^\circ}\Big(\Delta p_w - \frac{1}{2}\frac{(\Delta p_w)^2}{K_w} + \tau_w\,\Delta T_w\,\Delta p_w\Big) \\
&- \frac{1}{2}C_{pw}\,(\Delta T_w)^2 - \eta_w(p_w^\circ, T_w^\circ)\Delta T_w\,.
\end{aligned}
\tag{B.3}
$$

A slightly different form emerges if the temperature appaeraing below the heat capacity is kept at its current value, see Remark 3.3.1. The internal energy per unit mass of fluid

$\epsilon_W = \mu_w + T_w\,\eta_w - p_w/\rho_w$ becomes

$$\begin{aligned}\epsilon_W(p_w, T_w) &= \epsilon_W(p_w^\circ, T_w^\circ) + \frac{1}{\rho_w^\circ}\Big(\frac{\Delta p_w}{2\,K_w} p_w^\circ - \tau_w\,(p_w\,T_w - p_w^\circ\,T_w^\circ)\Big) \\ &+ \frac{C_{pw}}{T_w^\circ}\frac{\Delta T_w}{2}(T_w + T_w^\circ) - \frac{s\,\zeta^{(m)}}{\rho^w}\frac{\partial T^{(m)}}{\partial T_w} T_w\,.\end{aligned} \tag{B.4}$$

Appendix B.2 **Energies for the air phase.**

The air phase is assumed to behave as a perfect gas. Its total pressure P_a is linked to its temperature T_a and intrinsic density ρ_a through the universal gas constant R and its molar mass M_a,

$$P_a = p_a + p_{atm} = \frac{RT_a}{M_a}\,\rho_a\,. \tag{B.5}$$

Consequently, the bulk modulus and coefficient of thermal expansion are variable, namely

$$\frac{\Delta\rho_a}{\rho_a} = \frac{\Delta p_a}{K_a} - \tau_a\,\Delta T_a, \quad K_a \equiv P_a = p_a + p_{atm}, \quad \tau_a \equiv \frac{1}{T_a}\,. \tag{B.6}$$

As previously, using the fact that the expression defining the chemical potential is an exact differential, the change in entropy $\Delta\eta_a$ can be cast in at least two formats,

$$\begin{aligned}\Delta\eta_a \equiv \eta_a(p_a, T_a) - \eta_a(p_a^\circ, T_a^\circ) &= -\frac{R}{M_a}\,\mathrm{Ln}\frac{P_a}{P_a^\circ} + C_{pa}\,\mathrm{Ln}\frac{T_a}{T_a^\circ} - \frac{s\,\zeta^{(m)}}{\rho^a}\frac{\partial T^{(m)}}{\partial T_a}\,, \\ &= -\frac{R}{M_a}\,\mathrm{Ln}\frac{\rho_a}{\rho_a^\circ} + C_{va}\,\mathrm{Ln}\frac{T_a}{T_a^\circ} - \frac{s\,\zeta^{(m)}}{\rho^a}\frac{\partial T^{(m)}}{\partial T_a}\,.\end{aligned} \tag{B.7}$$

The second expression above involves the heat capacity C_{va} *at constant volume* per unit mass of gas defined as

$$C_{va} = C_{pa} - \frac{R}{M_a}\,. \tag{B.8}$$

Consequently, the chemical potential obtained by integration of his two partial derivatives is

$$\mu_a(p_a, T_a) = \mu_a(p_a^\circ, T_a^\circ) + \frac{R}{M_a}\,T_a\,\mathrm{Ln}\frac{P_a}{P_a^\circ} - C_{pw}\,T_a\,(\mathrm{Ln}\frac{T_a}{T_a^\circ} - 1) - \eta_a(p_a^\circ, T_a^\circ)\,\Delta T_a\,. \tag{B.9}$$

The resulting internal energy per unit mass of the gas $\epsilon_a = \mu_a + T_a\,\eta_a - P_a/\rho_a$ would depend as expected only on temperature, if the meniscus effect were neglected,

$$\epsilon_A(p_a, T_a) = \epsilon_A(p_a^\circ, T_a^\circ) + C_{va}\,T_a - \frac{s\,\zeta^{(m)}}{\rho^a}\frac{\partial T^{(m)}}{\partial T_a} T_a\,. \tag{B.10}$$

ADVANCED NUMERICAL MODELLING OF GRANULAR SOILS

Peter Keith Woodward

Department of Civil and Offshore Engineering, Heriot-Watt University, Edinburgh, EH14 4AS, U.K.

Abstract: This chapter initially presents a brief review of the different types of constitutive soil models used in geomechanics and illustrates the implementation of a typical isotropic model. The constitutive behaviour of a double-hardening isotropic model is also highlighted as part of this review. The majority of the chapter is however concerned with the description and implementation of a multi-surface kinematic elasto-plastic constitutive soil model into a finite element program. The calibration of the model to a particular sand under monotonic and cyclic loading and the results of typical boundary values problems under monotonic and dynamic loading are presented. The ability of the model to simulate earthquake induced liquefaction, through work published from the VELACS project, is also shown.

1 Granular Soil Models

In reviewing the literature one can easily see that many different types of constitutive granular soil models have been developed over the last 20-30 years. Some of these models have been developed into finite element models for the solution of real problems, whereas others have been developed largely for triaxial plane simulations. The most popular models for monotonic loading conditions are probably those based on bounding surface and classical plasticity. In cyclic loading conditions, the same types of popular models have been used together with nested surface models. In this section we will discuss the bases of the different forms of constitutive models proposed for modelling granular soils. The application of a specific soil model for simulating granular soils under monotonic and cyclic loading will be the main focus of the work presented in this chapter.

A comprehensive source of information on different granular soil models can be found in Saada & Bianchini (1987) and Arulanandan & Scott (1994).

1.1 Stress Invariants Used in Constitutive Model Description

The usual soil mechanics sign convention is adopted with compression positive. The isotropic stress s and the isotropic strain v can be calculated from the following expressions

$$s = \frac{1}{\sqrt{3}} \sigma_{ij} \delta_{ij} \qquad (1)$$

$$v = \frac{1}{\sqrt{3}} \varepsilon_{ij} \delta_{ij} \qquad (2)$$

where, σ_{ij} is the Cauchy stress, ε_{ij} is the strain and δ_{ij} is the Kronecker delta. The deviatoric stress t and the deviatoric strain e can be obtained from the following expressions

$$t = \sqrt{(t_{ij} t_{ij})} \tag{3}$$

$$e = \sqrt{(e_{ij} e_{ij})} \tag{4}$$

where t is the deviatoric stress tensor and e is the deviatoric strain tensor given by

$$t_{ij} = \sigma_{ij} - \frac{1}{3}\sigma_{kl}\delta_{kl}\delta_{ij} \tag{5}$$

$$e_{ij} = \varepsilon_{ij} - \frac{1}{3}\varepsilon_{kl}\delta_{kl}\delta_{ij} \tag{6}$$

As discussed by Dafalias (1994), soil models can broadly be split into three main groups.

1.2 Empirical Models

These models are developed based on 'analytical and/or experimental' observations of material behaviour. A direct tensorial relationship between stress and strain does not usually exist hence there is often no overall framework of elastoplasticty. In the strictest sense, they are not 'constitutive' soil models. They are often only applicable to a given boundary value problem (e.g. a retaining wall of given geometrical dimensions and idealised soil properties). If applied to other types of problems, for which they were not originally developed, they would require modifications and a new set of laboratory and/or field observations. For example, these types of models may have a direct relationship between shear modulus and shear strain, which maybe related to the mean effective stress. The development of 'plastic' strains or excess pore water pressures are usually due to empirical deviatoric or isotropic strain relationships or through a 'damage' quantity. These types of models are very popular in soil mechanics programs due to their simplicity. Typical examples include Inel et al (1993) and Finn et al (1977). The latter model is called TARA-2 in later publications.

1.3 Micromechanical Models

In these models, summing up the response of smaller 'units' simulates the overall material response. Plasticity is usually incorporated very simply at the unit level, and then transferred to global state as an average of the unit responses. The material parameters are usually related to the behaviour of the unit mechanism as well.

An example of these types of models is the work of Cundall and Strack (1979) and Thornton (1989) whereby the granular soil is modelled as an assemblage of unequal spherical particles. Movement of the particles is usually simulated through Newton's Laws and Conservation of Momentum. As the particles interact, shear and normal forces are generated leading to sliding and rotation, with peak shear stresses estimated using a Mohr-Coulomb criterion. The development of excess pore pressures is however often difficult to incorporate in these types of models.

Another example of a micromechanical model is the laminate model developed by Pande & Sharma (1983). In this model, the unit mechanism is defined on a laminate plane with normal and shear components. Standard plasticity formulations are then used for the remainder of the model.

1.4 Plasticity Models

By far the most extensive of soil models used in general engineering are the 'plasticity' based models. These models make use of classical plasticity theory (e.g. loading, unloading, hardening, softening, isotopic plasticity, kinematic plasticity, associative or non-associative flow rules, potential surfaces etc.).

Generalized plasticity models. Generalised plasticity models have generally come about through the work of Zienkiewicz et al (1985). Although many variants on the model have been proposed, it is generally based on the concept of a 'loading surface', whereby plastic deformation is generated by the direction of loading in principal stress space. The magnitude of the plastic modulus is related to the distance of the current stress ratio to that of the failure stress ratio.

Bounding surface models. Bounding surface models for soil are primarily based on the work of Dafalias and Popov (1975) and Krieg (1975). In these models a bounding (or ultimate) surface is first determined in principal stress space. For stress states inside the bounding surface plastic deformation can occur, the level of which is set by the 'distance' of the stress point from an image point on the bounding surface. This distance sets the magnitude of the plastic modulus at that point. The models can incorporate non-associated flow rules and hardening due to plastic isotropic and deviatoric shear strains. Plastic deformation generated during cyclic loading can also easily be incorporated. Many of the models are hypoplastic in that the strain-rate direction depends on the stress-rate direction. This means that the models can account for plastic deformation for a stress increment in any direction and can therefore account for the effects of principal stress rotation. An example of a bounding surface model for sands is the two-surface (yield and bounding) plasticity model described by Manzari and Dafalias (1997). In this model, the concepts of critical state soil mechanics, and a sand state parameter (Been and Jefferies, 1985), are used as a general framework for modelling the stress-strain behaviour of soil.

Nested-surface models. In the original description of these models (Mroz, 1967 and Iwan, 1967) a set of yield surfaces exist in principal stress space. Each of these surfaces has its own plastic modulus and hence plastic deformation occurs as the stress point pushes one surface towards the conjugate point on the next surface (kinematic hardening). The conjugate points ensure that the normal directions of loading are the same. The surfaces do not intersect, so once the next surface is reached it becomes the new active one (known as Mroz kinematic hardening).

Prevost (1985) uses an extension to this type of kinematic hardening, using conical yield surfaces in the deviatoric plane. These surfaces do intersect with each other during hardening to form a set of nested kinematic surfaces representing the stress history of the soil.

Classical plasticity models. The final group of models are those which fall under the 'classical' plasticity heading. These models may have yield and failure surfaces based on criteria like Mohr-Coulomb, Drucker-Prager, Lade etc. Isotropic hardening is achieved through expansion of the yield surface about the isotropic axes and the plastic strain increment is determined through a flow rule, usually non-associated for granular soils (i.e. the existence of a separate plastic potential). The direction of the plastic strain increment is defined as the vector normal to this surface (although some models first decompose the strain increment into isotropic and deviatoric parts). Kinematic hardening is simulated through translation of the yield surface in principal stress space. Some models include both isotropic and kinematic hardening (expansion with translation) and simulate stress history through the generation of multiple kinematic yield surface. Examples of these types of models can be found in Zienkiewicz and Chang (1978) and Jefferies (1993).

1.5 Isotropic Hardening

The model presented in this chapter falls into the classical plasticity group as defined by Dafalias (1994). We shall therefore first consider the derivation of the stiffness matrix for an isotropic model, and then examine in more detail the structure of the combined isotropic and kinematic hardening soil model.

For an elasto-plastic soil the relationship between the incremental stress $\{d\sigma\}$ and strain $\{d\varepsilon\}$ is given by

$$\{d\sigma\} = \lfloor D^{ep} \rfloor \{d\varepsilon\} \tag{7}$$

where $\lfloor D^{ep} \rfloor$ is the elasto-plastic matrix. In order to determine this matrix we must consider properties like the work hardening of the soil. To do this we first decompose the strain increment into an elastic $\{d\varepsilon^e\}$ and plastic $\{d\varepsilon^p\}$ parts.

$$\{d\varepsilon\} = \{d\varepsilon^e\} + \{d\varepsilon^p\} \tag{8}$$

The elastic strain is simply computed from the elastic matrix $\lfloor D^e \rfloor$.

$$\{d\varepsilon^e\} = \left[D^e\right]^{-1} \{d\sigma\} \tag{9}$$

The plastic strain is determined through the flow rule

$$\{d\varepsilon^p\} = d\lambda \left\{\frac{\partial G}{\partial \sigma}\right\} \tag{10}$$

where $d\lambda$ is a scalar defining the magnitude of the plastic strain increment and $\left\{\frac{\partial G}{\partial \sigma}\right\}$ defines the direction of the plastic strain increment due to an increment of stress.

A vector normal to a potential surface determines this direction. When the potential surface is assumed to be equal to the yield surface, the flow rule is said to be associated. When a separate surface is considered, the flow rule is said to be non-associated. The potential surface can take many different forms, for example, it may be a smaller version of the deviatoric yield surface, or it may be ellipsoidal in shape. The potential surface is chosen so that it can accurately simulate the plastic strain behaviour of the soil. In the case of dense sands, the surface must be capable of simulating the initial contraction of the soil (i.e. the normal of the potential points to the right in *t-s* space). It must then predict zero volume change at the phase transformation point (the normal points vertically upwards) and then finally dilation (the normal points to the left). In granular soils, associated flow rules usually lead to excessive dilation and non-associated flow rules are preferred.

By substitution, the following equation for the strain can be obtained

$$\{d\varepsilon\} = \left[D^e\right]^{-1}\{d\sigma\} + d\lambda\left\{\frac{\partial G}{\partial \sigma}\right\} \tag{11}$$

Since we now can determine the direction of the plastic strain, we must now determine its magnitude $d\lambda$. The plastic strain is only calculated when the yield surface is active and is either expanding or contracting. The yield surface can be written as

$$\mathrm{F} = \mathrm{F}(\{\sigma\},\kappa) = 0 \tag{12}$$

where, F represents the yield surface and is a function of the stress state and a hardening parameter κ. When F=0 the stress state is on the yield surface, whereas negative values indicate that it is inside and positive that it is outside. This equation can be written in an alternative form

$$\mathrm{F} = \mathrm{F}[\{\sigma\}] - f[\kappa] = 0 \tag{13}$$

$\mathrm{F}[\{\sigma\}]$ is now termed the loading function and $f[\kappa]$ is called the hardening function. The hardening function defines the position of the yield surface and the rate at which it expands or contracts due to plastic straining. During plastic yielding the stress state must remain on the yield surface, therefore equation (13) can be written as

$$\mathrm{dF} = \left\{\frac{\partial F}{\partial \sigma}\right\}^T \{d\sigma\} + \frac{\partial F}{\partial \kappa} d\kappa = 0 \tag{14}$$

where, $\left\{\frac{\partial F}{\partial \sigma}\right\}$ is the gradient normal to the yield surface. The hardening modulus H is now introduced

$$\mathrm{H} = \frac{-1}{d\lambda}\left(\frac{\partial F}{\partial \kappa} d\kappa\right) \tag{15}$$

Equation (14) can therefore be written as

$$\left\{\frac{\partial F}{\partial \sigma}\right\}^T \{d\sigma\} - \mathrm{H} d\lambda = 0 \tag{16}$$

re-arranging equation (16) gives

$$d\lambda = \frac{1}{\mathrm{H}}\left\{\frac{\partial F}{\partial \sigma}\right\}^T \{d\sigma\} \tag{17}$$

The magnitude of the plastic strain is therefore inversely proportional to the plastic modulus, i.e. the amount of plastic straining increases as the plastic modulus reduces. Through substitution into equation (14) and multiplying through by $\left\{\frac{\partial F}{\partial \sigma}\right\}^T \left[D^e\right]$ the following equation can be written

$$\left\{\frac{\partial F}{\partial \sigma}\right\}^T \left[D^e\right]\{d\varepsilon\} = \left\{\frac{\partial F}{\partial \sigma}\right\}^T \{d\sigma\} + d\lambda \left\{\frac{\partial F}{\partial \sigma}\right\}^T \left[D^e\right]\left\{\frac{\partial G}{\partial \sigma}\right\} \tag{18}$$

if we now substitute for the hardening modulus we obtain

$$\left\{\frac{\partial F}{\partial \sigma}\right\}^T \left[D^e\right]\{d\varepsilon\} = \mathrm{H} d\lambda + d\lambda \left\{\frac{\partial F}{\partial \sigma}\right\}^T \left[D^e\right]\left\{\frac{\partial G}{\partial \sigma}\right\} \tag{19}$$

This give rise to another expression for $d\lambda$ that can be substituted into equation (11)

$$d\lambda = \frac{\left\{\frac{\partial F}{\partial \sigma}\right\}^T \left[D^e\right]\{d\varepsilon\}}{H + \left\{\frac{\partial F}{\partial \sigma}\right\}^T \left[D^e\right]\left\{\frac{\partial G}{\partial \sigma}\right\}} \tag{20}$$

The elasto-plastic stress-strain relationship expressed in equation (7) can be written as a combination of the elastic matrix $\left[D^e\right]$ and plastic matrix $\left[D^p\right]$

$$\{d\sigma\} = \left[\left[D^e\right] - \left[D^p\right]\right]\{d\varepsilon\} = \left[D^{ep}\right]\{d\varepsilon\} \tag{21}$$

The plastic matrix $\left[D^p\right]$ can therefore be written as

$$\left[D^p\right]=\frac{\left[D^e\right]\left\{\frac{\partial G}{\partial \sigma}\right\}\left\{\frac{\partial F}{\partial \sigma}\right\}^T\left[D^e\right]}{H+\left\{\frac{\partial F}{\partial \sigma}\right\}^T\left[D^e\right]\left\{\frac{\partial G}{\partial \sigma}\right\}} \tag{22}$$

Early models (like Cam clay etc.) used 'density hardening' which means that plastic hardening was a function of the accumulated plastic volumetric strain (changes in specific density). However, for granular material this was found to be inadequate and hardening was then related to the accumulated plastic deviatoric strains. Some isotropic hardening models have both a volumetric 'cap', which can expand along the hydrostatic axis to simulate plastic volumetric compression during isotropic stress increases, and a deviatoric yield surface that can expand in the deviatoric plane to simulate shear behaviour. An example of this type of soil model is the double-hardening constitutive soil model MONOT (Molenkamp, 1981) shown pictorially in Figure 1. This type of model is limited to the solution of monotonic loading only as either yield surface can only be pushed out once. Purely elastic deformation will now be simulated during unloading and reloading to the current yield point.

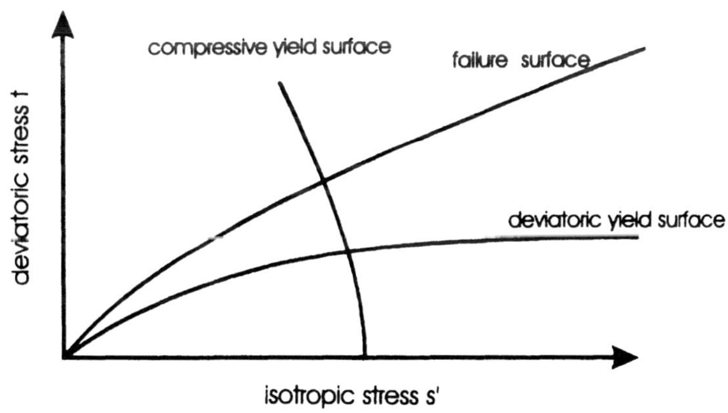

Figure 1. Illustration of the double hardening isotropic model MONOT

The model has been used successfully by Hicks (1990, 1995) to model a wide range of stress paths. The behaviour of the model is summerised in Figures 2 & 3 for the associated plastic compressive yield surface and the non-associated deviatoric yield surface. The ability to push out either yield surface for a variety of different stress paths is shown.

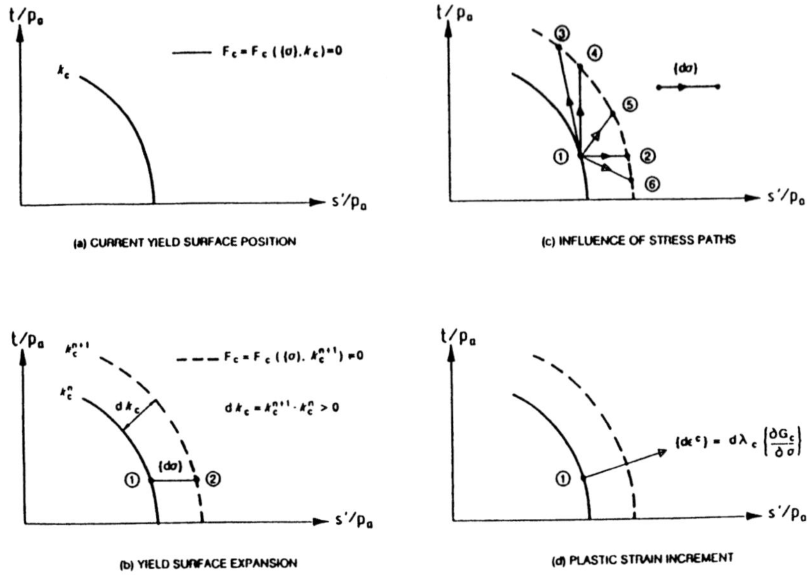

Figure 2. Behaviour of the plastic compressive yield surface in MONOT (after Hicks, 1990)

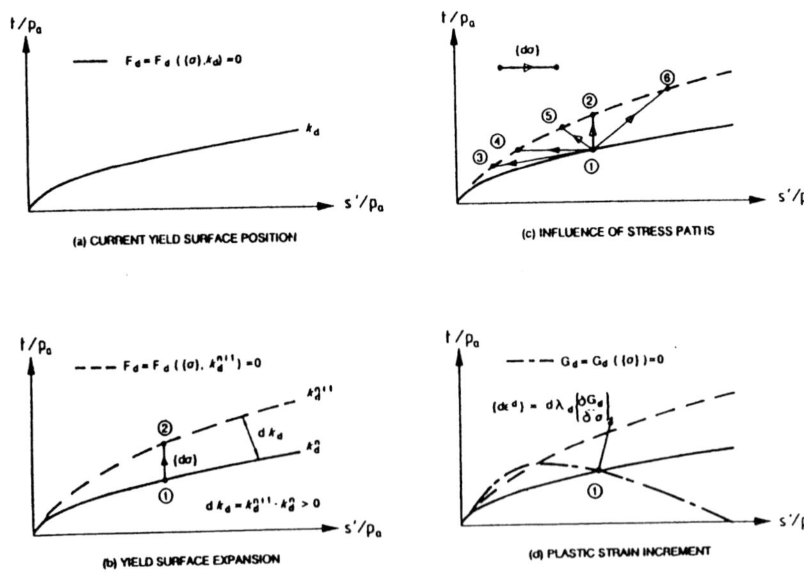

Figure 3. Behaviour of the plastic deviatoric yield surface in MONOT (after Hicks, 1990)

2. Kinematic Hardening Model ALTERNAT

In this section we will look in some detail at the multi-surface kinematic elasto-plastic constitutive soil model ALTERNAT (Molenkamp, 1982) as this model can simulate both cyclic and montonic loading.

2.1 Definition of a Co-Rotational State

It is necessary to first consider the geometrical basis, or *frame of reference*, used in the model when calculating the kinematic plastic properties. During continued alternating loading the soil deforms and rotates to a new state. To enable the plastic properties of the model to be determined, with respect to this new state, a co-rotational frame of reference is used (Molenkamp, 1986). This means that the plastic properties are determined with respect to a set of axes that rotate with the material during continued alternating loading. These axes can be seen in Figure 4 with respect to the fixed frame of reference.

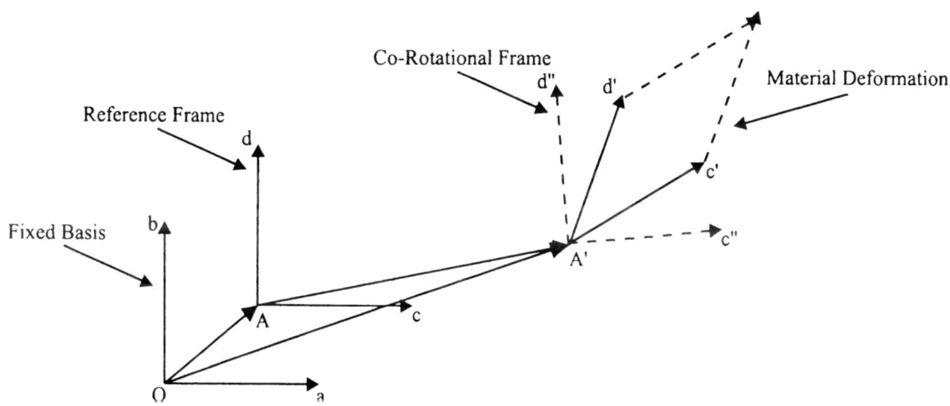

Figure 4. Definition of the co-rotational frame of reference

For this transformation to occur it is necessary to polar decompose the deformation gradient matrix F_{ij} into a symmetric stretch matrix V_{ik}, from which the magnitude and direction of the principal strains can be determined and an anti-symmetric rotation matrix R_{kj} where

$$F_{ij} = V_{ik} R_{kj} \tag{23}$$

The purpose of the rotation matrix R_{kj} is to convert from the reference state to the co-rotational state. In the co-rotational state pure deformation can then be considered. To model the changing state of deformation during loading the rate of change of the deformation gradient matrix $\dot{F}_{ij}$ is obtained from

$$\dot{F}_{ij} = L_{ik} F_{kj} \tag{24}$$

Here, L_{ik} is the velocity gradient matrix and for 'small' deformations can be calculated from

$$L_{ik} = \dot{V}_{ik} + \dot{R}_{ij} R_{jk}^{T} \tag{25}$$

where, $\dot{V}_{ik}$ is the symmetric stretch rate matrix and $\dot{R}_{ij} R_{jk}^{T}$ is the anti-symmetric spin matrix.

In the finite element implementation of the model, the initial stress state and the excess plastic stress correction procedure are with respect to the fixed frame of reference. However, during plastic deformation it is the co-rotational stress that must be used to determined the magnitude of the plastic strains. A more detailed description of the co-ordinate system and the deformation and rotation rates used in the model can be found in Molenkamp (1986, 1987).

2.2 Non-Linear Elastic Component of the Model

The behaviour of the soil during initial loading and immediately after a stress reversal is assumed to be non-linear elastic. This type of behaviour can be seen in Figure 5. Here, the isotropic stress is plotted against the isotropic strain during an isotropic triaxial consolidation test.

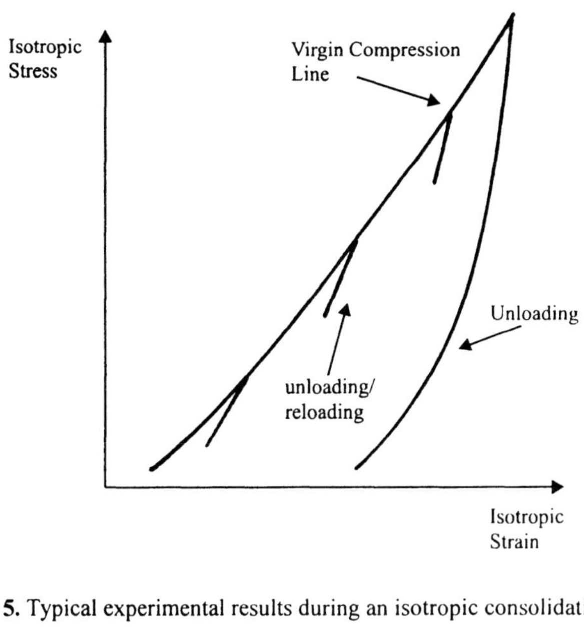

Figure 5. Typical experimental results during an isotropic consolidation test

The non-linear elastic isotropic unloading/ reloading curves can be simulated using the following equation (Lade, 1977 & Vermeer, 1980)

$$v = A\left(\frac{s}{p_a}\right)^{P} \tag{26}$$

To model non-linear elasticity Molenkamp (1988, 1992) applied the following complementary potential function

$$Q = A\frac{p_a}{(P+1)}\left(\frac{s}{p_a}\right)^{P+1}\left[1+\frac{3P(P+1)}{4R}\left(\frac{t}{s}\right)^2\right] \tag{27}$$

where, the scalars A, P and R are the elastic material parameters and p_a is the atmospheric pressure (assumed to be 100 kPa). The inverse of the elastic tangent stiffness matrix D_{ijkl}^{-1} is the second order partial derivative of equation (27)

$$D_{ijkl}^{-1} = \frac{\partial^2 Q}{\partial\sigma_{ij}\partial\sigma_{kl}} \tag{28}$$

The parameter R is the ratio between the shear modulus G and the isotropic bulk modulus K_{iso}

$$R = \frac{G}{K_{iso}} \tag{29}$$

where,

$$G = \frac{Rp_a}{3AP}\left(\frac{s}{p_a}\right)^{1-P} \tag{30}$$

and

$$K_{iso} = \frac{1}{3}\frac{p_a}{AP}\left(\frac{s}{p_a}\right)^{1-P} \quad \text{for } \left(\frac{t}{s}\right) = 0 \tag{31}$$

A detailed description on the influence of the 3 material parameters on the non-linear elastic response of the model can be found from Molenkamp (1992).

2.3 Kinematic Yield Surface

In ALTERNAT a pseudo stress T_{ij} is introduced to represent the mobilised friction with respect to the neutral axis.

$$T_{ij} = \sigma_{ij} - \frac{1}{3}\sigma_{kl}\delta_{kl}\xi_{ij} \tag{32}$$

where, σ_{ij} is the 'active' stress and ξ_{ij} is a non-dimensional deviatoric tensor of anisotropy. Each kinematic yield surface has its own hardening modulus. This enables the magnitude of the plastic strain increment to change during continued cyclic loading. The hardening modulus of the soil will change as the generated or activated surfaces expand and move in the deviatoric plane. The shape of the yield surface in this plane is given by the Lade & Duncan (1975) and Lade (1977) smooth conical surface.

$$F\left[I\left(T_{ij}\right)\right]-f(\chi)=\frac{I_1^3}{I_3}-27-f(\chi)=0 \tag{33}$$

where, χ is the hardening parameter and $f(\chi)$ is the size of the yield surface. I_1 and I_3 are the first and third invariants of the pseudo-stress respectively. It should be noted that the hardening modulus symbol κ has been replaced with the symbol χ so as not to confuse it with the ALTERNAT densification parameter κ.

The kinematic rule applied in ALTERNAT to control the distribution and movement of the yield surfaces in principal stress space is based on the assumption that kinematic yield surfaces remain tangent to each other at the stress reversal points σ_{ij}^r after Lade (1979). A consequence of this, is that the deviatoric components of the partial derivatives of the yield surfaces must remain proportional to each other after the stress reversal. This enables an algorithm to be developed to compute the increments in the tensor of anisotropy $d\xi_{ij}$, the pseudo-stress dT_{ij} and the hardening parameter $d\chi$.

2.4 Kinematic Modelling of Cyclic Mobility

The kinematic rule applied in ALTERNAT ensures that plastic deviatoric strains are developed during loading and unloading. To predict cyclic mobility, a modification to the hardening modulus is required to ensure that the absolute magnitude of the plastic devatoric strain increment during the loading phases are greater than the unloading phases. The level of cyclic mobility and densification can then be controlled through additional material parameters. The modified hardening modulus is obtained from the following equation

$$H_m = H \cdot X = H \exp(-\eta\psi) \tag{34}$$

Where, ψ is a scalar measure of the direction of loading. ψ is used to modify the hardening modulus H by the factor X, which also incorporates the material parameter η to control the amount of cyclic mobility. H_m is the modified hardening modulus. The cyclic properties of the model are illustrated in Figures 6 & 7 for cyclic triaxial stress and strain controlled loading of Levenseat sand at an initial void ratio of e_v=0.78, in both drained and undrained states. The ability of the model to simulate both cyclic densification and cyclic liquefaction is clearly seen.

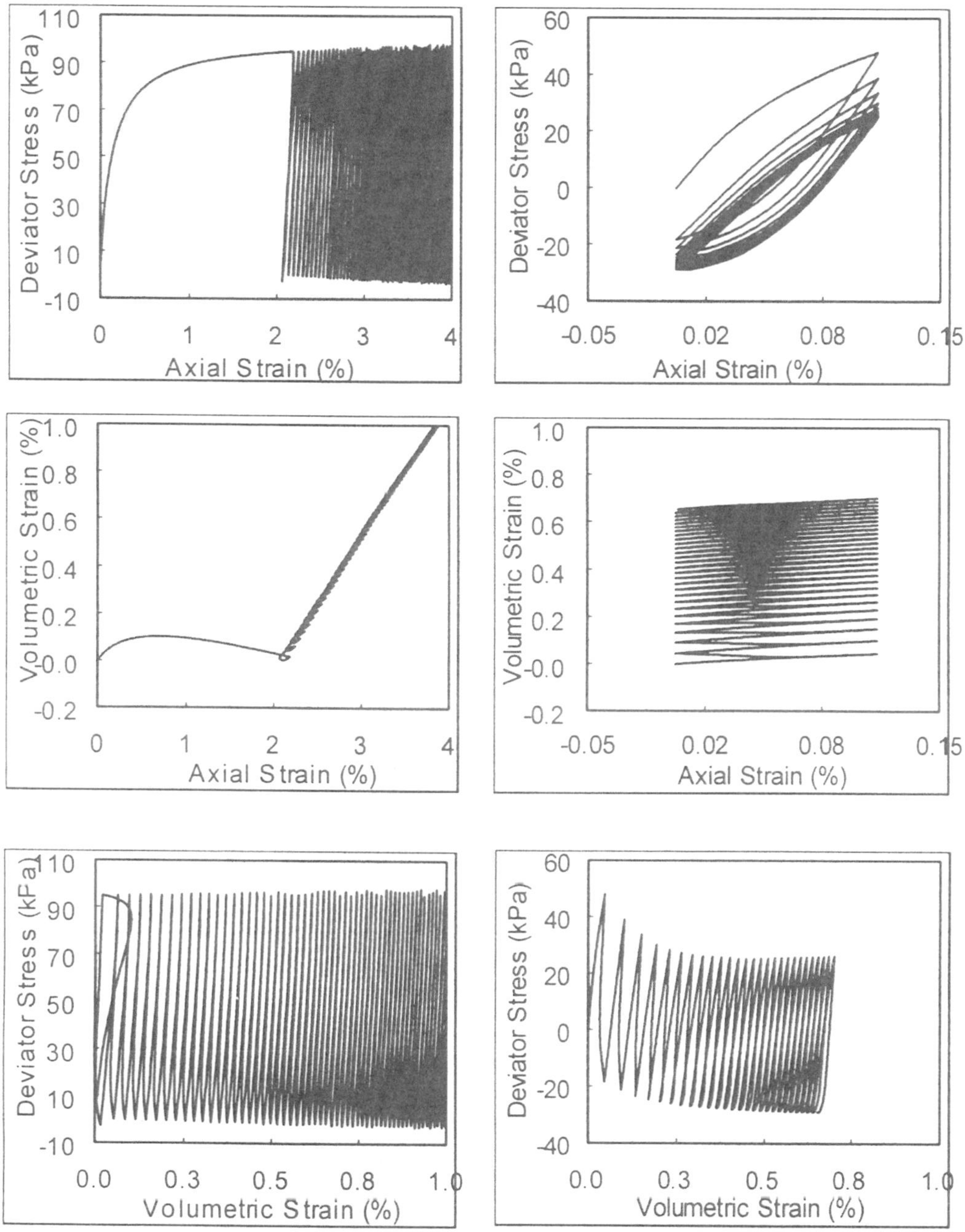

(a) Stress Controlled

(b) Strain Controlled

Figure 6. Typical results of simulated drained cyclic triaxial tests of Levenseat sand at an initial void ratio of e_v=0.78 using ALTERNAT illustrating cyclic densification properties of the model

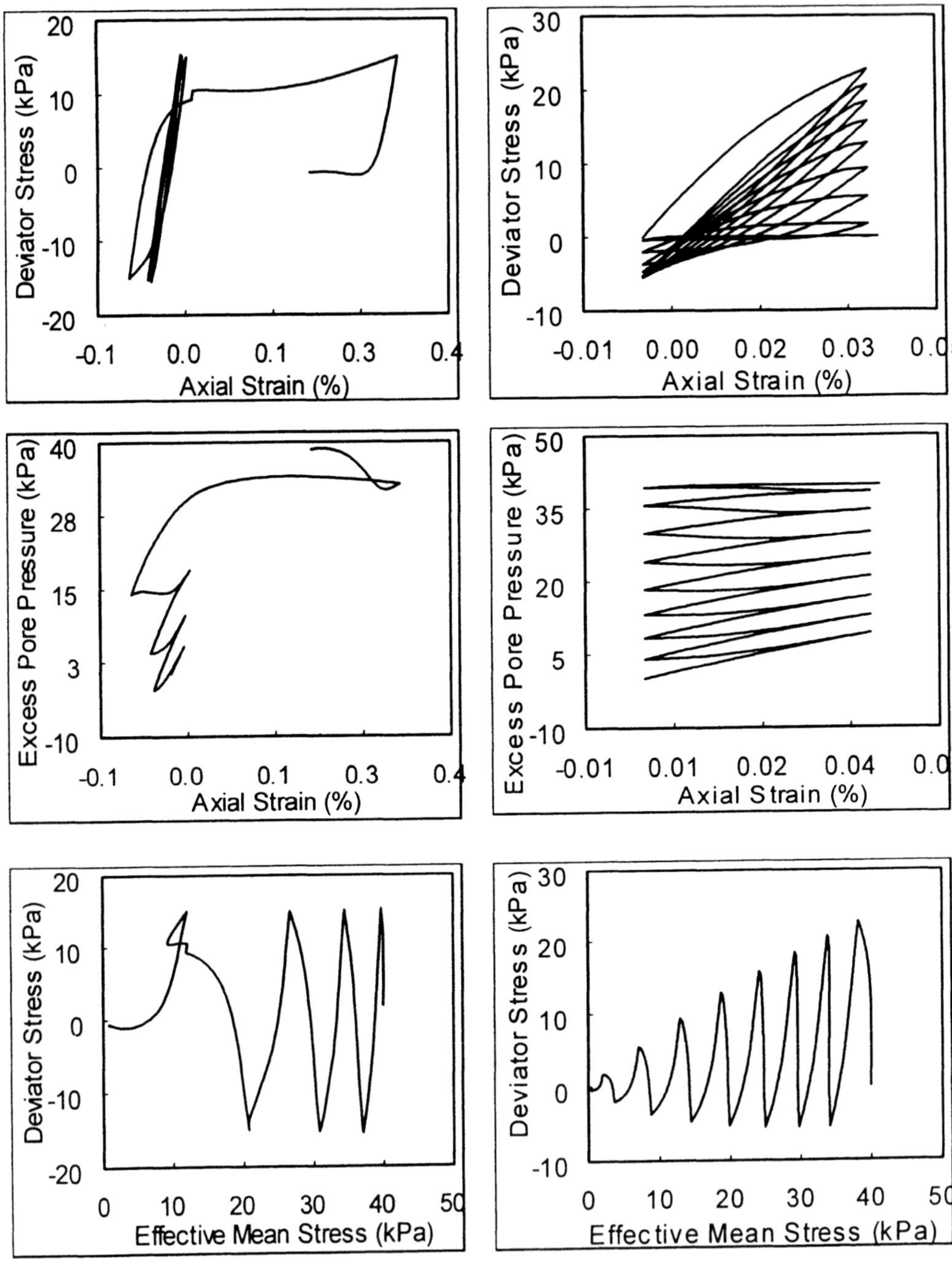

(a) Stress Controlled (b) Strain Controlled

Figure 7. Typical results of simulated undrained cyclic triaxial tests of Levenseat sand at an initial void ratio of e_v=0.78 using ALTERNAT illustrating cyclic liquefaction properties of the model

2.5 Plastic Potential

The flow rule used in ALTERNAT to determine the plastic strain increment $d\varepsilon_{ij}^{p}$ is composed of both an isotropic and a deviatoric component.

$$d\varepsilon_{ij}^{p} = \frac{\partial G}{\partial \sigma_{ij}} \lambda = \left(\alpha \delta_{ij} + \frac{\partial G^{d}}{\partial \sigma_{ij}} \right) \lambda \tag{35}$$

Isotropic component. The isotropic component $\alpha\delta_{ij}$ is based on Rowe's stress-dilatancy theory (Rowe, 1962, 1971 and Josseling de Jong, 1976) for granular soils and is obtained from the plastic dilatancy ratio

$$\frac{dv^{p}}{de^{p}} = \alpha\sqrt{3} \tag{36}$$

Molenkamp (1982) derived the following expressions for the plastic dilatancy ratio for loading in triaxial compression and extension

$$\frac{dv^{p}}{de^{p}} = \frac{-\sqrt{2}(1-K) - (2+K)\left(\frac{t}{s}\right)}{(1+2K) + \sqrt{2}(1-K)\left(\frac{t}{s}\right)} \qquad \text{loading in triaxial compression}$$

$$\frac{dv^{p}}{de^{p}} = \frac{\sqrt{2}(1-K) - (1+2K)\left(\frac{t}{s}\right)}{(2+K) - \sqrt{2}(1-K)\left(\frac{t}{s}\right)} \qquad \text{loading in triaxial extension}$$

(37a&b)

where,

$$K = \tan\left(45 + \frac{\phi_{\mu}}{2}\right)^{2} \tag{38}$$

and ϕ_{μ} is the true interparticle friction angle. A function is then used to compute the plastic dilatancy ratio for any Lode angle θ based on the above two cases. The point of zero dilatancy is determined from the true interparticle friction angle ϕ_{μ}, which is assumed to depend on the void ratio e_{v} and the isotropic stress $\phi_{\mu} = \phi_{\mu}\left(e_{v}, \frac{s}{p_{a}}\right)$. The shear stress level $\left(\frac{t}{s}\right)_{\mu c}$ at zero dilatancy can be obtained from the following

$$\left(\frac{t}{s}\right)_{\mu c} = \frac{2\sqrt{2}\sin\phi_{\mu}}{3 - \sin\phi_{\mu}} = C\left[(1+\xi) - \xi\exp\left(-\theta_{p}\left(\frac{s}{p_{a}}\right)\right)\right]\left[1 - \exp(-\varsigma e_{v})\right] \tag{39}$$

where, C, ξ, ζ and θ_p are material parameters. Dilation of the material will therefore begin once this shear stress level has been exceeded during shearing. Figure 8 shows that decreasing the value of C will reduce the shear stress level after which dilation will occur for a given void ratio. This simulates the increase in dilation with density of a granular material.

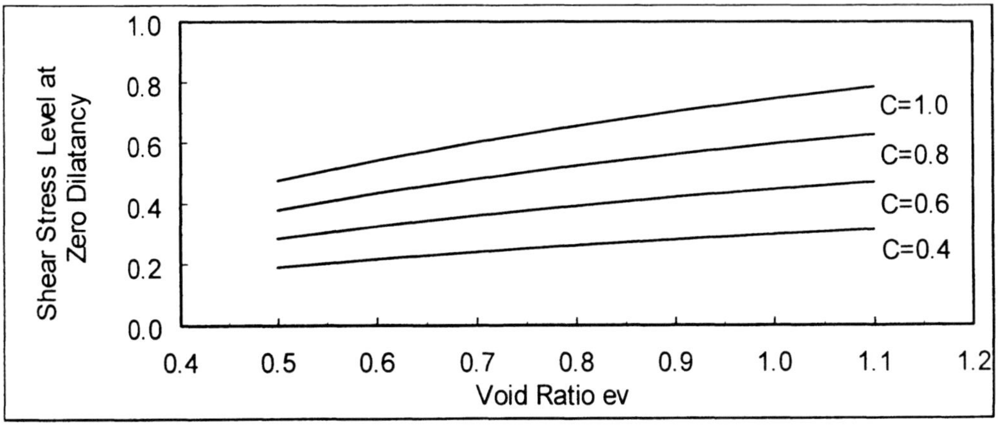

Figure 8. Variation of the parameter C on the shear stress level at zero dilatancy (s/p_a=1.0, ξ=0.12, θ_p=2.0 ζ=1.13)

Figure 9 shows the effect of the isotropic stress level $\left(\frac{s}{p_a}\right)$. As the isotropic stress increases the shear stress level at which dilation will occur also increases, simulating the reduction in dilation with high confining pressures.

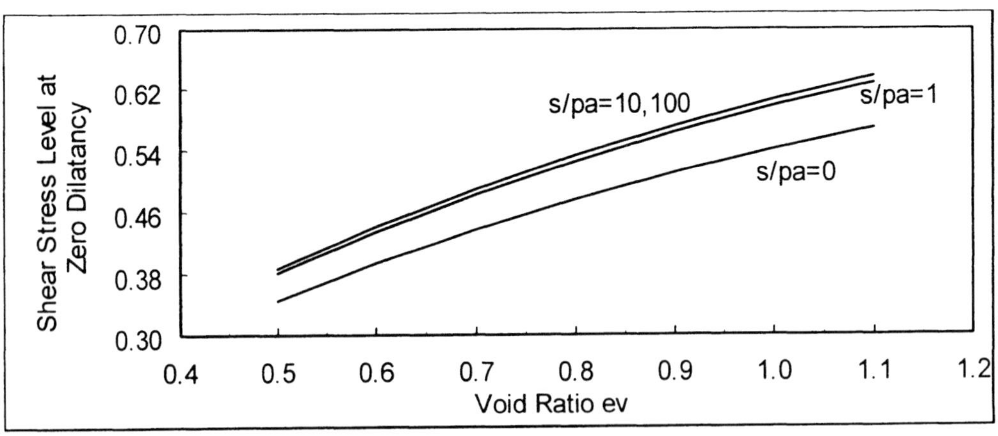

Figure 9. Variation of the parameter C on the shear stress level at zero dilatancy (C=0.8, ξ=0.12, θ_p=2.0 ζ=1.13)

Deviatoric component. The deviatoric component of the plastic potential $\left(\frac{\partial G^d}{\partial \sigma_{ij}}\right)$ is based on the expression used to determine the yield function (as normal for potential surfaces). To make the surface more circular in the deviatoric plane, a reduced pseudo-stress T^* is applied. The level of pseudo-stress reduction is determined by an additional material parameter R_p. However, T^* has the same tensor of anisotropy and isotropic component as the pseudo stress. The shape of the potential and yield surfaces in the deviatoric plane are shown in Figure 10, together with the directions of both yield surface expansion and the increment in the plastic deviatoric strain.

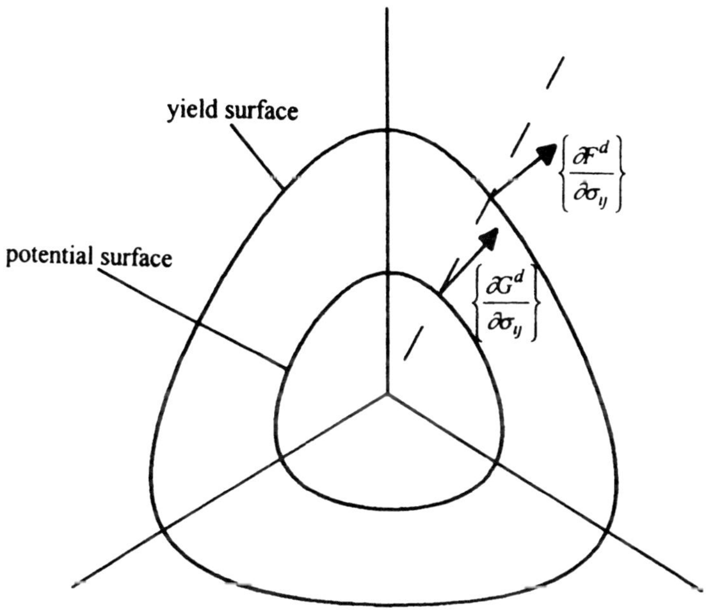

Figure 10. Illustration of yield and potential surfaces in the deviatoric plane

2.6 Hardening Parameter

The relationship between the increment in the plastic deviatoric strain (de^p) and the hardening parameter ($d\chi$) is given below

$$d\chi = de^p \frac{\left(\dfrac{n_l - n_d}{n_l - n_d - \Sigma_c \Delta\varepsilon^p_{vol,c}}\right)^K}{k(\chi,\theta)\left(\dfrac{s}{p_a}\right)^L} \frac{\dfrac{\partial F}{\partial T_{kl}} dT_{kl}}{\dfrac{\partial F}{\partial \sigma_{mn}} d\sigma_{mn}} \tag{40}$$

where $k(\chi, \theta)$ is a function which describes the distribution of the plastic stiffness around the circumference of the yield surface with hardening parameter χ and Lode angle θ, and is used to ensure that for drained cyclic deviatoric loading, at constant isotropic stress, a closed form stress strain loop is simulated. The function $k(\chi, \theta)$ is given as

$$k(\chi,\theta) = -\left(1 + S(1+\cos 6\theta)\frac{\dfrac{\partial I_3(T^c)}{\partial T_{kl}} T^{de}_{kl}}{\dfrac{\partial I_3(T^e)}{\partial T_{mn}} T^{de}_{mn}}\right) \tag{41}$$

where, S is a material parameter, the superscripts c and e refer to the third invariant of the triaxial compression and extension pseudo-stresses respectively and T^{de} is the deviatoric component of the pseudo-stress in triaxial extension. This ensures that cyclic mobility is a function of η only.

n_i and n_d are the initial (before shearing) and final (at maximum densification) porosities and $\Sigma_c \Delta\varepsilon^p_{vol,c}$ is the accumulated plastic volumetric strain due to alternating loading. L is a material parameter. The term with the porosities simulates plastic densification due to cyclic loading and ranges from unity at the initial isotropic state to infinity at the maximum densification. The modelling of cyclic densification was previous shown in Figure 6 with respect to the cyclic mobility. However, an interesting analysis is shown in Figure 11 and demonstrates the level of simulation possible. Here the cyclic mobility parameter is switched off (set to zero) and the soil is cycled. The figure shows that as the material densifies the gradient of the stress strain curve increases (i.e. the shear modulus increases) reducing the magnitude of the axial strain increment and volumetric strain increment.

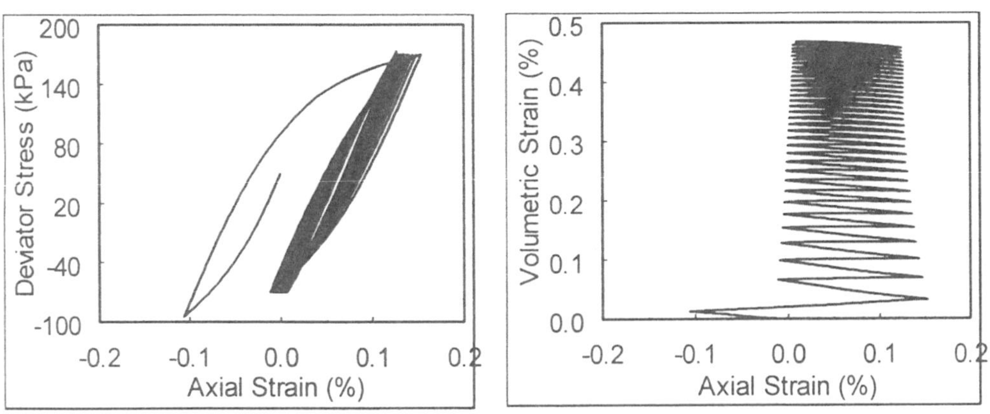

Figure 11. Effect of cyclic densification parameter κ assuming zero cyclic mobility

2.7 Hardening Modulus

The model has been formulated in such a way as to enable it to be applied to many different stress paths. Often models are formulated for specific types of analysis, e.g. triaxial compression tests, cyclic liquefaction studies etc. Often circular yield and failure surfaces are used in kinematic models as they tend to be easier to implement and cause less numerical problems. However, difficulties are then encountered when they are used to simulate monotonic failures with varying Lode angles.

The deviatoric hardening modulus H is used to describe the plastic stiffness of the soil (i.e. it relates deviatoric strain increments to stress increments) and can be represented by the gradient of the curve in Figure 12. This gradient is dependent on the position of the yield surfaces in principal stress space. The hardening modulus is modified by taking into account the cyclic stress history of the soil as discussed previously (i.e. it increases with density).

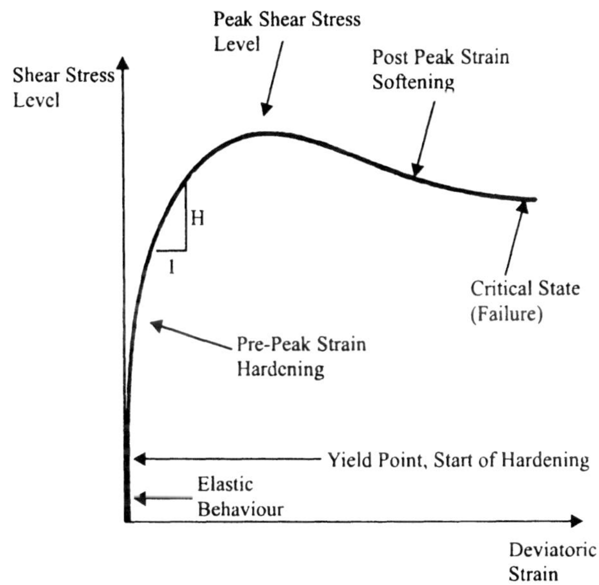

Figure 12. Illustration of shear stress level with material deformation

Figure 12 shows that the hardening modulus is related to the pre-peak strain hardening and post-peak strain softening of the material. The hardening modulus H in ALTERNAT is found from the following equation,

$$H = \frac{df}{d\left(\frac{t}{s}\right)} \frac{d\left(\frac{t}{s}\right)}{d\chi} \frac{\left(\frac{n_l - n_d}{n_l - n_d - \Sigma_c \Delta\varepsilon^{p}_{vol,c}}\right)^{\kappa}}{\kappa(\chi,\theta)\left(\frac{s}{p_a}\right)^{L}} \tag{42}$$

where,

$$\frac{df}{d\left(\frac{t}{s}\right)} = 81\frac{\left(\frac{t}{s}\right)}{\left[1+\sqrt{2}\left(\frac{t}{s}\right)\right]^2\left[1-\frac{1}{\sqrt{2}}\left(\frac{t}{s}\right)\right]^3} \tag{43}$$

This equation is obtained by rearranging equation (33) and substituting for the stress invariants I_1 & I_3 and the shear stress level $\left(\frac{t}{s}\right)$ for triaxial compression. It should be noted at this point that the size of the yield surface $f(\chi)=0$ if $\left(\frac{t}{s}\right)=0$ and $f(\chi)\cong 100$ as $\left(\frac{t}{s}\right)=1$.

The $\frac{d\left(\frac{t}{s}\right)}{d\chi}$ component of equation (42) is obtained by considering the strain hardening and softening behaviour of the material. A combined equation (equation 44) is used to describe the transition from pre-peak strain hardening to post peak strain softening, in terms of the shear stress level in triaxial compression as a function of the plastic deviatoric strain. The pre-peak shear stress level is termed Y_1 and the post-peak shear stress level is termed Y_2.

$$\left(\frac{t}{s}\right) = \frac{Y_1 Y_2}{\left(Y_1^n + Y_2^n\right)^{\frac{1}{n}}} \tag{44}$$

where n is a material parameter and can be chosen to ensure that the correct peak shear stress level is observed. Y_1 is obtained from the following

$$Y_1 = \left[\frac{e^p}{\left(\frac{s}{p_a}\right)^L \frac{(1+e_v)E}{(Q-e_v)^2}}\right]^{\frac{1}{D}} \tag{45}$$

and Y_2 from

$$Y_2 = M\left\{(1-\upsilon)+\upsilon.\exp\left(-\psi\left(\frac{s}{p_a}\right)\right)\right\}\exp(-\beta e_v) \tag{46}$$

where, E, Q and D are the parameters for the pre-peak strain hardening range after Hardin and Hardy (1968) and Tatsuoka et al (1979). M, υ, ψ and β are the parameters for the post-peak strain softening range. The effect of Y_1 and Y_2 on the $\frac{d\left(\frac{t}{s}\right)}{d\chi}$ component is illustrated in Figure 13. Equation 44 ensures a smooth transition from pre-peak strain hardening to post-peak strain softening and eventual critical state.

The effect of the two main parameters E and M are shown in Figures (14) & (15). In Figure (14a) the pre-peak shear stress level Y_1 is plotted against the plastic deviatoric strain e^p.

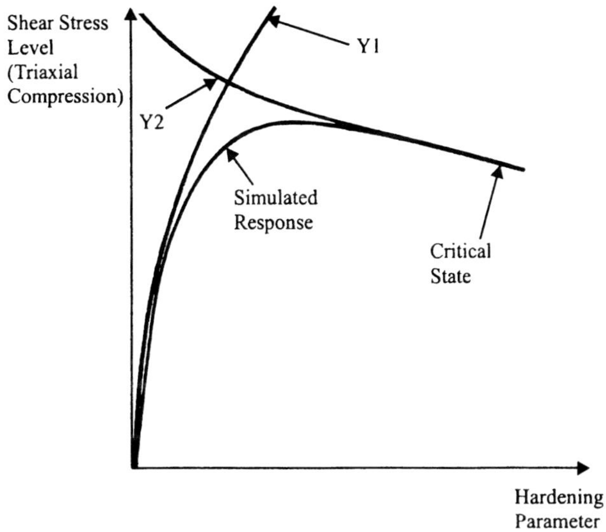

Figure 13. Illustration of shear stress level in triaxial compression with hardening parameter ($\frac{d\left(\frac{t}{s}\right)}{d\chi}$ component)

As E decreases the shear stress level $\left(\frac{t}{s}\right)$ increases for any value of e^p and any void ratio e_v.

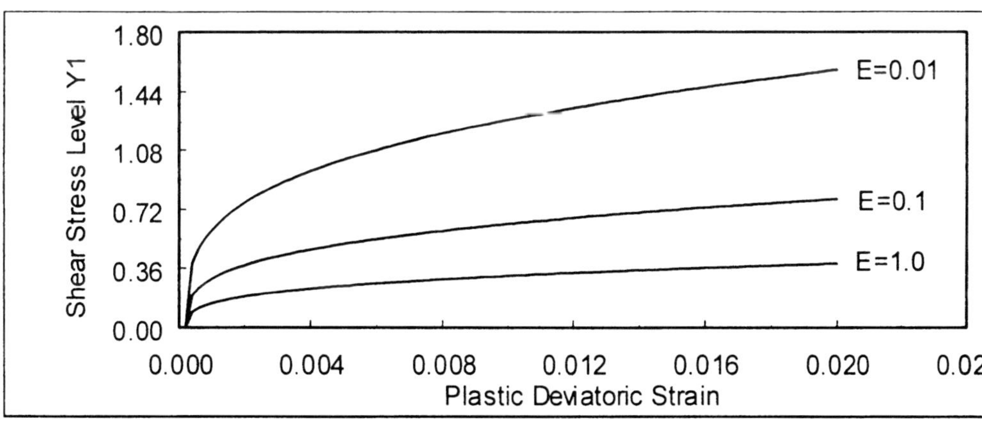

Figure 14(a). Pre-peak strain hardening shear stress level $Y1$ with plastic deviatoric strain e^p (s/p_a=1, e_v=0.6, D=3.3 & Q=2.5)

Figure (14b) shows that decreasing values of e_v generate increases in $\left(\frac{t}{s}\right)$ with e^p simulating increasing pre-peak strain hardening with increases in the relative density. In Figure (15a) the post-peak strain softening shear stress level Y_2 is plotted against e_v to demonstrate the influence of the main material parameter M. The predicted post-peak shear stress level decreases with M

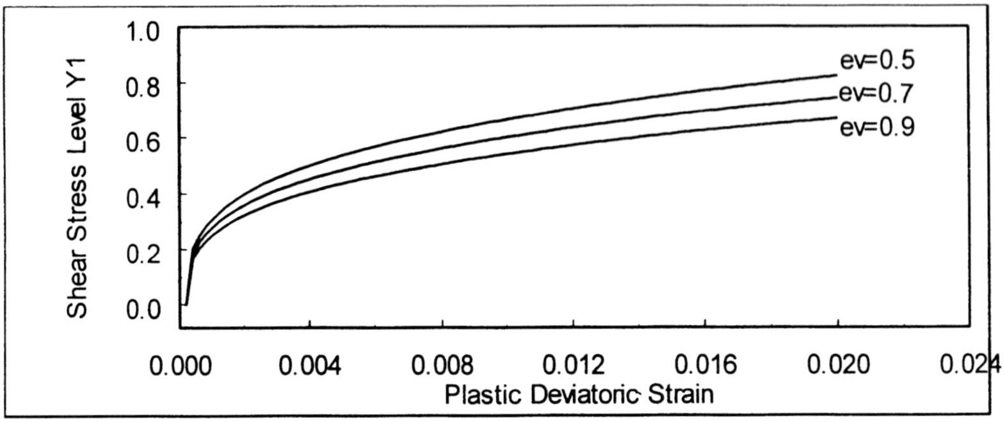

Figure 14(b). Pre-peak strain hardening shear stress level *Y1* with plastic deviatoric strain e^P (s/p_a=1, E=0.1, *D*=3.3 & *Q*=2.5)

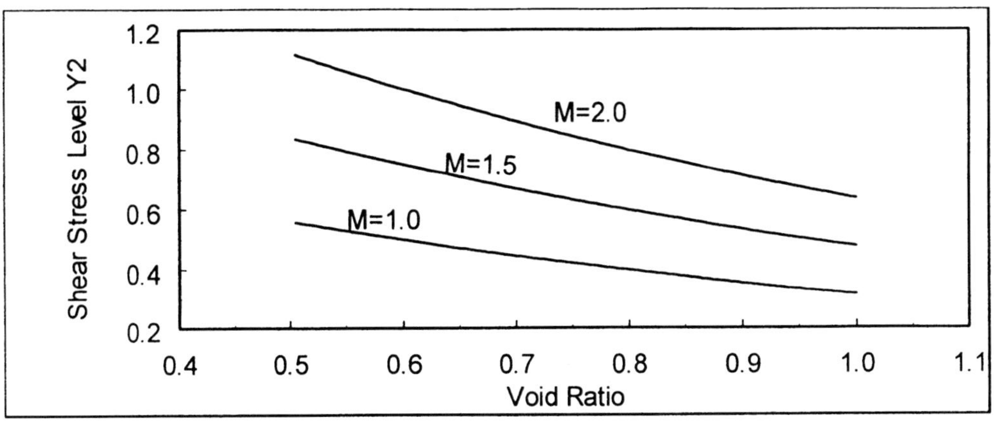

Figure 15(a). Post-peak strain softening shear stress level *Y2* with void ratio (s/p_a=1, ν=0.1, β=1.13 & ψ=0.2)

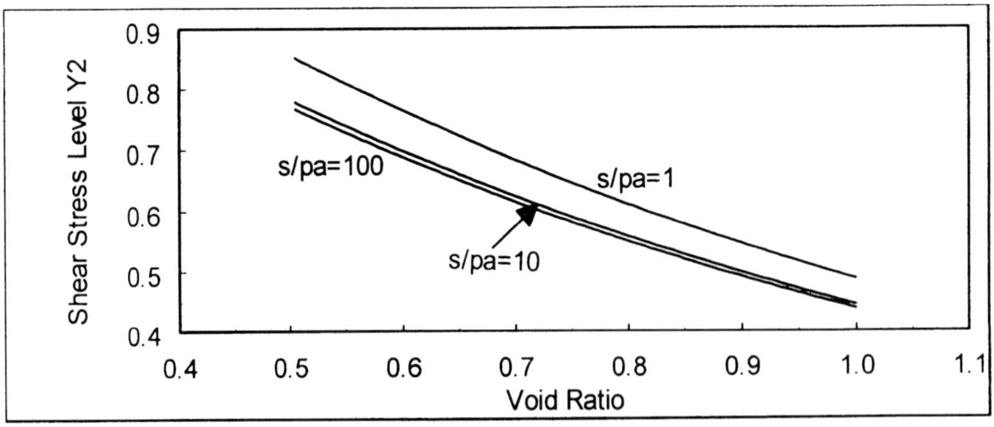

Figure 15(b). Post-peak strain softening shear stress level *Y2* with void ratio (*M*=1.5, ν=0.1, β=1.13 & ψ=0.2)

and with increases in e_v. Figure (15b) shows Y_2 will also decrease with increasing isotropic stress level $\left(\frac{s}{p_a}\right)$. To obtain the $\frac{d\left(\frac{t}{s}\right)}{d\chi}$, equation (46) is differentiated to give

$$\frac{d\left(\frac{t}{s}\right)}{d\chi} = \frac{1}{\left(Y_1^n + Y_2^n\right)^{\left(\frac{1}{n}-1\right)}}\left(Y_1^{n+1}\frac{dY_2}{d\chi} + Y_2^{n+1}\frac{dY_1}{d\chi}\right) \tag{47}$$

2.8 Critical State

The critical state (Casagrande, 1936 and Schofield and Worth, 1968) of a granular material is achieved when changes in the stress and the isotropic strain, with increasing deviatoric strain, become equal to zero. This is illustrated in Figure 16 where several samples of the same sand, having different initial densities but the same isotropic stress level, are subjected to triaxial compression.

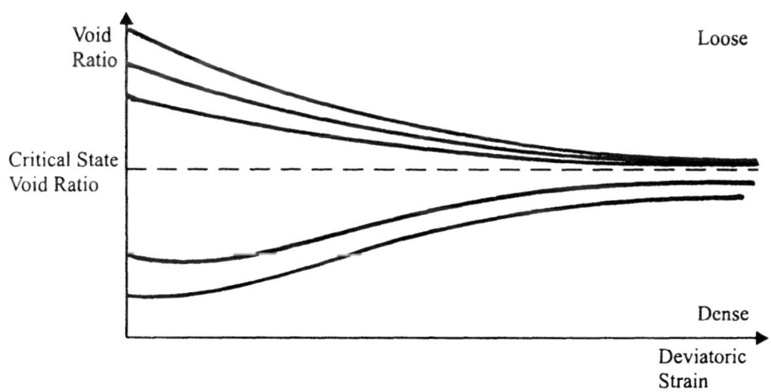

Figure 16. Illustration of granular soil behaviour during shearing towards the critical state void ratio (isotropic stress constant)

If no localisation occurs during shearing then their final densities will be approximately the same (i.e. their critical state void ratios).

If during a triaxial test, a dense sand is confined by a relatively low cell pressure the stress-strain curve will peak during shear and then decrease towards the critical state and the sand will dilate. If the same sand is subjected to a much higher initial confining pressure and sheared towards the critical state there may be no observable peak. This shows that the confining pressure has a significant effect on the rate of dilation of granular soils. Loose sands tend to contract during shear with the stress strain curve rising monotonically towards the critical state void ratio. This type of behaviour is illustrated in Figure 17 where the critical state line of a idealised granular soil is plotted with respect to the isotropic stress. The plastic volumetric behaviour of the granular soil at various initial densities is very much dependent on its position with respect to the critical state line.

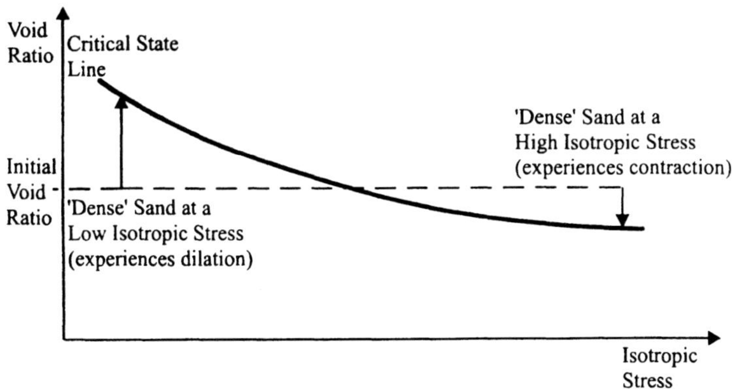

Figure 17. Illustration of granular soil behaviour during shearing at constant isotropic stress towards the critical state line

At large strains the friction angle corresponding to shearing at constant volume (ϕ_{cv}) should in fact be equal to the true interparticle friction angle (ϕ_{μ}). This is because it is only at this friction angle that deviatoric straining at constant volume can occur.

To account for this phenomenon the critical state component of the model is formed by equating the post-peak shear stress level to the shear-stress level at zero dilatancy to give

$$\exp(-\beta e_v) = \left[M\left\{(1-\upsilon)+\upsilon\exp\left(-\psi\left(\frac{s_c}{p_a}\right)\right)\right\} + C\left\{(1+\xi)-\xi\exp\left(-\theta_p\left(\frac{s_c}{p_a}\right)\right)\right\}\right]^{-1} \times C\left\{(1+\xi)-\xi\exp\left(-\theta_p\left(\frac{s_c}{p_a}\right)\right)\right\} \tag{48}$$

The void ratio at the critical state of the material is therefore a function of the isotropic stress level at the critical state $\left(\frac{s_c}{p_a}\right)$, the post-peak strain softening parameters and the true interparticle friction angle parameters.

2.9 Summary of Current Model Capabilities and Future Research

The current capabilities of the model are summarized below

1. Non-linear elasticity
2. Pre-peak strain-hardening
3. Post-peak strain-softening
4. Dilatancy based on Rowe's Stress-Dilatancy Theory

5. Critical state
6. Cyclic mobility
7. Cyclic densification
8. Tensile strength

Tensile strength is defined by setting a minimum value of the isotropic stress.

Although the model can simulate many different stress paths and has been shown to be able to analyse a wide variety of problems, development of the constitutive relationships is still ongoing. For example, the following areas are currently being researched for inclusion into the model

1. Development of an isotropic loading cap yield surface (as in the Monot model).
2. Simulation of an immediate elasto-plastic response to small principal stress rotation.
3. A material length formulation to simulate post-localisation behaviour.
4. Simulation of creep at large shear stress levels.
5. Formulation of a plastic spin to ensure a constant anisotropy at the critical (steady / ultimate) state for continuos simple shear.

2.10Summary Material Parameters

Table 1 shows all the material parameters used in the formulation of ALTERNAT with reference to their constitutive purpose and typical values. These parameters represent typical values for a granular soil; modified parameters will be presented later for different granular soils. Although the list of parameters seems very long, many of them can be considered as constant (only requiring further calibration for very 'sensitive' calibration studies). It is important to mention here that the list includes all possible variations with the constitutive laws (i.e. all power and scaling terms). When examining constitutive models, based on their ease of use compared to their predictive capabilities, one should look at all the scalars (whether fixed or variable) involved in the definition of the constitutive laws. Many authors' claim that their models have only a small number of parameters that require calibration, however, when all the scalars are included (as in Table 1) the 'parameter' list increases significantly. By adopting a transparent approach to the fixed scalars, the user of the model would be able to verify the significance of their variation if in any doubt.

In ALTERNAT, the important parameters required for the simulation of granular material behaviour, i.e. those which should be calibrated to experimental data, are highlighted by the star in Table 1. From the original 24 material parameters, only 6 for monotonic loading and 8 for cyclic loading are required to be calibrated for general use. This can be reduced if only the drained response of the soil is required. The number of material parameters required for calibration is therefore dependent on the stress paths under consideration, the available test data and the accuracy of solution required. For example, accurate cyclic densification studies would require e_{vd} to be calibrated; this is easily done if the maximum dry density is known from standard classification tests. It should be remembered that inconsistencies in experimental data often exist, particularly in cyclic undrained tests. The user of the model must therefore be able to interpret the data and calibrate the model to those stress paths that are most likely to reflect the problem at hand. It should also be remembered that to use a simple elastic perfectly plastic linear Mohr-Coulomb constitutive model 5 material parameters are required: Young's Modulus E, Poisson's Ratio ν, effective cohesion c', effective friction angle ϕ', and the dilation angle ψ.

Parameter	Description	Typical Value
A^*	Non-linear elastic model	0.0009
P^*		1.0
R		0.6
E^*	Pre-peak strain hardening for triaxial compression	0.02
D		3.3
Q		2.5
N		3.0
L		0.3
M^*	Post-peak strain softening for triaxial compression	1.5
υ		0.1
ψ		0.2
β		1.13
C^*	Dilatancy and Critical State	0.96
ξ		0.12
θ_p		2.0
ζ		1.13
η^*	Cyclic mobility	4.0
e_{vi}^*	Initial void ratio	0.675
e_{vd}	Void ratio at the maximum densification	0.52
κ^*	Plastic densification parameter	1.0
R_p	Plastic potential in the π-plane	0.3
S	Hardening in the π-plane	0.0
σ_t/Pa	Normalised tensile strength	1.0E-5
c/Pa	Normalised cohesion	0.0

Table 1: ALTERNAT material parameters *required for general calibration (after Woodward & Molenkamp, 1999)

3 Implementation of the ALTERNAT Soil Model

Over the last few decades there has been a significant increase in the number and sophistication (i.e. level of simulation possible) of constitutive soil models. However, the implementation of these models into general-purpose finite element programs for the solution of real problems has not. In reviewing the literature one can easily see that the solution of many real problems is still performed using relatively simple constitutive laws and models. For example, the vast majority of work performed on the numerical modelling of granular foundations has used simple elastic-perfectly plastic linear Mohr-Coulomb models. Although these types of models can give an insight into the performance of these foundations there are unable to give accurate descriptions of settlement behaviour. Yet, in the case of granular foundations, settlement behaviour is of prime importance in the design. The application of the ALTERNAT model to granular foundation analysis will be discussed later in this chapter. In cyclic problems simple models cannot be applied if the analysis of cyclic densification or liquefaction is required. It would seem that the development of constitutive models is only useful if they are accompanied by research into their implementation. The rest of this chapter is concerned with the implementation of the model, its calibration to real data and a series of boundary value simulations for validation and demonstration are presented.

3.1 Subroutine Implementation

To implement the full constitutive model into a finite element program Molenkamp (1990) has written 5 main subroutines; ELAST: non-linear elastic matrix, INITST: generation of the initial state, DSPVOL: current value of plastic isotropic strains, PLASCY: generation of plastic properties and SALTER: determination of improved estimate in stress due to a strain increment. There are also around 26 lower order subroutines.

The model has been implemented into several different finite element programs by different authors and used to examine many different types of problems. The subroutines are written in such away as to enable either 2-dimensional or 3-dimensional finite element analyses to be performed, but the authors have tended to implement the model into separate codes rather than a single one.

In this chapter, two finite element 'programs' are used, a 2-dimensional 'static/ dynamic' program and a 2-dimensional fully coupled 'implicit dynamic' program. The first program (called ALTICA) was developed by the Woodward (1993) and is uncoupled (Naylor's, Method, 1974, is used for excess pore water pressure generation). The second program has been developed by Kaddouri (2000) to study earthquake induced liquefaction and uses BIOT's Theory coupled with the dynamic UP formulation to simulate the interaction between the pore fluid and solid skeleton. The coupled program is an extension of the uncoupled program and hence the two programs have many common features. For the purposes of this chapter, the ALTICA program will be discussed with respect to its structure, followed by a more general discussion of the coupled program. Although only 2-dimensional problems are considered, the programs have been developed for 3-dimensional analysis as well.

The name ALTICA (ALTernat Incremental Computer Algorithm) refers to several finite element codes developed for the solution of non-inertial and inertial problems in both 2 and 3 dimensions using a relatively straightforward, but reliable, stress correction algorithm.

When illustrating the ability of sophisticated constitutive models many papers only present the results of single stress paths responses, sometimes using only one finite element. However, when real boundary value problems are analysed many algorithms break down when large numbers of finite elements are used, due to the large number of differing stress-paths occuring at the same load increment. This creates convergence problems due to the development of out of balanced plastic stresses (or 'bodyloads'). The algorithm used in ALTICA determines both 'bodyloads' and the error in the 'bodyloads'. This procedure not only helps the convergence process with each load increment, and subsequent iterations, but also decreases computational times in many applications.

3.2 General ALTICA Program Structure

ALTICA uses 8-noded quadrilateral finite elements for two-dimensional analysis or 20-noded brick elements for three dimensional analysis. In the coupled analysis, a further 4-noded quadrilateral element, or an 8-noded brick element, are used to represent the pore water pressure. A 2x2 integration rule is used in the example 2-D 'static' problems presented in this chapter. The algorithm is based on the work of Zienkiewicz et al (1969) using the Initial Stress Method. A constant stiffness matrix is adopted as the reference and an iterative procedure is used to correct the unbalanced plastic forces with a note kept on which yield surface is currently active for stress redistribution.

In the dynamic part of the program the load increment loop becomes the time loop and the equation of motion is solved at every time step using direct integration methods, usually via Newmark type schemes. The new velocities and accelerations are calculated at the end of every time loop.

3.3 Initial State

The internal parameters and arrays of the model are first set by the initial state INITST routine. Here it is assumed that the soil is in a condition immediately after sedimentation with both horizontal stresses σ_h equal and related to the vertical stress σ_v through the earth pressure at rest coefficient K_o.

$$\sigma_h = K_o \sigma_v \tag{49}$$

To simulate other initial stress states, for example the shear stresses in a slope, the model would first have to be used to simulate the construction or excavation of the slope from level ground. This can be achieved by adding or removing finite elements during an initial analysis. It is also possible to determine the equivalent additional bodyload forces representing the difference between K_o stresses and 'gravity turn on' stresses, i.e. the first step in the analysis would be to self equilibrate between internal and external forces. Displacements generated during these phases would usually be set to zero. Figure 18 shows that at the initial state three yield surfaces are assumed to exist (although if $K_o \cong 1$ the second yield surface would coincide with the third).

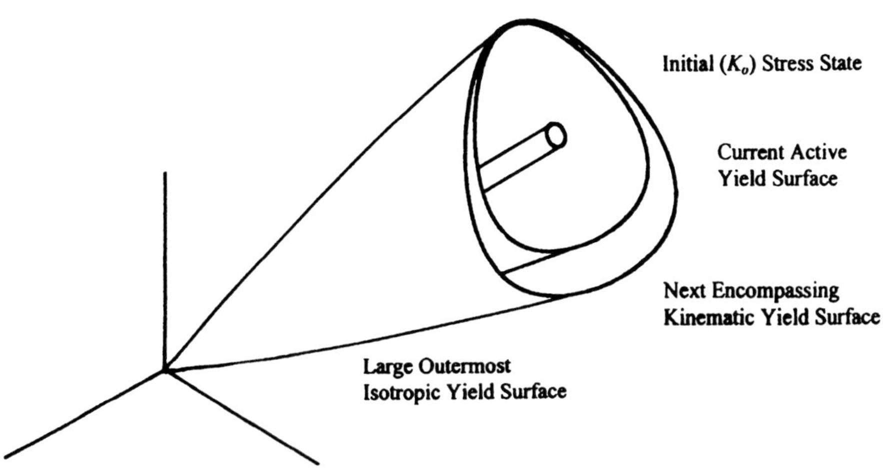

Figure 18. Illustration of the three yield surfaces at the initial state

The 'outermost' yield surface can only expand about the isotropic axis, which ensures that at 'large' deformations the material behaviour simulated is isotropic (soils tend to forget their stress history at large strains).

3.4 Initial Movements of the Yield Surface

When the soil is loaded from the initial state (with or without an initial anisotropy) its initial response is assumed to be elastic (the stress point is within the first yield surface). When the active stress crosses this initial yield surface the surface will expand and the material will experience elasto-plastic behaviour. For this first initial yield surface it is assumed that its anisotropy does not change during loading. When this yield surface reaches the next encompassing yield surface it is erased and kinematic elasto-plastic behaviour is simulated. During continued monotonic loading (e.g. a conventional triaxial test) the kinematic yield surface would continue to expand until it touched the outermost yield surface.

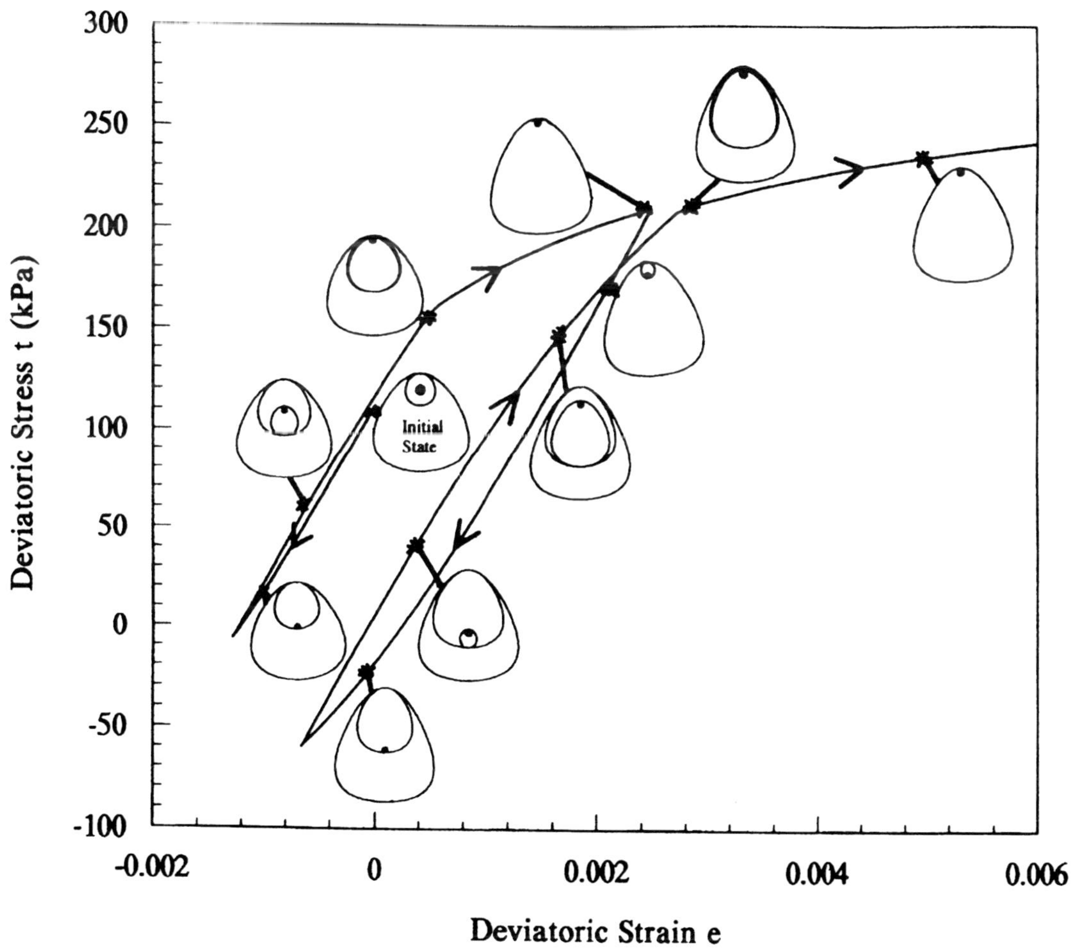

Figure 19. Illustration of the yield surface movements during stress reversals (after Woodward & Molenkamp, 1999)

At this point, the kinematic yield surface would be erased and the isotropic yield surface would then expand towards the peak shear stress level and eventually the critical state. If the material experiences a stress reversal, a new yield surface is generated at the stress reversal point at a predefined size f_{min} . The initial behaviour would be non-linear elastic then elasto-plastic as the minimum size is surpassed and the kinematic yield surface starts to expand.

Typical movements of the yield surfaces are shown in Figure 19. Here the soil is first unloaded from the initial state then loaded towards failure. Just before the peak resistance is achieved, the soil experiences a stress reversal. During this loading and unloading sequence the creation, expansion and destruction of kinematic yield surfaces can clearly be seen. A further kinematic rule can be applied if the size of the current surface tries to expand beyond the next encompassing one (if tensors of anisotropy do not coincide).

3.5 Numerical Implementation

The algorithm used in ALTICA to determine the next state of stress due to the application of a stress increment is of the initial stress type. It should be noted however that many other types of iterative schemes could be used, for example, Newton-Raphson, Broydens Method, BFGS iterations etc. The initial stress algorithm used in ALTICA is shown in Figure 20 and has been shown to be reliable and time efficient.

In order to use a symmetric stiffness matrix the plastic deviatoric component is omitted during the formation of the element stiffness matrix. The stiffness matrix is therefore based on the non-linear elastic component only (updated when necessary during load application). Although using the elastic component only tends to lead to more elasto-plastic iterations it is more stable during strain softening and storage requirements are significantly reduced. The initial stress iterations performed within ALTICA are similar to the iteration process used by Hicks (1990). Outside of SALTER the iterations can be summarized by

$$\{dR_I\}_n^m = K_{IJ}^{-1}\{dF_J\}_n^{m-1} \tag{50}$$

where, dR_I is the global incremental displacements vector, dF_J is the global increment loads vector, K_{IJ}^{-1} is the inverse of the global non-linear elastic stiffness matrix, and the subscript n and superscript m refer to the increment and iteration numbers respectively. The vector dF_J is composed of the global incremental external loads vector dF_J^{ext} , the error in the global incremental bodyloads vector dF_J^{err} and the global incremental bodyloads vector dF_J^{body} given by

$$\{dF_J\}_n^{m-1} = \{dF_J^{ext}\}_n^0 + \{dF_J^{err}\}_n^0 + \{dF_J^{body}\}_n^{m-1} \tag{51}$$

Once the displacements dR_I have been found, each Gauss point is addressed in turn and the Gauss point strains $d\varepsilon_I$ and incremental rotation matrix ΔR_{IJ} and elastic stress increment $d\sigma_I$ are determined.

Declarations and input solution parameters

For each Gauss point derive the initial state using INITST

For each load/ displacement/ time increment

Form and reduce global non-linear elastic stiffness matrix K_{ij}

For each stress redistribution iteration

Form the global incremental loads vector $\{dF_j\}_n^{m-1}$

Compute the displacement increment

$$\{dR_i\}_n^m = K_{ij}^{-1}\{dF_j\}_n^{m-1} \quad \text{(Non-inertial)}$$

$$\{\Delta R_i\}_{t+\Delta t}^m = K_{ij}^{-1}\{\Delta F_j\}_{t+\Delta t}^{m-1} - M_{ij}\{\Delta \ddot{R}_i\}_{t+\Delta t}^m - C_{ij}\{\Delta \dot{R}_i\}_{t+\Delta t}^m \quad \text{(Inertial)}$$

Null the bodyloads vector dF_{body}

Check convergence (equation 40)

Compute the incremental strains, form the incremental rotation matrix and determine the non-linear elastic stress increment $dF_j^{elastic}$

Using SALTER compute the new total stress F_j^{total} (after conversion from the co-rotational frame of reference) and determine the new material behaviour and yield surface positions

Determine the bodyloads vector $\{dF_j^{body}\}_n^m$ (equation 38)

Determine the error in the bodyloads $\{dF_j^{err}\}_{n+1}^0$ (equation 39)

$$\{dF_j^{body}\}_{n+1}^0 = \{dF_j^{body}\}_n^m$$

Update displacements (Non-inertial)
Update displacements, velocities and accelerations (Inertial)

Output

Figure 20. Structure of the ALTICA program.
(Equation numbers refer to the original paper:
Woodward & Molenkamp, 1999)

The subroutine SALTER is then called and the new stress state is determined using the previously described constitutive relationships. In order to ensure that the magnitude of the strain increment is not too large, a sub-increment system exists in SALTER to reduce the strain increment to smaller sub-increments. This is essential to prevent any numerical drift due yield surface expansion and translation. Once the improved stress state $\left\{F_J^{total}\right\}_n^m$ has been determined it is necessary to redistribute any excess stresses through the updated bodyloads vector dF_J^{body}

$$\left\{dF_J^{body}\right\}_n^m = \left\{dF_J^{elastic}\right\}_n^m + \left\{F_J^{total}\right\}_n^0 - \left\{F_J^{total}\right\}_n^m \tag{52}$$

where, $\left\{dF_J^{elastic}\right\}_n^m$ is the non-linear elastic stress increment and $\left\{F_J^{total}\right\}_n^0$ is the previous incremental stress state. The error in the bodyloads $\left\{dF_J^{err}\right\}_n^o$ is determined when convergence has been achieved.

$$\left\{dF_J^{err}\right\}_{n+1}^0 = \left\{dF_J^{body}\right\}_n^m - \left\{dF_J^{body}\right\}_n^{m-1} \tag{53}$$

Convergence is assumed to occur when changes in the incremental displacements, normalised with respect to the maximum absolute value Smith & Griffiths (1988), fall below a threshold value given by TOL.

$$\frac{\left|\left\{dR_I\right\}_n^m - \left\{dR_I\right\}_n^{m-1}\right|_{\max}}{\left|\left\{dR_I\right\}_n^m\right|} \langle \text{ TOL} \tag{54}$$

Since $\left\{dF_J^{err}\right\}_n^o$ is passed onto the next load increment higher values of TOL, than at first thought, can be used in certain problems. The effect of this is to reduce computational times in real boundary value problems.

In dynamic analyses the Newmark (1965) implicit time stepping algorithm is used and equation (50) is replaced by the equation of motion

$$M_{IJ}\left\{\Delta\ddot{R}_I\right\}_{t+\Delta t}^m + C_{IJ}\left\{\Delta\dot{R}_I\right\}_{t+\Delta t}^m + K_{IJ}\left\{\Delta R_I\right\}_{t+\Delta t}^m = \left\{\Delta F_J\right\}_{t+\Delta t}^{m-1} \tag{55}$$

where M_{IJ} is the mass matrix, C_{IJ} is the viscous damping matrix, $\Delta\ddot{R}$ and $\Delta\dot{R}$ are the incremental accelerations and velocities. Damping is primarily hysteretic, but the viscous damping matrix C_{IJ} can be used if viscous damping is to be simulated as well. The standard Newmark relationships are then used to determine new displacements, velocities and accelerations at the next time step.

3.6 The SALTER Routine

The SALTER routine is summarised in Figure 21 for a typical sub-increment.

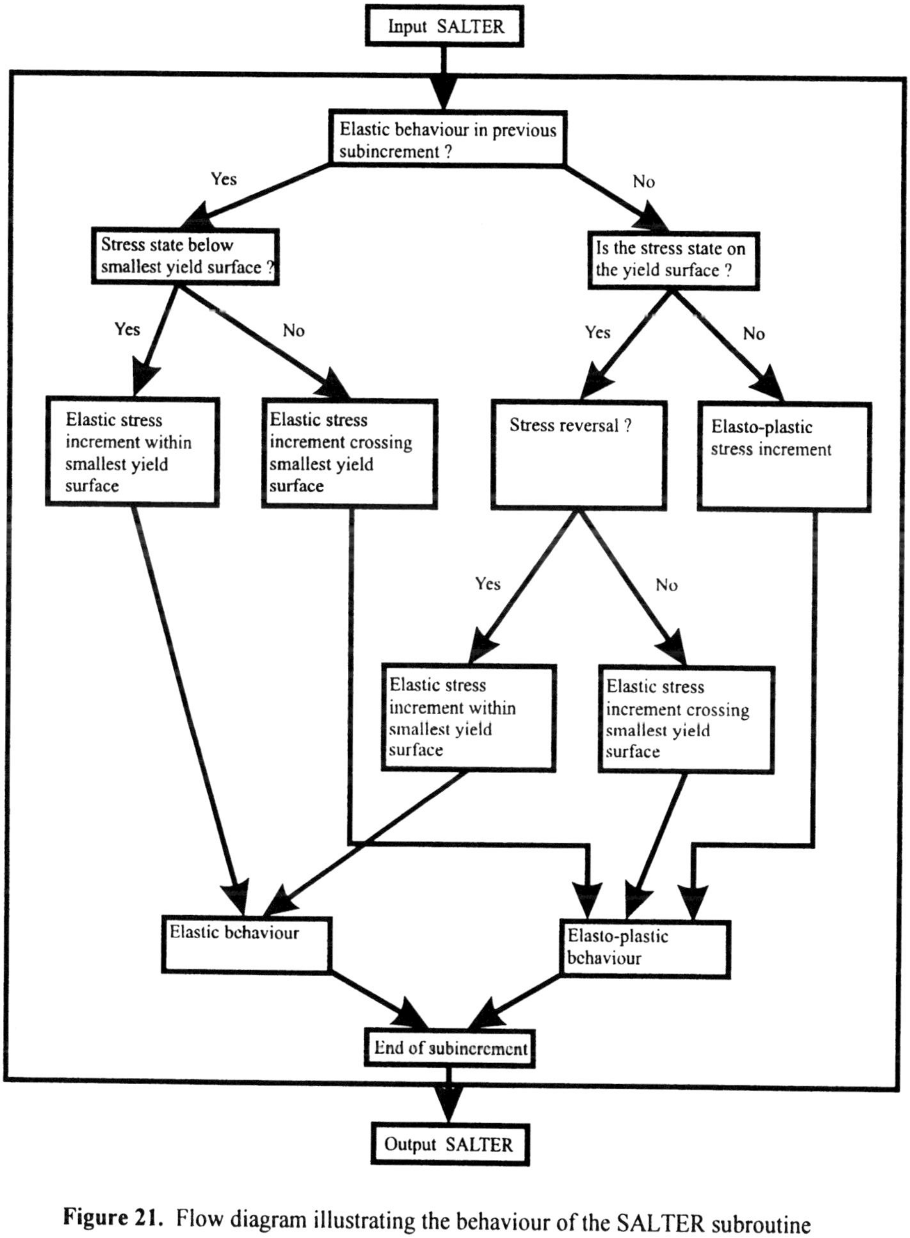

Figure 21. Flow diagram illustrating the behaviour of the SALTER subroutine

The output of the routine depends on a series of possibilities with respect to material behaviour at the previous sub-increment and the new predicted behaviour based on the current stress sub-increment. If required, new yield surfaces are created or destroyed to follow the soils anisotropy during loading. The total number of yield surfaces allowed at any Gauss point are predetermined by the user, the SALTER routine automatically erases the least important surfaces once this limit has been reached.

3.7 Coupled Analyses

The kinematic properties of the model have been developed so that the model can be used to look at problems involving cyclic and dynamic loading. However, ALTICA can only look at problems in which the soil is either in a drained or undrained state. This type of program is often called a 1-phase analysis as the soil and pore water are assumed to move together within the soil matrix. To simulate problems like earthquake induced liquefaction or dynamic consolidation, where the solid skeleton and the pore water may not move together, additional numerical computations are required. To study this type of problem Kaddouri & Woodward (1998) and Kaddouri (2000) have developed ALTICA into a coupled UP formulation. The starting point of the formulation is BIOT's 2-phase theory (BIOT, 1962).

Linear momentum equilibrium of the bulk soil. Neglecting mass exchanges and accounting for inertia effects, the following 'mixture' equation can be written

$$\sigma_{ij} + \rho b_i = \rho \ddot{u}_i + \rho_f \ddot{w}_i \tag{56}$$

where, σ_{ij} is the stress tensor, ρ & ρ_f are the mass densities of the mixture and fluid respectively, b_i is the body force vector and u_i is the solid displacement vector.

$w_i = n(U_i - u_i)$, where n=porosity and U_i is the fluid displacement vector.

Linear momentum equilibrium of the fluid. The linear momentum of the fluid is given by

$$-p_i + \rho_f b_i = \rho_f \left(\ddot{u}_i + \frac{1}{n} \ddot{w}_i \right) + \frac{\dot{w}_i}{k} \tag{57}$$

where, p is the pore water pressure, k is the isotropic permeability (k_{ij} in case of anisotropy).

Fluid mass conservation. The fluid mass conservation is given by

$$-\dot{w}_{ii} = \dot{\varepsilon}_{ii} + \frac{p}{Q^{*}} \tag{58}$$

where, $\dot{\varepsilon}_{ii}$ is the rate of volumetric deformation of the skeleton and

$$\frac{1}{Q^*} = \frac{n}{K_f} + \frac{1-n}{K_s} \tag{59}$$

where, K_f is the bulk modulus of the fluid phase and K_s is the bulk modulus of the solid phase. The usually effective stress principal is used as

$$\sigma_{ij} = \sigma'_{ij} - \delta_{ij} p \tag{60}$$

where, σ_{ij} & σ'_{ij} are the total and effective stress tensors. In the above equation compression is taken as negative.

Space discretisation. Most earthquakes have low frequency contents and so $\ddot{w} = 0$. This means that the analysis reduces to two variables only, namely the displacement of the solid u, and the pore fluid pressure p. After procedures like Galerkin's method in the finite element discretisation a set of governing equations can be written. These governing equations are summarised in the following matrices

$$\begin{bmatrix} M & 0 \\ 0 & 0 \end{bmatrix} \begin{pmatrix} \ddot{u} \\ \ddot{p} \end{pmatrix} + \begin{bmatrix} C & 0 \\ 0 & S \end{bmatrix} \begin{pmatrix} \dot{u} \\ \dot{p} \end{pmatrix} + \begin{bmatrix} K & -Q^T \\ 0 & H \end{bmatrix} \begin{pmatrix} u \\ p \end{pmatrix} = \begin{pmatrix} f_s \\ f_p \end{pmatrix} \tag{61}$$

where, M is the mass, K is the stiffness, C, S & H are coupling and damping matrices etc. In the finite element code a Newmark algorithm is used for u and a backward difference for p. In the 2-dimensional program, the solid skeleton is modelled using 8-noded quadrilateral elements and the pore water pressure is modelled using 4-noded quadrilateral elements. The elements are superimposed, so that 3 degrees of freedom exist at the element corners of the global mesh.

4 Calibration

One of the most important areas in the modelling of real boundary value problems is the calibration of the constitutive model. Constitutive models must be able to be calibrated to tests that can routinely be performed in the geotechnical laboratory. The material parameters, that need to be calibrated to triaxial test data, were highlighted in Table 1. This section discusses the calibration of the model to a Nevada sand in a more detail. Often it is at this stage were implementation of constitutive models finish. Although the ability to simulate triaxial data must be a fundamental requirement, the simulation of real problems must be the ultimate goal of the model developer. However, we will first start by discussing, in general terms, how to calibrate the ALTERNAT model to the most important parameters.

Non-linear elastic. To calibrate A & P the isotropic strain versus the isotropic stress for an isotropic triaxial test needs to be performed. This can be achieved by simply increasing the cell pressure and measuring the isotropic strains induced for several cycles. Equation (26) can then be used to determine A & P.

Pre-peak strain hardening. The parameter E is obtained by isolating the plastic deviatoric strains during initial pre-peak strain hardening and fitting the relevant terms of equation (45).

Post-peak strain hardening. The most accurately way to calibrate M is to perform at least 3 triaxial compression tests at constant isotropic stress. The peak shear stress level from each test is plotted against the isotropic stress level. An exponential is then fitted and the peak shear stress level at zero isotropic stress found. Equation (46) can then be used directly to find M.

Dilatancy and critical state. The most accurately way to calibrate C is from at least 3 triaxial compression tests at constant isotropic stress. The peak ratio of σ_1/σ_3 is plotted against $-2\varepsilon_3/\varepsilon_1$ and the value of K (equation, 38) found. The shear stress level at zero dilatancy is then found and plotted against the isotropic stress level. An exponential curve can then be fitted and C found from equation (39).

Cyclic parameters. The cyclic parameters are best obtained from drained cyclic compression tests, ideally at constant isotropic stress.

Where possible, monotonic and cyclic undrained triaxial tests should also be performed.

4.1 Calibration of Nevada Sand

To illustrate some of the abilities of the model to simulate real material behaviour the results of calibrations to Nevada sand used in the VELACS (1992) project (Verification of Liquefaction by Centrifuge Studies) is presented using a single finite element in axi-symmetry (Figure 22).

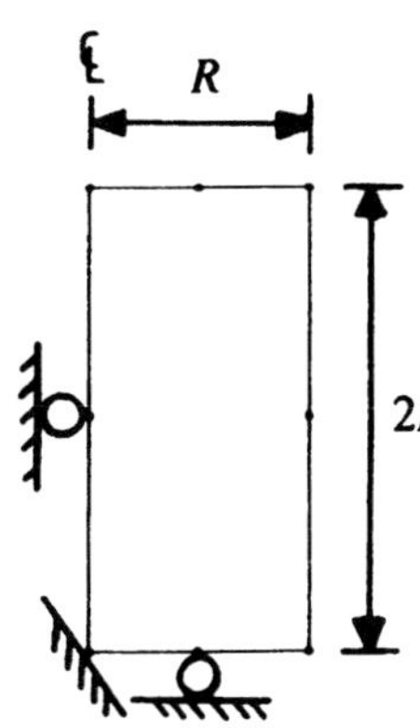

Figure 22. Single finite element in axi-symmetry

A significant problem occurred during the calibration process while attempting to determine the non-linear elastic parameters A and P, as purely isotropic versus volumetric strain relationships (measuring increases in volumetric strain with increasing confining pressure) were not performed. The calibration process was made more difficult due to the typical scatter associated with undrained cyclic tests. In some instances, samples of the sand at similar initial void ratios, confining pressures and cyclic deviator stresses showed marked differences in response. Table 2 gives the set of material parameters determined from the calibration procedure for both initial densities, from the available VELACS data.

Parameter	Dr=40%	Dr=60%
A	0.0012	0.002
P	0.6	0.52
E	0.016	0.016
M	1.62	1.72
C	0.93	1.01
$\mathbf{e_{vi}}$	0.73	0.662

Table 2. Modified ALTERNAT material parameters for Nevada sand (after Woodward, 1993 and Woodward & Molenkamp, 1999)

Additional parameters like the void ratio at maximum densification for Nevada sand could be found (perhaps for dynamic compaction studies) from equation (62)

$$e_{vd} = G_s \frac{\gamma_w}{\gamma_{\max}} - 1 = 0.511 \qquad (62)$$

where, γ_w=unit weight of water and γ_{max}=unit weight at maximum density.

Figure 23 shows the experimental (solid line) and simulated (dashed line) responses for the isotropically-consolidated drained compression tests at constant confining pressures of σ_m =40,80 & 160 kPa. Figures 24 and 25 show typical responses to the isotropically-consolidated undrained cyclic compression tests at 1 Hz using a non-inertial displacement-controlled simulation. In the first of the cyclic tests the initial effective confining pressure is σ_m=160kPa and the amplitude of the cyclic deviator stress is q_d=48.4kPa, with a q_{off}=20.7kPa offset. The second test has a initial effective confining pressure of σ_m=40kPa, with q_d=18.6kPa and q_{off}=0.0kPa. In both cases, the number of cycles required to generate material instability and eventual liquefaction is simulated. The plots show that the butterfly effect at liquefaction is simulated. This phenomenon is due to the stress path crossing the shear stress at zero dilatancy line (i.e. the material switches from contractive to dilative tendencies). In the undrained environment this dilative tendency generates negative increases in the excess pore water pressure. As the stress path crosses this line it therefore starts to bend to the right as it heads towards the peak resistance. As the load reverses the sand still has some effective isotropic stress and therefore some strength. The shear stress ratio is high however hence significant permanent strain occurs. In the experiment total liquefaction occurred (i.e. no butterfly effect) due to the experiment being load control. This results in total collapse of the material at the onset of liquefaction.

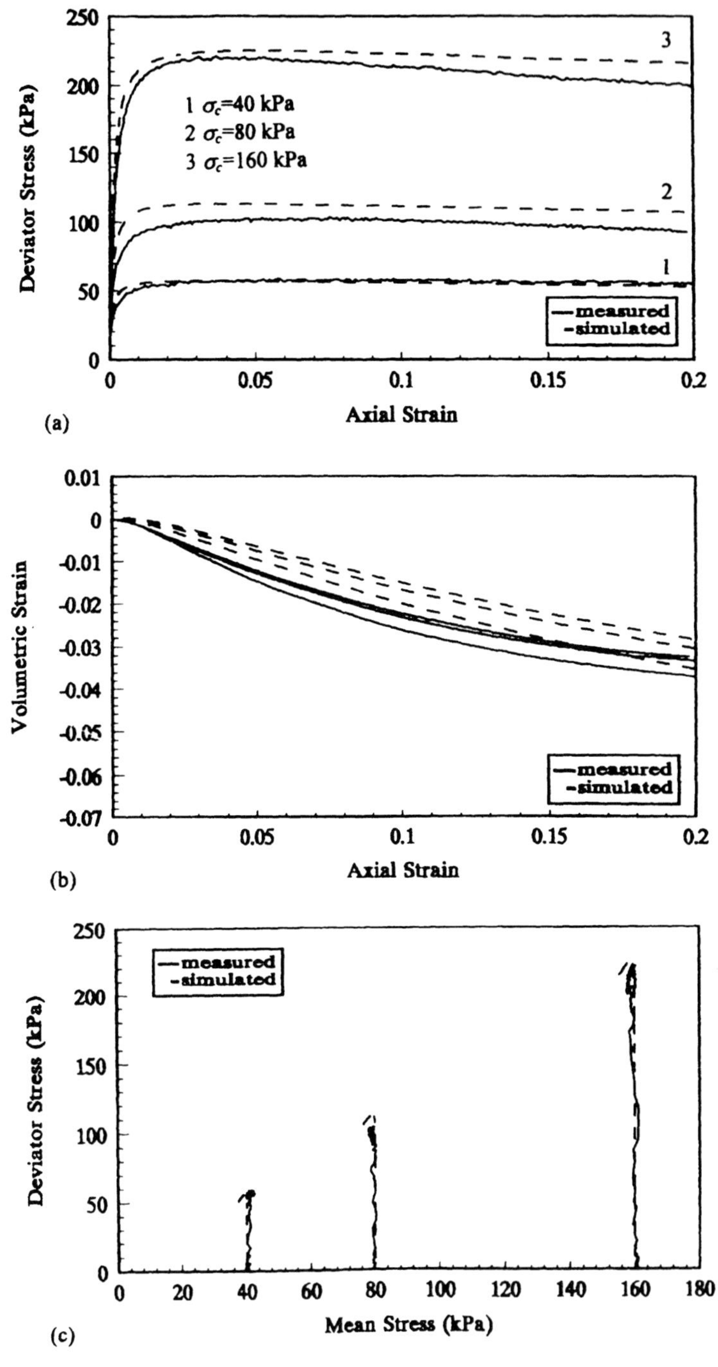

Figure 23. Calibration of ALTERNAT to Nevada sand in drained triaxial compression (constant s) with e_v=0.73 (after Woodward & Molenkamp, 1999)

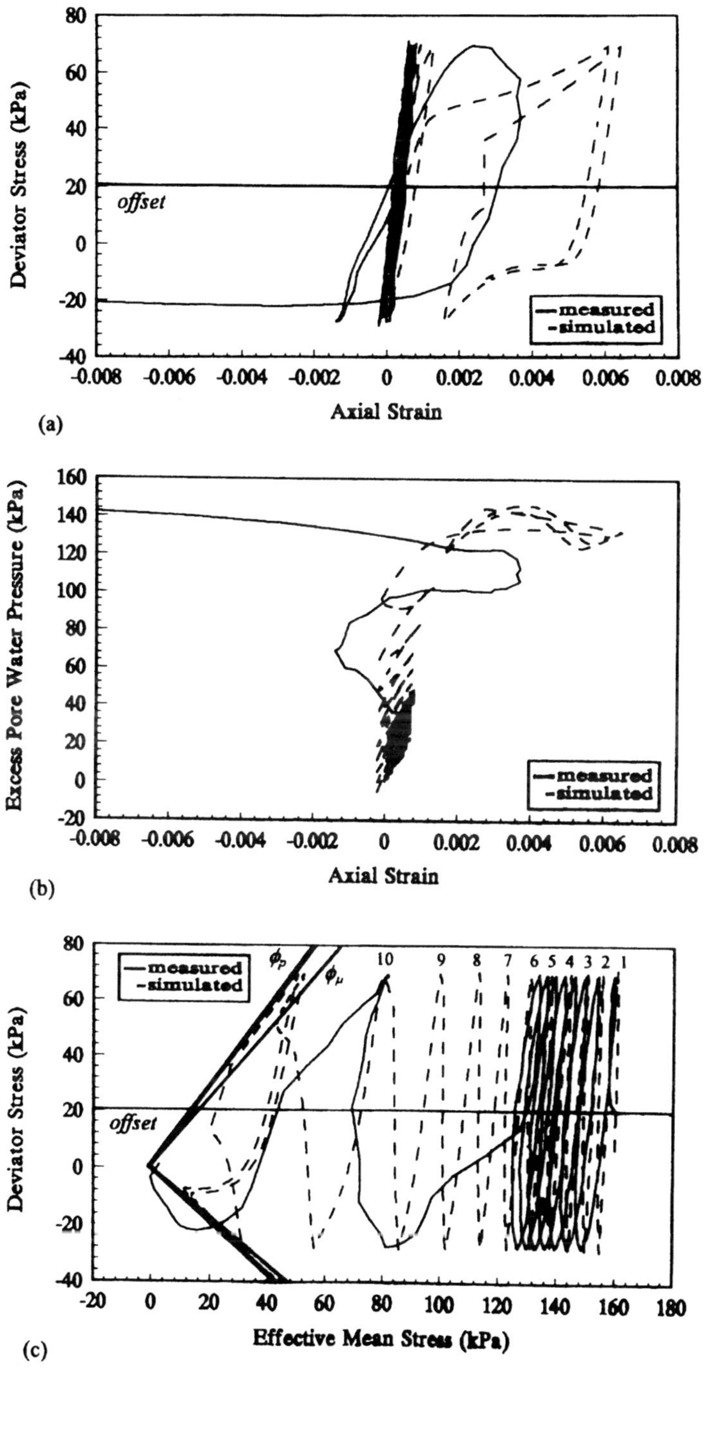

Figure 24. Calibration of ALTERNAT to Nevada sand in undrained cyclic triaxial compression at 1 Hz with e_v=0.73, q_d=48.4 kPa and q_{off}=20.7 (after Woodward & Molenkamp, 1999)

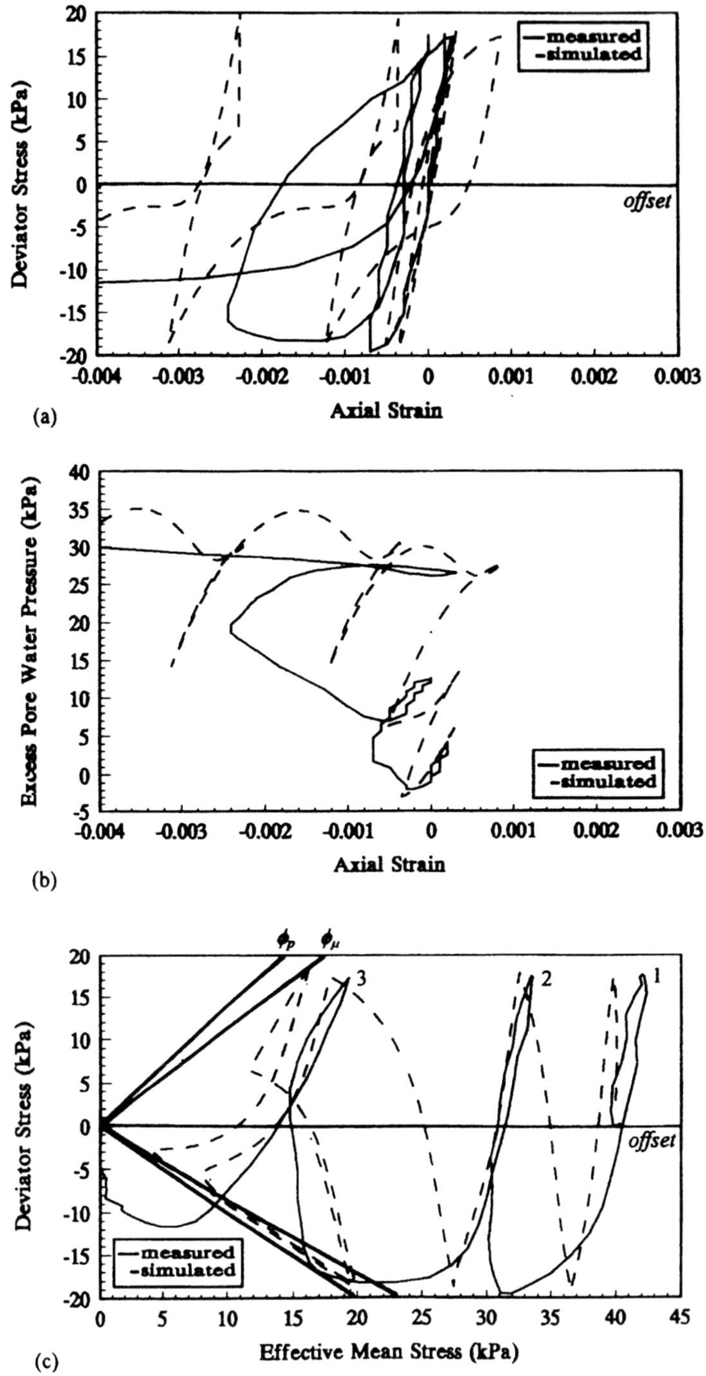

Figure 25. Calibration of ALTERNAT to Nevada sand in undrained cyclic triaxial compression at 1 Hz with e_v=0.73, q_d=18.6 kPa and q_{off}=0.0 (after Woodward & Molenkamp, 1999)

Calibration of the model to these types of stress paths could of course be performed using load control (as discussed below). However, it is in the interests of the user to calibrate the model to as many different stress paths as possible using a variety of different analysis procedures and techniques. In the author's experience small changes in one particular material parameter may not affect the response of the model to the stress path being considered, but may dominate the response of the model to another stress path. Also, the affects of time must not be neglected in cyclic simulations. For example, a calibration may be performed via a load or displacement controlled simulation using a 'static' program. Although a good response may be predicted, this may not be the case when a 'dynamic' program is used with the same material parameters, with respect to the build up in pore pressure with time. It is of paramount importance that the calibration is performed well if any confidence in the results is to be assured. If in doubt, the model should always be calibrated to those stress paths that are most likely to occur in the actual soil structure. It should be remembered however, that the stress paths being simulated are in the triaxial plane only. They are also undrained, whereas in reality the soil maybe consolidating while loading takes place, depending on the soil's permeability. An accurate knowledge of the drainage properties of the real structure is therefore of paramount importance.

Figure 26 shows the calibration of the model to the second undrained cyclic test using the Newmark algorithm. In this analysis, the Newmark parameters are α=0.5 and β=0.25, with Δt=5.0e-4secs. As with the strain-controlled results, the correct number of cycles to material instability is simulated. However, this time, as the material becomes unstable it rapidly collapses as the material heads towards zero effective confinement. The program was set to remove the external load at an effective stress of σ_m=3kPa. To continue the analysis further it is necessary to switch off the model close to zero effective mean stress (the frictional model ALTERNAT is not defined at the apex of the yield surfaces at zero isotropic stress) and use (perhaps) a viscous type constitutive law (1987).

At the liquefied state, sand has little strength in shear and extremely large shear strains develop as shown in Figure 27, which shows the full deviator stress versus axial strain record for the second cyclic test. Currently, it is not possible to continue the analysis to any degree of accuracy as good post liquefaction constitutive models are still at the development stage. Switching the analysis to simulate viscous fluid movement may be the answer, although some dilation of the material is still observed during continued loading. The post liquefaction stress path is significantly influenced by the testing machine being used, which adds to the difficulty in studying the fluid like properties. As the post liquefaction flow starts to arrest, perhaps due to a combination of dilation and consolidation affects, the material will start to gain strength. This will require the viscous model to be switched off and the original constitutive model to be switched back on. An example of this type of analysis is the simulation of lateral spreading of bridge piers due to earthquake induced liquefaction. During earthquake loading the lower layers of a sand or silt bearing stratum may liquefy resulting in very high excess pore water pressures (often seen as sand fountains) and very little shear strength. The material may then start to flow in the direction of the propagating stress wave. Any overlying layers of soil will then be 'dragged' with the material. If this then impinges on a structure (for example a bridge column), a passive wedge will develop in the overlying layer as it tries to 'flow' past the structure. If sufficiently high, failure of the column will occur. This means that the bridge column must be designed to withstand the earthquake forces and the post-liquefaction conditions of the soil. Simply switching the model off and keeping the excess pore pressure constant is not realistic.

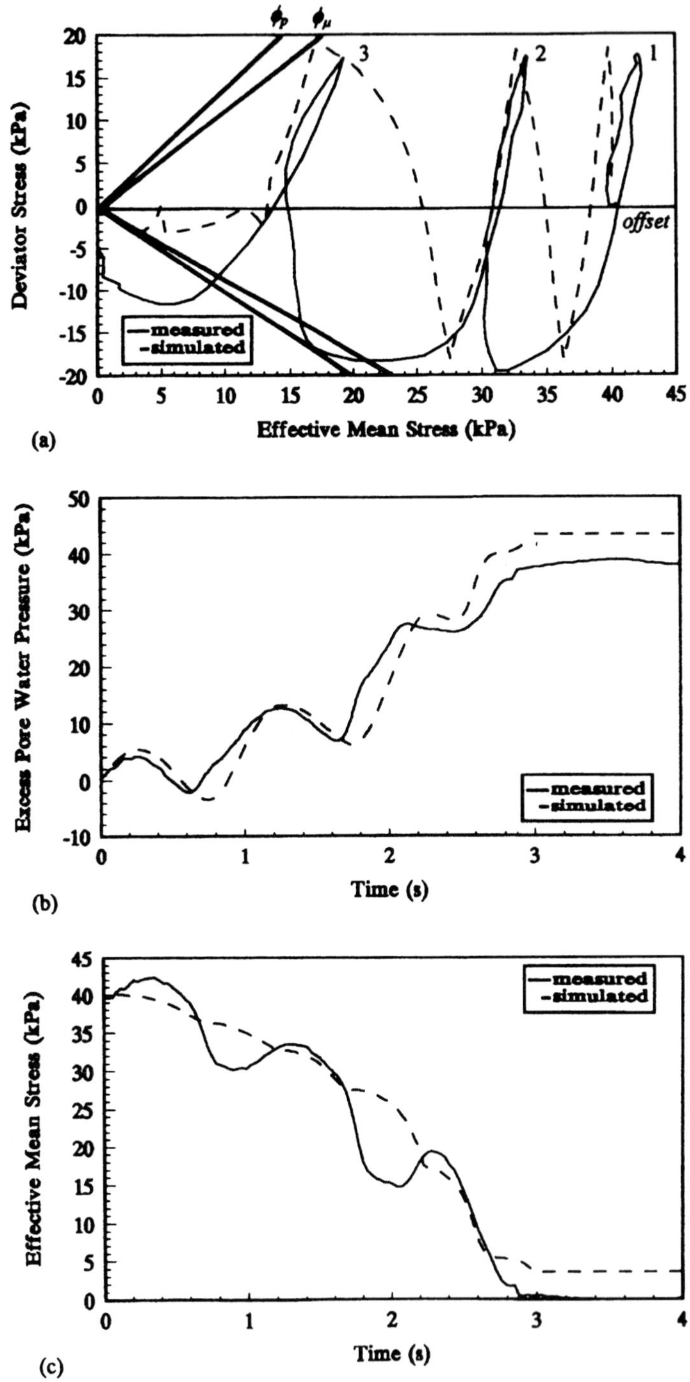

Figure 26. Calibration of ALTERNAT to Nevada sand in undrained dynamic triaxial compression at 1 Hz with e_v=0.73, q_d=18.6 kPa and q_{off}=0.0 (after Woodward & Molenkamp, 1999)

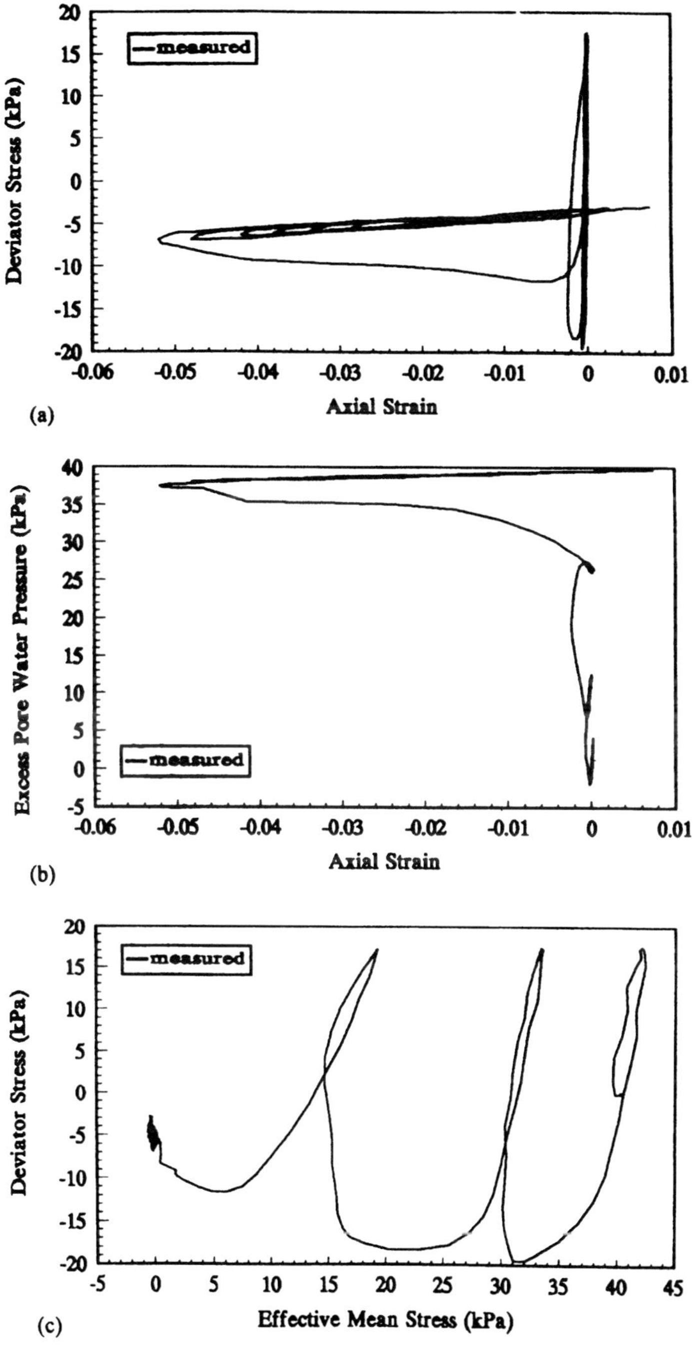

Figure 27. Full stress path plots of Nevada sand in undrained dynamic triaxial compression at 1 Hz with e_v=0.73, q_d=18.6 kPa and q_{off}=0.0

5 Examples

5.1 Example 1: Bearing Capacity Analysis

Although the bearing capacity of a foundation on the surface of granular soil is a traditional problem in geotechnical engineering, it is a good example of the advantages of using an advanced soil model. The ability of the model to not only predict the failure of the foundation, but also accurately describe its behaviour prior to failure, is of paramount importance. This is particularly the case with granular soils, as the displacement of the foundation under working load is usually more important than its ultimate bearing capacity. The prediction of the allowable bearing capacity is therefore of primary concern.

The allowable bearing capacity is usually determined by considering a suitable factor of safety with respect to its peak resistance, and the maximum displacement allowed by foundation under working loads. However, before we determine the allowable bearing capacity, it is first useful to review the literature on foundations to demonstrate the importance of using models like ALTERNAT in this area.

Traditionally the analysis of footings on the surface of granular soils has been one of the most difficult finite element simulations to be attempted (de Borst and Vermeer, 1984 and Griffiths, 1982). This is particularly noticeable by a general lack of published work in foundation analysis using finite elements in the literature, especially for 'rough' footings situated on the surface of sands with ϕ>35°. Recent attempts include Simonini, (1993), Manoharan & Dasgupta (1995), Frydman & Burd (1997) and Woodward & Griffiths (1998). In all of these studies a linear-elastic perfectly-plastic linear Mohr-Coulomb constitutive soil model was used. Simple models of this type do not predict reductions in N_γ with foundation width and certainly would not be able to predict deformation behaviour very accurately. This is because they cannot simulate the reduction in the friction angle and dilatancy properties with confining pressure. Determination of the allowable bearing pressures is also very difficult to obtain accurately. Most analytical solutions do not take into account material dilatancy or confining pressure effects. In fact, there is no unique solution to the problem and many solutions have been proposed. For example,

$N_\gamma = 1.8(N_q - 1)\tan\phi$ (Brinch Hansen) $N_\gamma = \left(N_q - 1\right)\tan\left(1.4\phi\right)$ (Meyerhof)

$N_\gamma = 2(N_q + 1)$ (Caquot & Kerisel) $N_\gamma = 2(N_q - 1)\tan\phi$ (Vesic)

$$N_q = \exp\left(\pi \tan\phi\right)\tan^2\left(45 + \frac{\phi}{2}\right) \qquad \text{(Terzaghi)}$$

Given the complexity of the problem it seems that accurate predictions of both the settlement behaviour and the ultimate bearing capacity of foundations resting on granular soils can only be achieved using more advanced constitutive relationships. A significant advantage of using kinematic yield surfaces is the ability to look at effects such as the over consolidation ratio as well. In this example, we will look at the collapse of a circular foundation on the surface of Nevada sand using the material parameters determined earlier in the calibration section. Figure 28 shows the finite element mesh used in the analysis (322 8-noded quadrilateral elements). The footing is assumed to be rough and rigid and is subjected to equal vertical displacement increments applied at the vertical degrees of freedom at the contact nodes. The rough assumption is simulated by ensuring no movement of the horizontal degrees of freedom during load application. The footing width B=2m and the overall mesh dimensions are 10B in the vertical and horizontal directions. Since the footing is symmetrical about the centreline only one half needs to be analysed using rollers at the sides and assuming a fixed base. The analysis was performed on a Pentium II, 450 MHz computer with 256MB of memory.

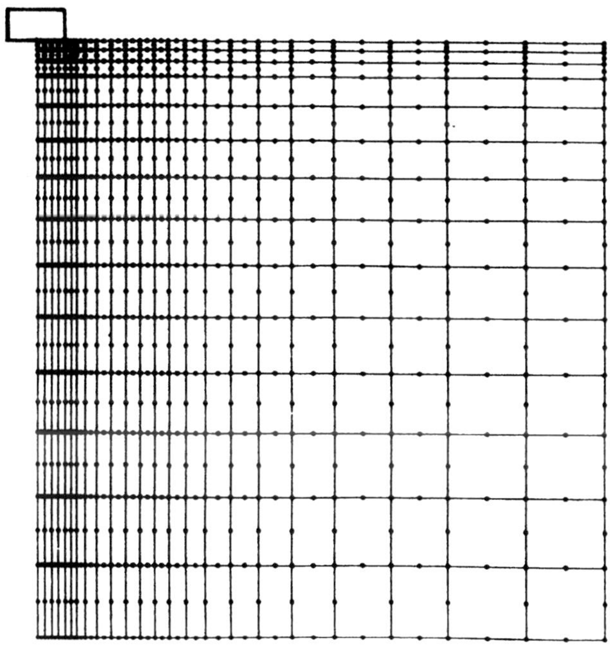

Figure 28. Mesh used in foundation example

The bearing stress q is determined by first defining contact elements underneath the footing. The nodal forces are then calculated for these elements from the following equation

$$F = \int_V B^T \sigma \, d\Omega - \int_V N^T N \rho g \, d\Omega \tag{63}$$

where, N are the elemental shape functions and are used to calculate the direct effect of the soil's weight and B^T is the transpose of the strain-displacement matrix. The vertical component of the nodal forces are accumulated and divided by the footing area over one radian to give the bearing stress q. The bearing capacity factor N_γ is then found from the peak value of the bearing stress q_p

$$N_\gamma = \frac{2q_p}{\gamma B} \tag{64}$$

At the initial state the peak triaxial compression friction angle ϕ_p=36.35°, the friction angle at the critical state ϕ_{cv}=31.3° and the true interparticle friction angle ϕ_μ=28.0°. Figure 29 shows the results of the analysis, presented as normalised bearing capacity $N_\gamma S_\gamma$ against normalised vertical displacements δ/B.

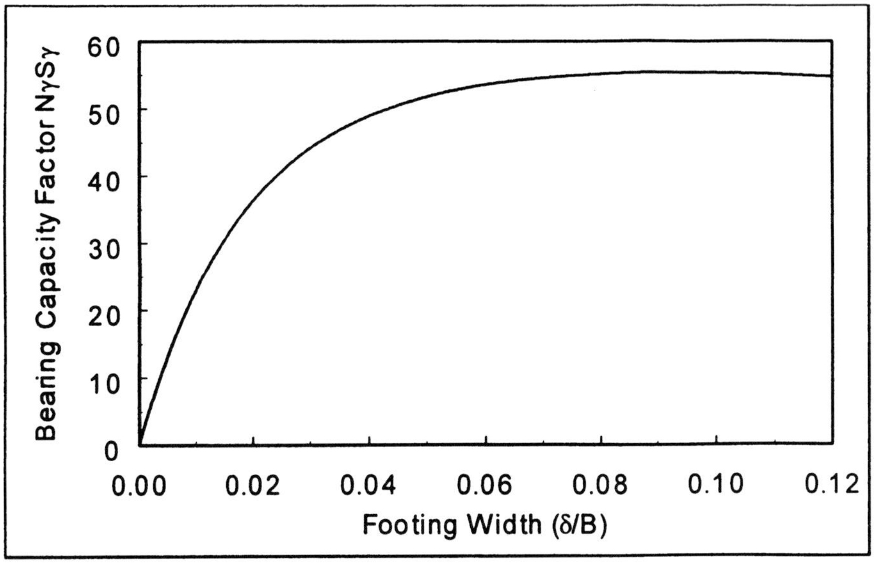

Figure 29. Normalised bearing capacity $N_\gamma S_\gamma$ against normalised vertical displacements δ/B

No numerical solution of this problem has been reported in the literature. The only available comparison is to the analytical values published by Bolton & Lau (1993). For a friction angle of ϕ=36° they quote N_γ=101. If we assume that S_γ=0.6, then from Figure 29 N_γ=92. Although the results appear to be similar, it should be remembered that the finite element results are based on the response of a real sand and are therefore a function of stress path, Lode angle (Griffiths, 1990), density, dilatancy, mean stress etc. For example, Figure 30 shows the predicted reduction in the friction angle with confining pressure for Nevada sand and other granular materials. This type of dependency is not simulated in the analytical solutions. The value of N_γ is also related to parameters like the size, shape and roughness of the foundation as well as the soil parameters. Figures 31 and 32 shows this dependency on the footing's physical dimensions, by illustrating the reduction in the bearing capacity factor and the allowable bearing capacity with foundation width for a rough footing on several different granular soils, including Nevada sand, as computed by ALTICA.

Figure 33 completes the summary of the analysis by showing a displacement vector plot at the end of loading. The inserts have a geometrical magnification of 3. The formation of a general shear failure mechanism can be seen.

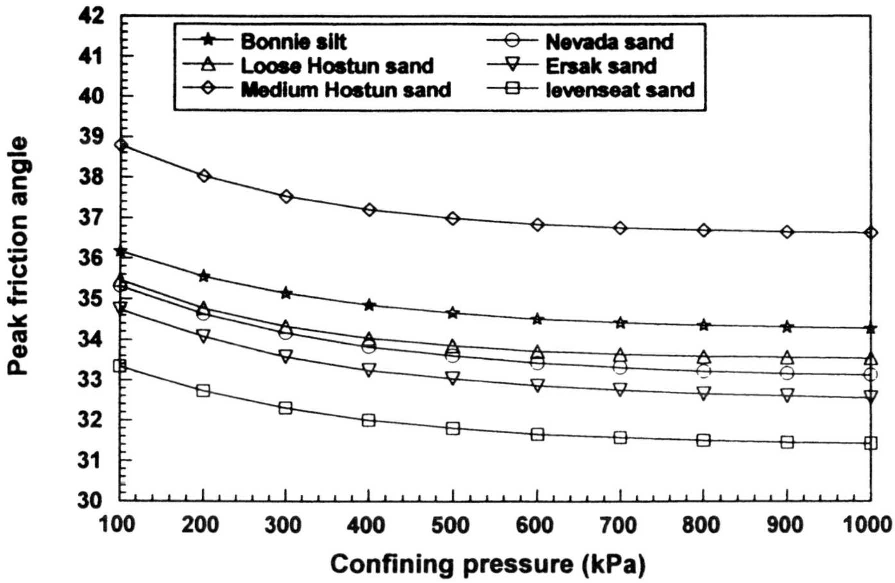

Figure 30. Variation of the peak friction angle with confining pressure for several different granular soils (after Woodward & Nesnas, 2000)

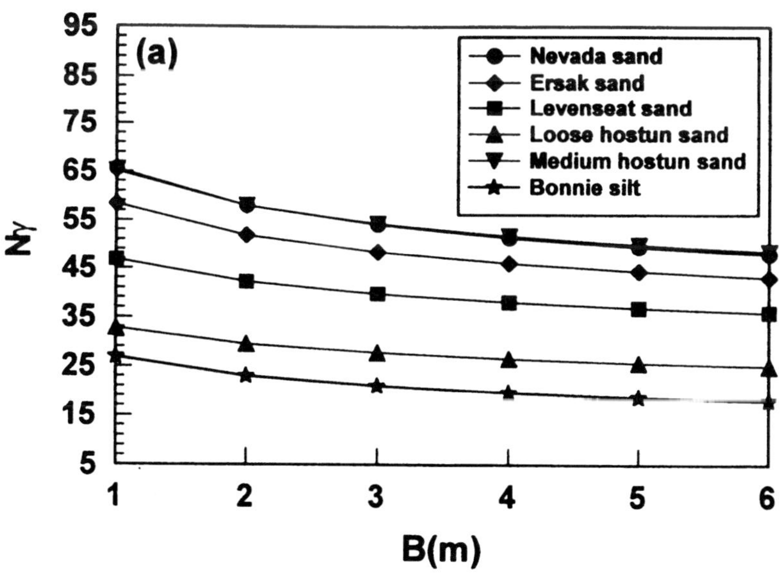

Figure 31. Variation of the bearing capacity factor with width of a rough circular footing for six granular materials (after Woodward & Nesnas, 2000)

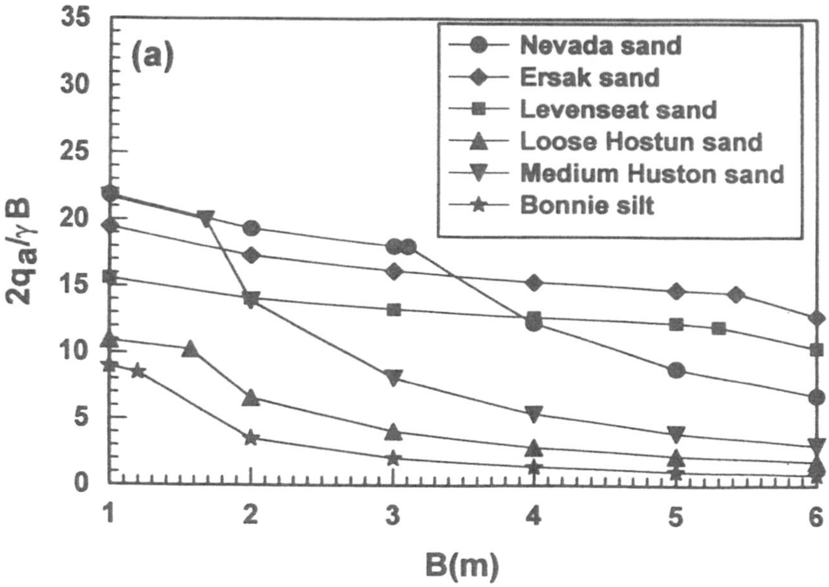

Figure 32. Variation of the allowable bearing capacity factor with width of a rough circular footing for six granular materials (after Woodward & Nesnas, 2000)

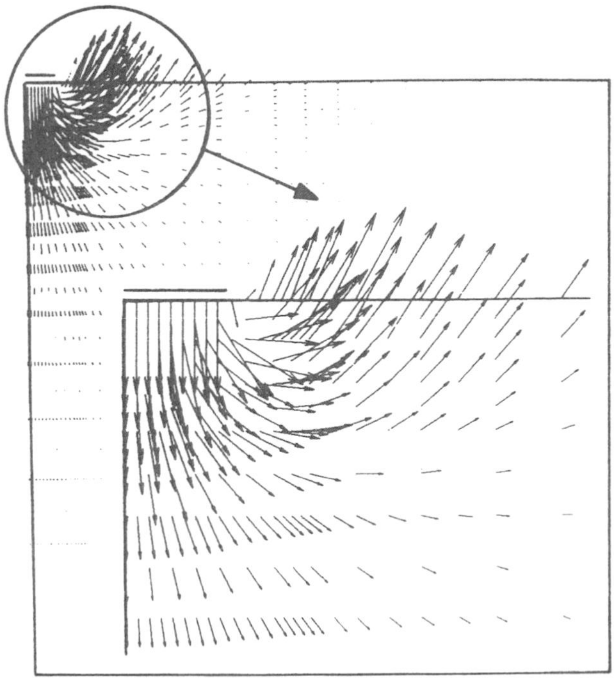

Figure 33. Displacement vectors at the peak resistance of a rough circular footing on Nevada sand

5.2 Example 2: Simulation of Earthquake Induced Liquefaction

The second example considered is the centrifuge earthquake simulation of Model No. 1 from the VELACS project shown in Figure 34.

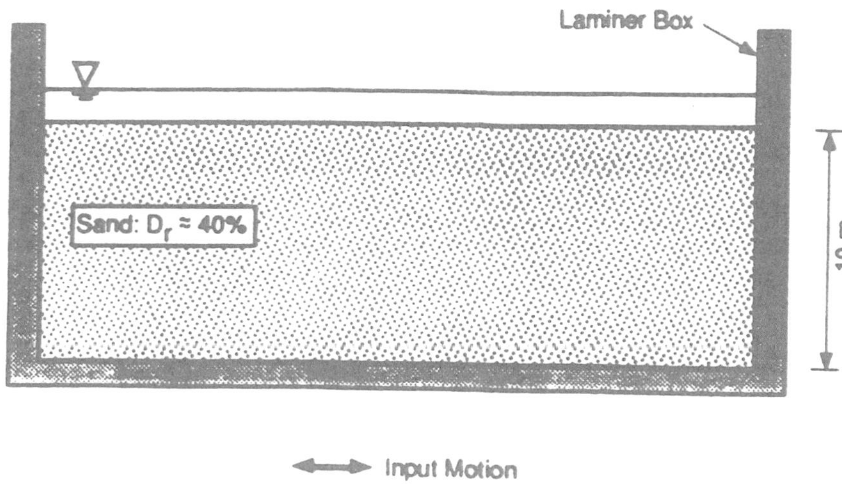

Figure 34. VELACS test model No. 1

The material is saturated Nevada sand at a relative density of D_r=40% and initial void ratio e_i=0.73. The box is subjected to a base acceleration (shown in Figure 35) with a peak amplitude of 0.24g.

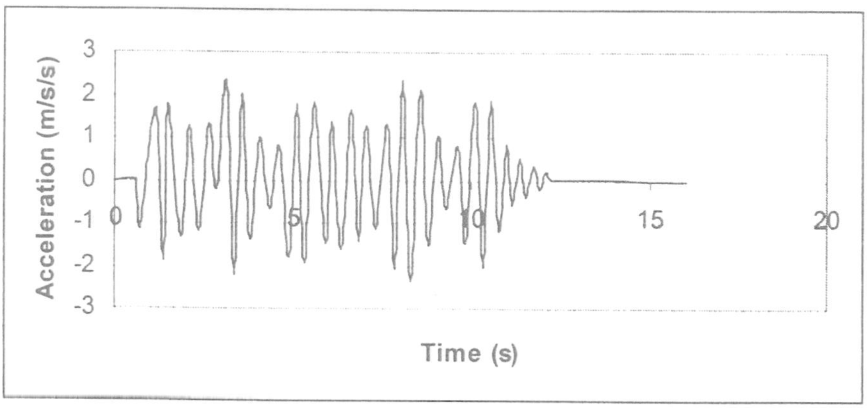

Figure 35. Input base acceleration time history

The finite element mesh used in the analysis is shown in Figure 36. It should be noted that the ALTERNAT model is no longer valid when liquefaction is reached (as discussed previously). The analysis can only therefore be simulated accurately up to the point of liquefaction, after this point the model must be switched off. The coupled UP formulation was used, however difficulties exist with respect to which permeability to use in the analysis. In centrifuge tests the fluid velocities do not scale and so the value of the 'correct' permeability to use between the full scale and the model is debatable. To illustrate this, the analysis is performed assuming full scale conditions and compared directly to the centrifuge results.

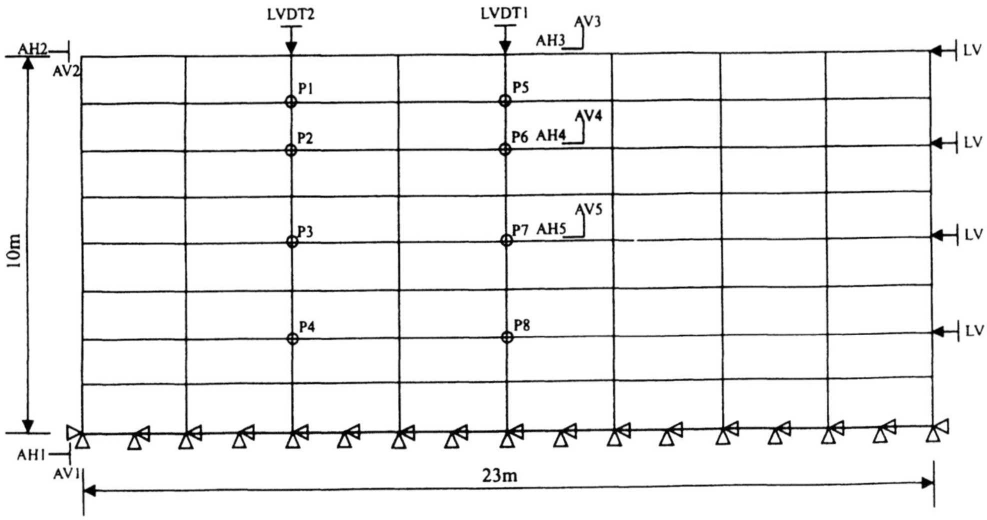

Figure 36. Finite element mesh used in analysis

The results of the analysis for position P5 are shown in Figure 37, in terms of excess pore water pressure generation. The figure shows that the sand in the model liquefied at this position after approximately 6 secs. The model predicts that liquefaction occurs at this position after only 1.2 secs. As illustrated, decreasing the numerical permeability of the sand will increase the time to liquefaction and generate a better 'fit' to the model results.

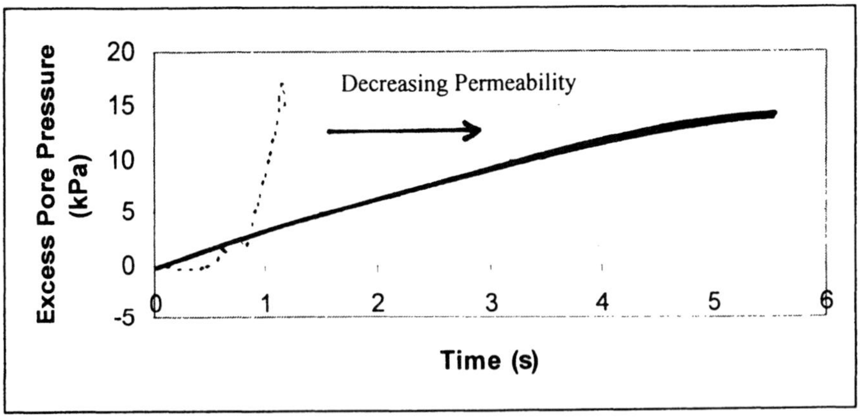

Figure 37. Excess pore water pressure up to liquefaction (solid experimental, dashed numerical)

7 Concluding Remarks

In this chapter an advanced constitutive soil model was presented. The soil model makes use of combined isotropic and kinematic hardening and can simulated the cyclic densification and liquefaction phenomena of sand through a series of multiple yield surfaces.

Using a single finite element the model was calibrated to Nevada sand using both a 'static' and 'dynamic' finite element program. The model was then used to look at the response of two boundary value problems using the single element calibration data. The first of these was the response of a surface footing. The reduction in the peak friction angle and bearing capacity factor N_γ with increasing footing width was shown. In the second problem a coupled finite element approach (UP formulation) was used to examine the buildup in excess pore water pressure with time for Model No. 1 of the VELACS project. The difficulties in predicting the correct value of the permeability with respect to the prototype, model and numerical analysis were highlighted.

The development of a complete post liquefaction constitutive soil model for earthquake induced liquefaction studies is currently being researched at Heriot-Watt University. This is being combined with a large displacement analysis procedure and should lead to the complete pre and post liquefaction flow analysis of granular soil under earthquake loading. This approach is preferred to that of simply switching off the soil model, but keeping the excess pore pressure constant during further shaking until consolidation is assumed to occur (i.e. after shaking has stopped) as it will allow post liquefaction interaction studies to be modelled.

8. References

Anandarajah, A. (1994). Procedures for elasto-plastic liquefaction modeling of sands. *J.of Engineering Mechanics*, Vol 120, No. 7, 1563-1587.

Arulanandan, K. and Scott, R. (1994). *Verification of numerical procedures for the analysis of soil liquefaction problems*. Balkema, Rotterdam.

Arulmoli, K., Muraleetharan, K.K., Hossain, M.M. and Fruth, L.S. (1992) VELACS, Verification of liquefaction analysis by centrifuge studies, laboratory testing program, soil data report, The Earth Technology Corporation, 13900 Alton Parkway, Suite 120, Irvine, CA.

Been, K. and Jefferies, M.G. (1985). A state parameter for sands. *Geotechnique* 35, No. 2, 99-112.

Bolton, M.D. and Lau C.K. (1993). Vertical bearing capacity factors for circular and strip footings on Mohr-Coulomb soil. *Canadian Geotechnical Engineering Journal*, 30, 1024-1033.

de Borst, R. & Vermeer, P.A. (1984). Possibilities and limitations of finite elements for limit analysis, *Geotechnique*, 34, No. 2, 199-210.

Casagrande, A. (1936) Characteristics of cohesionless soils affecting the stability of slopes and earth fills, *J. Boston Soc. Civ. Engrs.*, 257-276.

Crouch, R.S. & Wolf, J.P. (1994). Unified 3D critical state bounding-surface plasticity model for soils incorporating continuous plastic loading under cyclic paths. Part I: Constitutive relations. *Int. J. for Num. Meth. in Geomech.*, Vol. 18, No. 11, 735-758.

Crouch, R.S. & Wolf, J.P. (1994). Unified 3D critical state bounding-surface plasticity model for soils incorporating continuous plastic loading under cyclic paths. Part II: Calibration and simulations. *Int. J. for Num. Meth. in Geomech.*, Vol. 18, No. 11, 759-784.

Cundall, P.A. and Strack, O.D.L. (1979). A discrete numerical model for granular assemblies. *Geotechnique* 29, No. 1, 47-65.

Dafalias, Y.F. and Popov, E.P. (1975). A model of nonlinearly hardening materials for complex loading. *Acta Mechanica* 21. 173-192.

Dafalias, Y.F. (1994). Overview of constitutive models used in VELACS. *Verification of numerical procedures for the analysis of soil liquefaction problems*. Balkema, Rotterdam, 1293-1303.

Finn, W.D.L., Lee, K.W. and Martin, G.R. (1977). An effective stress model for liquefaction. *Journal Geotechnical Engineering Division* 103, ASCE, 517-533.

Frydman, S. and Burd, H.J. (1997) Numerical studies of bearing capacity factor N_γ, *J. Geo. Eng., ASCE*, Vol 123, No. 1, 20-29 (1997).

Griffiths, D.V. (1982) Computation of bearing capacity factors using finite elements, *Geotechnique*, 32, No. 3, 195-202.

Griffiths, D.V. (1990). Failure criteria interpretation based on Mohr-Coulomb friction, *J. Geo. Eng., ASCE*, Vol 116, No. 6, 986-999.

Hardin, B.O. and Hardy, W.L. (1968). Vibration modulus of normally consolidated clay, *J. of Soil Mech. Div., ASCE,* SM 2, 353-369.

Hicks, M. (1990) Numerically modelling the stress strain behaviour of soils, *Doctural Thesis*, University of Manchester.

Hicks, M. (1995). A computer algorithm for solving boundary value problems using the double-hardening constitutive law MONOT: I. Algorithm development, *Int. J. for Num. Meth. in Geomech.*, Vol. 19, 1-27 (1995).

Hicks, M. (1995). A computer algorithm for solving boundary value problems using the double-hardening constitutive law MONOT: II. Algorithm validation, *Int. J. for Num. Meth. in Geomech.*, Vol. 19, 29-57 (1995).

de Josselin de Jong, G. (1976) Rowe's stress-dilatancy relation based on friction, *Geotechnique* 26, No. 3, 527-534.

Inel, S., Roth, W.H. and de Rubertis, C. (1993). *Nonlinear dynamic effective-stress analysis of two case histories*. 3rd Int. Conf. On Case Histories in Geot. Eng., St Louis, MO, June.

Iwan, W.D. (1967). *On a class of models for the yielding behaviour of continuous and composite systems*. ASME J. Appl. Mech. 34, 612

Jefferies, M.G. (1993). *Nor-sand: a simple critical state model for sand.* Geotechnique 43, 91-103.

Kaddouri, A. and Woodward, P.K. (1998). Advanced numerical modelling of earthquake induced liquefaction. *Proceedings 6th Conference of SECED on Seismic Design Practice into the Next Century*: University of Oxford, UK. 213-218.

Kaddouri, A. (2000). Title to be set. *Doctural Thesis*, Heriot-Watt University.

Krieg, R.D. (1975). A practical two-surface plasticity theory. *ASME J. Appl. Mech.* 42. 641-646.

Lade, P.V. and Duncan, J.M. (1975). Elasto-plastic stress-strain theory for cohesionless soils, *J. Geotech. Eng. Div. ASCE.* 107, GT10, 1037-1053.

Lade, P.V. (1977). Elasto-plastic stress strain theory for cohesionless soil with curved yield surfaces, *Int J. Solids Struct.* 13, 1019-1035.

Lade, P.V. (1979). Three-dimensional stress-strain behaviour and modelling of soils, *Bochum, Ruhr-Univ., Schriffenreihe des Inst. F. Grundb., Wasserwesen und Verkehrswesen, Serie Grundbau*, Heft 4, Aug.

Lindenberg, J. and Koning, H.L. (1981). Critical density of sand, *Geotechnique*, 31, No. 2, 231-245.

Manoharan, N. and Dasgupta, S.P. (1995). Bearing capacity of surface footings by finite elements, *Comp. Struct*, 54, No.4, 563-586.

Manzari, M.T. and Dafalias, Y.F. (1997). A critical state two-surface plasticity model for sands. *Geotechnique* 47, No. 2, 255-272.

Molenkamp, F. (1981). Elasto-plastic double hardening model MONOT, *LGM Report* CO-218595, Delft Geotechics.

Molenkamp, F. (1982) Kinematic model for alternating loading ALTERNAT, *LGM Report* CO-218598, Delft Geotechnics.

Molenkamp, F. (1986). Limits to the Jaumann stress rate, *Int. J. for Num. Meth. in Geomech.*, Vol. 10, 151-176.

Molenkamp, F. (1987). Elasto-plastic model for the simulation of liquefaction under alternating loading, *Workshop on constitutive laws for the analysis of fill retention structures, Workshop Report*, Department of Civil Engineering, Univeristy of Ottawa, July, 60-194.

Molenkamp, F. (1988). A simple model for isotropic non-linear elasticity of frictional materials, *Int. J. for Num. Meth. in Geomech.*, Vol. 12, 467-475.

Molenkamp, F. (1990) Reformulation of ALTERNAT to minimise numerical drift due to cyclic loading, *University of Manchester Internal Report*. UK.

Molenkamp, F. (1992). Application of non-linear elastic model, *Int. J. for Num. Meth. in Geomech.*, Vol. 16, 131-150.

Mroz, Z. (1967). On the description of anisotropic workhardening. *J. Mech. Physics Solids* 15, 612.

Naylor, D.J. (1974). Stresses in nearly incompressible materials by finite elements with applications to the calculation of excess pore pressures, *Int. J. Num. Meth. Eng.*, 8, 443-460.

Newmark, N.M. (1965). Effects of earthquakes on dams and embankments, *Geotechnique* 15, No. 2, 139-160.

Pande, G.N. and Sharma, K.C. (1983). Multi-laminate model of clays – a numerical evaluation of the influence of rotation of the principal stress axes. *Int. J. Num. Anal Meth. Geomechanics*. 7. 397-418.

Poulos, S.J. (1981). The steady state of deformation, *J. of Geotech. Eng., ASCE*, 107, 5, 553-562.

Prevost, J.H. (1977). Mathematical modelling of monotonic and cyclic undrained clay behaviour, *Int. J. for Num. Meth. in Geomech.*, Vol. 1, No. 2, 195-216.

Prevost, J.H. (1982) Two-surface vs multi-surface plasticity theories, *Int. J. for Num. Meth. in Geomech.*, Vol. 6, 323-338.

Prevost, J.H. (1985). A simple plasticity theory for frictional soils. *Soil Dynamics and Earthquake Engineering*, 4, 9-17.

Prevost, J.H. & Keane, C.M. (1990). Multimechanism elasto-plastic model for soils. *J. of Engineering Mechanics*, Vol. 116, No. 9, 1924-1945.

Rowe, P.W. (1962). The stress-dilatancy relation for static equilibrium of an assembly of particles in contact, *Proc. Royal Soc.*, Vol. 269, 500-527.

Rowe, P.W. (1971). Theoretical meaning and observed values of deformation parameters for soil, *Proc. Roscoe Mem. Symp.*, Cambridge, 143-194.
Saada, A. and Bianchini, G. (1988). *Constitutive equations for granular non-cohesive soils*. Balkema, Rotterdam.
Schofield, A.N. and Wroth, C.P. (1968). *Critical state soil mechanics*, McGraw-Hill, London.
Simonini, P. (1993) Influence of relative density and stress level on the bearing capacity of sands, *Int. J. for Num. Meth. in Geomech.*, Vol. 17, 871-890.
Smith, I.M. and Griffiths, D.V. (1988). *Programming the finite element method*, 2nd Edn, Wiley, Chichester.
Tatsuoka, F., Iwasaki, T., Yoshida, S., Fukushima, S. and Sudo, H. (1979). Shear modulus and damping by drained tests on clean sand specimens, reconstituted by various methods, *Soils and Foundations*, Vol. 19, No. 1, 39-54.
Thorton, C. (1989). A direct approach to micromechanically based continuum models for granular material. Satake M., ed., *Mechanics of Granular Materials, Report of ISSMFE Technical Committee on Mechanics of Granular Materials*, Japanese Society of Soil Mechanics and Foundation Engineering, 145-150.
Vermeer, P.A. 'Formulation and analysis of sand deformation problems'. *PhD Thesis*, Delft University of Technology, 1980.
Woodward, P.K. (1993). Earthquake engineering and advanced constitutive modelling in geomechanics, by finite elements, *Doctural Thesis*, University of Manchester (1993).
Woodward, P.K. (1995). Application of a kinematic elasto-plastic soil model to monotonic and cyclic loading, *Proc the Sixth Int. Con. on Civil and Struct. Eng. Comp.*, St Johns College Cambridge, England 229-237.
Woodward, P.K. and Griffiths, D.V. (1998) Observations on the computation of the bearing capacity factor N_γ by finite elements, *Geotechnique*, 48, No. 1, 137-141.
Woodward, P.K. and Molenkamp, F. (1999). Application of an advanced multi-surface kinematic constitutive soil model, *Int. J. for Num. Meth. in Geomech.*, Vol. 23, 1995-2043.
Woodward, P.K. and Nesnas, K. (2000) Analysis the behaviour of footings on sand using an advanced constitutive soil model. *EPSRC Final Research Report*, UK.
Zienkiewicz, O.C., Valliappan, S. and King, I.P. (1969). Elasto-plastic solutions of engineering problems. Initial-stress, finite element approach, *Int. J. for Num. Meth. in Geomech.*, 1, 75-100.
Zienkiewicz, O.C. and Chang, C.T. (1978). Nonlinear seismic response and liquefaction. *Int J. Num. Anal. Meth. In Geomechanics*, 2, 381-404.
Zienkiewicz, O.C., Leung, K.H. and Pastor, M. (1985). A simple model for transient soil loading in earthquake analysis. I: Basic model and its application, and II: Non-asociative models for sands. *Int. J. Num. Anal. Meths. Geomechanics*. 2. 381-404.

Appendix: Notation

A	material parameter
B	width of footing
B^T	transpose of the strain-displacement matrix
b_i	body force vector
C	material parameter
C_{ij}	global viscous damping matrix

D	material parameter
D_{ijkl}	elastic tangent stiffness matrix
E	material parameter
e	measure of the deviatoric strain
e_{ij}	components of the deviatoric strain tensor
e_i	initial void ratio
e_v	current void ratio
e_{vd}	void ratio at the maximum densification
d	increment in
F_{ij}	deformation gradient matrix
$\dot{F}_{ij}$	deformation gradient rate matrix
F_j	total global loads vector
g	acceleration due to gravity
G	shear modulus
G	plastic potential function
G_s	specific gravity
H	hardening modulus
H_w	height of retaining wall
I_1	first invariant of the pseudo-stress T_{ij}
I_3	third invariant of the pseudo-stress T_{ij}
K_{iso}	elastic bulk modulus for isotropic unloading
K_o	earthpressure coefficient at rest
K_p	passive earthpressure coefficient
K_{ij}	global non-linear elastic stiffness matrix
L	material parameter
L_{ik}	velocity gradient matrix
M	material parameter
M_{ij}	global mass matrix
N	element shape functions
N_γ	bearing capacity factor due to soil self weight
N_q	bearing capacity factor due to surcharge
n	material parameter
n_i	initial porosity
n_d	final porosity
P	material parameter
P_p	total passive force
p_a	atmospheric pressure
p	pore pressure
Q	complementary potential function per unit volume
Q	material parameter
q_p	bearing capacity of footing
R	material parameter
R_p	material parameter
R_{ij}	anti-symmetric rotation matrix
$\dot{R}_{ij}$	anti-symmetric spin matrix

R_i	global displacement vector
S	material parameter
s	measure of isotropic stress
T_{ij}	components of the pseudo-stress tensor
t	measure of the deviatoric stress
t_{ij}	components of the deviatoric stress tensor
U_i	fluid displacement vector
u_i	displacement vector
V_{ik}	symmetric stretch matrix
$\dot{V}_{ik}$	symmetric stretch rate matrix
v	measure of isotropic strain
X	a factor used to modify the hardening modulus to account for cyclic mobility
Y_1	pre-peak shear stress level
Y_2	post-peak shear stress level

α	isotropic component of plastic potential
β	material parameter
χ	hardening parameter of a kinematic model
$f(\chi)$	measure of the yield surface size
δ_{ij}	Kronecker delta
Δ	finite increment
ε_{ij}	components of the Eulerian (small) strain tensor
ϕ	Mohr-Coulomb friction angle
ϕ_μ	true interparticle friction angle
γ_d	dry unit weight
γ_{max}	maximum dry unit weight
γ_w	unit weight of water
η	material parameter
κ	material parameter or hardening parameter of an isotropic model
λ	magnitude of the plastic strain increment
θ	Lode angle
θ_p	material parameter
ρ	material density
ρ_f	fluid density
σ_{ij}	components of the Cauchy stress tensor
υ	material parameter
ξ	material parameter
ξ_{ij}	non-dimensional deviatoric tensor of anisotropy
ψ	scalar measure of the direction of loading
ψ	material parameter
ζ	material parameter

Superscripts

body	contribution to the total global loads vector due to excess plastic stress redistribution
elastic	elastic component
err	contribution to the total global loads vector due to errors in the excess plastic stress redistribution at the last load increment
ext	contribution to the total global loads vector due to the external loads
m	iteration number
p	plastic
total	improved stress state with respect to the reference state
0	quantity before the start of iterations
-1	inverse

Subscripts

vol	volumetric
n	load increment number
t	time step